Tubular Heat Exchanger Inspection, Maintenance, and Repair

Tubular Heat Exchanger Inspection, Maintenance, and Repair

Carl F. Andreone, P.E., FASME
Stanley Yokell, P.E., FASME

McGraw-Hill
New York San Francisco Washington, D.C. Auckland Bogotá
Caracas Lisbon London Madrid Mexico City Milan
Montreal New Delhi San Juan Singapore
Sydney Tokyo Toronto

Library of Congress Cataloging-in-Publication Data

Andreone, Carl F.
Tubular head exchanger inspection, maintenance, and repaire / Carl F. Andreone, Stanley Yokell.
p. cm.
Includes bibliographical references and index.
ISBN 0-07-001778-6
1. Heat exchangers—Inspection. 2. Heat exchangers—Maintenance and repair. I. Yokell, Stanley. II. Title.
TJ263.A54 1997
621.402'5—dc21 97-26077
CIP

McGraw-Hill
A Division of The McGraw-Hill Companies

1 2 3 4 5 6 7 8 9 0 DOC/DOC 9 0 2 1 0 9 8 7

ISBN 0-07-001778-6

The sponsoring editor for this book was Robert Esposito, the editing supervisor was David E. Fogarty, and the production supervisor was Sherri Souffrance. It was set in Century Schoolbook by Estelita F. Green of McGraw-Hill's Professional Book Group composition unit.

Printed and bound by R. R. Donnelley & Sons Company.

This book is printed on recycled, acid-free paper containing a minimum of 50% recycled, de-inked fiber.

Contents

Preface

We wrote this book with the intention of providing plant and maintenance engineers with a starting point for addressing heat exchanger inspection, maintenance, and repair tasks.

Engineering curricula for chemical and mechanical engineers provide students with a basic understanding of the thermodynamics, thermal design, fluid dynamics, and stress analysis of shell-and-tube heat exchangers. However, for the most part, they do not deal with the mechanical design and fabrication practices and procedures. Design and construction details, to which pressure safety codes do not apply, have always been developed by individual manufacturers. Except where they collaborated to produce agreed-on standards, the manufacturers have closely held such details as proprietary information. Only when some aspect of design or manufacture became a troublesome maintenance problem for owners or operators did user groups and manufacturers focus enough attention on heat exchanger inspection, maintenance, and repair problems. One of this book's goals is to point the way to help readers avoid the pitfalls of the past.

In 1978, Stanley Yokell sold his heat exchanger manufacturing company after being in business for 25 years. At that time he was persuaded to present a 4-day intensive continuing education course dealing with shell-and-tube heat exchangers; it was based on his extensive manufacturing experience, with emphasis on various aspects of construction, inspection, maintenance, and troubleshooting. The presentations of this course were well attended, and in 1982, in response to the many questions by early attendees about feedwater heaters and other power plant heat exchangers, he persuaded Carl Andreone to join in preparing another seminar devoted to feedwater heaters.

The shell-and-tube heat exchanger course became so well known through presentations in the United States, Canada, and Europe, that in 1990, Mr. Yokell prepared a much-expanded version of the course notes, published by McGraw-Hill Book Company under the title *A Working Guide to Shell-and-Tube Heat Exchangers*.

In presenting these courses, we realized that there was significant interest in the inspection, maintenance, and repair sections included in the curriculum. We prepared a third course speaking directly to life extension of heat exchangers; the goal was to shed light on inspection and maintenance practices and procedures that would help to prevent failures that result in costly unscheduled shutdowns. We included repair procedures and concepts we believed would alleviate design or manufacturing inadequacies. These included ones that could have resulted from insufficient design-basis information or changes in operating conditions not originally considered, poor design concepts, or faulty and inadequate manufacturing procedures.

The course notes for the third intensive course was the starting point for this book. We recognize that shell-and-tube exchangers seemingly have an almost infinite number of variables and consequently can fail in an infinite number of ways. But we hope the information in this work will point readers to solutions to their heat exchanger problems and ease their tasks of procuring suitable new and replacement equipment, maintain them in satisfactory condition, and safely alter them to meet new conditions as they arise.

We are indebted to our many friends and colleagues for their help and advice as well as useful information we have included in this text.

The book is dedicated to our wives, Ida Porcello Andreone and Edith Gersen Yokell, who have demonstrated once again the support, patience, and tolerance they have shown us for the past 50 years.

Carl F. Andreone
Stanley Yokell

Chapter

1

General Information

Introduction

In this work, the terms *heat exchanger, exchanger,* and *tubular* are used interchangeably to mean shell-and-tube heat exchangers, whether used to transfer sensible heat, condense vapors, boil or vaporize liquids, heat boiler feedwater, or condense steam. This book does not deal with the several other types of heat transfer equipment also generally called *heat exchangers,* for example, plate and frame equipment and double-pipe, compact, and air-cooled exchangers.

Throughout, we refer to the following organizations and publications by their abbreviations and acronyms:

American National Standards Institute	ANSI
American Petroleum Institute	API
American Petroleum Institute, *Pressure Vessel Inspection Code*	API 510 or the API Inspection Code[1]
American Petroleum Institute, *Inspection of Pressure Vessels, Recommended Practice 572*	API 572[2]
ASME Boiler and Pressure Vessel Code	ASME Code[3]
ASME Boiler and Pressure Vessel Code Committee	Code Committee
American Society for Nondestructive Testing	ASNT*

*See the supplement to this chapter for similar and associated societies in other countries.

ASTM (formerly named the American Society for Testing Materials, currently named ASTM)	ASTM
American Welding Society	AWS
Heat Exchange Institute of Cleveland, Ohio	HEI
Heat Exchange Institute, *Standards for Closed Feedwater Heaters*	HEI CFHS[4]
Heat Exchange Institute, *Standard for Power Plant Heat Exchangers*	HEI PPS[5]
Heat Exchange Institute, *Standard for Steam Surface Condensers*	HEI SSCS[6]
National Board of Boiler and Pressure Vessel Inspectors	National Board
National Board Inspection Code	National Board Code, also NBIC[7]
Standards of the Tubular Exchanger Manufacturers Association	TEMA Standards[8]
Tubular Exchanger Manufacturers Association	TEMA

We urge readers to maintain current editions of the following publications in their libraries:

1. *ASME Boiler and Pressure Vessel Code,* Section II, "Materials Specifications"; Section V, "Nondestructive Examination"; Section VIII, Division 1, "Pressure Vessels"; Section VIII, Division 2, "Pressure Vessels—Alternative Rules"; Section IX, "Welding and Brazing Qualifications," "Code Cases," and "Interpretations." We recommend that readers concerned with heat exchangers that handle radioactive materials have access to Section III, "Rules for Construction of Nuclear Power Plant Components" and Section XI, "Rules for In-Service Inspection of Nuclear Power Plant Components"*
2. The pressure-vessel codes prevailing in the countries of heat exchanger origin and installation
3. The National Board Inspection Code

*Although the rules apply for nuclear power plants, it is the authors' opinion that they should be applied when the risks from the failure of an exchanger handling radioactive water or process fluids are similar to the risks of a failure in a nuclear power plant component.

4. The inspection code prevailing in the countries of origin and installation.†
5. The American Petroleum Institute *Pressure Vessel Inspection Code,* API 510
6. The American Petroleum Institute *Inspection of Pressure Vessels,* API 572
7. The *Standards of the Tubular Exchanger Manufacturers Association*
8. The Heat Exchange Institute *Standard for Power Plant Heat Exchangers*
9. The Heat Exchange Institute *Closed Feedwater Heater Standards*
10. The Heat Exchange Institute *Steam Surface Condenser Standards*

Common usages for some of these publications are as follows. The terms *ASME Code* and *Code* are generally used to mean *The ASME Boiler and Pressure Vessel Code.* These terms may also be used to mean Section VIII, Division 1, the part of the code that applies to most tubular exchangers. The terms *National Board Code* and *NBIC* are used interchangeably to refer to the *National Board Inspection Code.* The term *API 510* is used for the American Petroleum Institute *Pressure Vessel Inspection Code,* API 510; the term *API 572* is used to refer to the American Petroleum Institute *Inspection of Pressure Vessels,* API 572. *TEMA Standards* is the short common term for the *Standards of the Tubular Manufacturers Association.* The terms *HEI Power Plant Heat Exchanger Standards, HEI Closed Feedwater Heater Standards,* and *HEI Steam Surface Condenser Standards* are used as the short ways to refer to the several Heat Exchange Institute standards.

To make sure heat exchangers subject to pressure are safely designed and constructed, political bodies require adherence to pressure safety codes. The inspections these codes require apply only to parts under pressure. Inspectors responsible for inspecting to ensure the safe construction of pressure vessels may have little or no knowledge of material quality and construction requirements of their non-

†To the best of the authors' knowledge, no other national inspection code equivalent to the National Board Inspection Code or the American Petroleum Institute's API 510 has been produced or is in use anywhere. The NBIC has formally been adopted in many countries.

pressure parts. Therefore, we distinguish between *pressure-vessel inspections* and pressure-vessel *Authorized Inspectors* on one hand and heat exchanger *mechanical quality inspections* and heat exchanger *mechanical quality inspectors* on the other hand.

Authorized Inspectors are commissioned or certified by organizations such as Technische Überwachunzs-Verein (TUV) in Germany and the Stoomwet in the Netherlands. In North America, Pressure-Vessel Inspectors are commissioned by states, commonwealths, and provinces. Most North American Pressure-Vessel Inspectors also hold commissions from the National Board. European and Japanese pressure-vessel Authorized Inspectors may hold National Board commissions in addition to the commissions of their home countries and other political jurisdictions. National Board commissioned Inspectors may inspect pressure vessels built to the ASME Code and registered with the National Board, and perform in-service inspection, and inspection of repairs and alterations of such pressure vessels. Individuals not familiar with the ASME Code and NBIC should be aware that the ASME Code applies only to new construction, and that the National Board Code applies only to in-service inspection and inspection of repairs and alterations of pressure vessels. Where a state or provincial authority accepts API 510, the Pressure-Vessel Inspector may work according to its rules. These codes designate the Pressure-Vessel Inspector as the *Authorized Inspector.* European and Japanese Authorized Inspectors work according to such codes as A. D. Merk-blatt (Germany), Stoomwesen (the Netherlands), and other European Union codes, The Pressure Vessel Code Dai Isshi Atsuryoku Youki Kousous Kikahu (Japan) and may also work according to the ASME Code and NBIC.

Throughout this work, we use the terms *Authorized Inspector (AI), Code Authorized Inspector,* or simply Inspector interchangeably to refer to commissioned inspectors concerned with construction, alterations, and repairs that conform with a pressure safety code or pressure-vessel repair code. The AI required qualifications, duties, and responsibilities unique to each pressure safety code are defined in the individual codes and are not repeated here.

We use the lowercase term *inspector* to mean purchasers', users' or architect-engineers' non–code inspectors. We use the lowercase term *inspection* for the inspection performed by the inspector. We define *Code Inspection* to describe the inspection performed by Authorized Inspectors. To avoid confusion, we restrict the use of *examination* and *examiner* to nondestructive examinations, for example, radiographic, ultrasonic, and eddy-current examinations and the individuals who perform such examinations. Readers are cautioned that the ASME

Code and the standards which it incorporates by reference sometimes use the term *inspection* when they mean *examination* and the term *examination* when *inspection* is more appropriate.

Inspector's Qualifications

Few heat exchanger users can justify maintaining a staff inspector qualified only to inspect tubular heat-transfer equipment and devoted to this purpose. Consequently, inspectors are seldom fully competent to inspect heat exchangers. Following are *desirable* attributes of inspectors responsible for inspecting tubulars:

- Experienced in heat exchanger construction
- Familiar with exchanger fabrication and repair-shop practices
- Trained in reading and understanding heat exchanger construction drawings
- Knowledgeable about the applicable pressure safety and inspection codes, API Inspection Code sections, and TEMA Standards and/or other trade organization standards customarily applied to the work and their acceptability criteria
- Familiar with and skilled in using inspection tools and instruments
- Trained to make, record, and interpret measurements
- Experienced with, and knowledgeable about, destructive and nondestructive testing and examination
- Knowledgeable about materials used in constructing heat exchangers and applicable materials specifications
- Familiar with erosion, corrosion, erosion-corrosion, and evidence of failure and deterioration of heat exchanger parts
- Trained and qualified by an inspection organization, for example, TUV or the National Board, in pressure-vessel inspection
- Certified or commissioned as a qualified inspector by a private or public agency in accordance with national, international, or industry standards
- Aware of hazards when inspecting in operating facilities and in heat exchanger factories
- Conscious of safety requirements when performing inspections
- Aware that inspection and inspection records are essential to proper maintenance, repair, and life extension of tubular heat exchangers

Roles of Code Authorized Inspectors and Non–Code Inspectors

The obligations of pressure-vessel code Authorized Inspectors are different from those of inspectors employed by purchasers, architect-engineer firms, and end users. Consequently, they have different perspectives. The following sections discuss the different roles pressure safety code Authorized Inspectors and users' inspectors play.

Responsibilities of Authorized Inspectors

Authorized Inspectors are bound by the responsibilities delineated in the applicable pressure safety construction and inspection codes. (The ASME Code also states the users' and manufacturers' responsibilities; the National Board Inspection Code lists the responsibilities of repair/alteration organizations.) Their primary concern is the safety of the pressure-containing parts of the exchanger and the welds of any nonpressure bearing parts to the pressure envelope.

The responsibilities of Authorized Inspectors, listed in the ASME Code and the NBIC, are typical of most pressure-safety codes. Some of these are:

- Verifying that design calculations have been made in accordance with the code rules or that user's and manufacturer's design reports are on hand and are being followed.
- Making sure that all pressure parts, parts welded to them, and welding materials conform with the prevailing pressure vessel code requirements.
- Confirming that only qualified personnel perform welding and that the welding conforms with qualified procedures.
- Ascertaining that all manufacturing, repair, or alteration methods accord with the pressure code or inspection code rules.
- Determining that all nondestructive examinations are performed and interpreted by qualified examiners and accepting or rejecting work examined in accordance with the code criteria.
- Verify that the manufacturer follows the quality assurance and quality control manuals submitted to the certifying authority.
- Inspect fit-ups and pressure welds.
- Witness specified leak and pressure tests required by the code.
- Sign the appropriate report form to indicate acceptance of the newly constructed pressure vessel. In the United States and Canada, this is the ASME Code's Manufacturer's Data Report.

- Sign the appropriate report form to indicate acceptance of a repaired or altered pressure vessel. In the jurisdictions where the National Board Code applies, this is one of the R report forms for alterations or repairs.

Authorized Inspectors do not concern themselves with any specification or manufacturers standard to which the pressure safety or inspection code does not refer. They pay scant attention to deviations from approved drawings that are not code violations—for example, cocked or slightly misplaced nozzles, misaligned support boltholes, and under- or oversized baffles and supports. Furthermore, they are not obliged to verify that a repair or alteration will not adversely affect an exchanger's operation.

AIs are *not* obliged to determine that an exchanger is well constructed for its performance requirements. They need not ascertain whether the clearance between replacement bundles and existing shells is smaller or larger than in the original exchanger. They may not consider that if clearance is too small, the bundle may be damaged during insertion or realize that if it is too large, it will permit more-than-designed quantities of shell fluid to flow through the space between the shell and the baffles, thereby reducing performance. They are not required to verify that replacement exchangers are so constructed that they perform as well as the originals.

Responsibilities of non–code inspectors (inspectors)

Responsibilities of inspectors may partially overlap those of Code Authorized Inspectors. Their primary duties are to

- Verify that the work meets the purchaser's specifications. This is so even when the specifications do not include construction and stamping in accordance with the construction code or shop repairs or alterations in accordance with the repair code.
- Make sure that all dimensions and fits of parts conform with agreed-on specifications, standards, and approved drawings.
- Make sure that the manufacturer stores and handles materials and parts in ways that will not shorten exchanger life.
- Be present at critical times such as tubing the cage and stabbing the bundle into the shell.
- Deal with inadvertent factory errors.
- Expedite materials procurement and production (sometimes).

Hold points and coordination with the shop. The users's or purchaser's inspector must coordinate with the manufacturer the hold points required by the purchase order and inspection brief. Inspectors have to work with the manufacturer to keep track of the approach of hold points and to avoid delaying the work. On the other hand, shops must stop construction at hold points that the Authorized Inspector designates. They must arrange their schedules to coordinate with the AI's schedule.

Few inspectors have the qualifications and training listed above. Consequently, it is important to prepare guidelines before sending them into the field. Even those with training, instruction, qualification, and certification-by-test in inspection techniques should be provided with instructions for doing their work. In the next chapter, we discuss the inspection brief, which is such a for-the-purpose document.

Preinspection Conferences

We recommend holding preinspection conferences to prepare inspectors and make sure that they understand the guidelines embodied in the inspection brief. Conference subjects depend on what is to be inspected: new equipment or replacement parts, mothballed or used units, or exchangers undergoing scheduled or unplanned outages.

The aim of preinspection conferences for new exchangers and replacement parts is to ensure that all concerned individuals are fully aware of and understand the brief. Each participant should understand what is to be inspected, at what stage of production it is, and the acceptance and rejection criteria. In addition, use the conference to review the procedure to be followed for final documentation and release for shipment and be sure to review special shipping requirements, such as monitoring of impacts during transit.

For mothballed or used tubulars, the goal is to make sure that the participants know the considerations for restoring a unit to service, putting a mothballed one into a new service, or procuring a used exchanger as a replacement. The latter may be for temporary use or for a new service when production requirements do not allow time to procure a new one. Jointly discussing the brief can forestall confusion by emphasizing what must be done and what is acceptable. Have available and review the brief's reference documents. Avoid disorder by highlighting applicable and important provisions. Make sure that the inspector understands their content.

Inspection Tools and Instruments

Most tools and instruments used for inspecting tubulars and replacement parts under construction are also used for inspecting operating ones or disused units being considered for service. The tools and instruments used during construction are simple. Manufacturers use them to make sure that dimensions, thicknesses, locations, construction, etc. conform to the approved drawings and agreed-on specifications and standards. However, maintenance inspecting during outages or surveying mothballed or used equipment requires more elaborate tools and methods to assess the condition of the parts and to measure thicknesses. Here, the aim is to try to determine the present condition of previously accepted equipment which has deteriorated because of its service or exposure to the elements.

Table 1.1 suggests parts to be inspected and tools and instruments that should be available to the inspector. Those that apply mostly to shop inspections are marked S; those that apply mostly in the field are marked F; the ones that apply to both are marked SF.

TABLE 1.1 Parts to Examine vs Tools and Instrument to Use

	Thing to examine	What to examine it with
S	Baffle cuts	Scale
SF	Bolt elongation	Micrometer calipers; sonic micrometer
S	Bolthole orientation	Bolts used as pins in boltholes; level, square, and straight-edge
SF	Channel cover flatness	Straight-edge with mounted dial indicator (Fig. 1.1)
SF	Tubesheet flatness	
SF	Flange flatness	

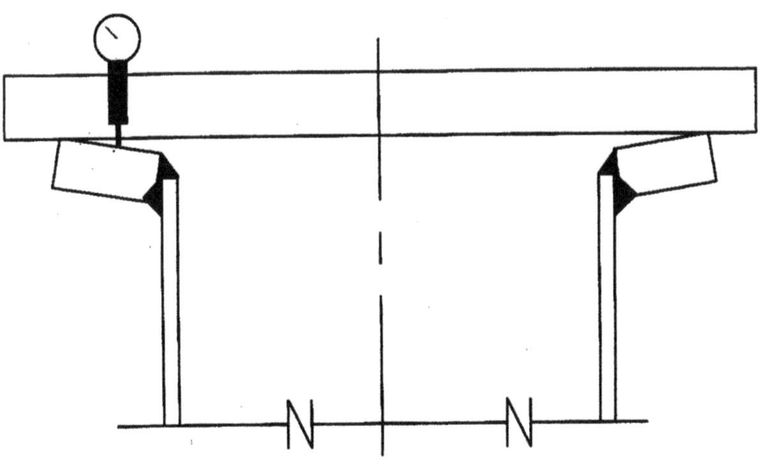

Figure 1.1 Measuring flatness. Illustration for flange cupping due to welding.

TABLE 1.1 Parts to Examine vs Tools and Instrument to Use (*Continued*)

	Thing to examine	What to examine it with
S	Surface finishes	Tactile comparison samples; tactile comparison gage; optical comparator
SF	Flange scalloping (distortion between bolt-holes)	Feeler gages for assembled flanges; straight-edge with mounted dial indicator for disassembled flanges
SF	Tubesheet ligament distortion	Internal dial calipers; strain gages
S	Nozzle and support angularity	Level, straight-edge, or square
F	Pass-partition bending	Scale, straight-edge with mounted dial indicator
SF	Tube bowing	High-intensity light
F	Tube condition	Borescope with videorecorder; eddy-current test system; magnetic resonance test system; pulsed ultrasonic beam inspection system; Internal Rotary Inspection Scan (IRIS)
SF	Tube diameter, ovality, and thickness	Hole gage; internal dial calipers; outside micrometer calipers
S	Tube holes in tubesheets	Go/no-go gage (Fig. 1.2)
SF	Tube-to-tubesheet joints	10-power magnifying glass; 30-power illuminated pocket microscope; bubble test kit; fluid penetrant examination kit; halogen leak sniffer (halogen diode tester) and accessories; helium leak sniffer (mass spectrometer) and accessories
S	Tubesheet-to-tubesheet parallelity	Steel measuring tape; strung piano wire; transit; split-image transit
S	Tubesheet rotation about longitudinal axis	Level, straight-edge, plumb line, and protractor
S	Tubesheet waviness	Straight-edge with mounted dial indicator
F	Tube-wall thickness	*Before installation*—micrometer calipers; *after installation*—IRIS, eddy-current tester
SF	Shell interior condition	Borescope and videorecorder; scrapers; pit gage
F	Shell expansion joint angular deflection and squirm	Trammels and scale, level and straight-edge
S	Weld size	Welding gage
SF	Nozzle weld condition	Magnetic particle examination kit

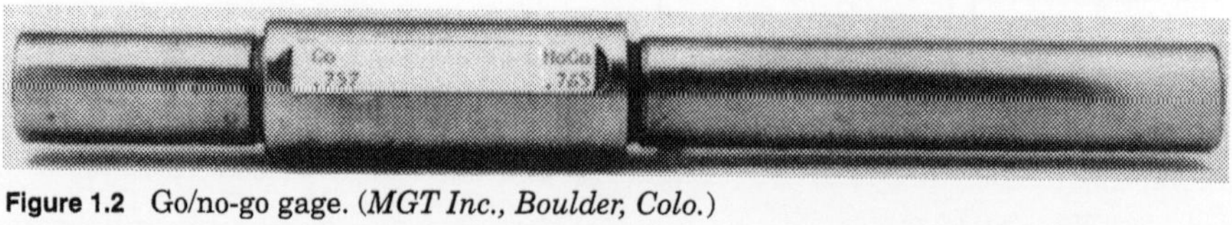

Figure 1.2 Go/no-go gage. (*MGT Inc., Boulder, Colo.*)

Table 1.2 is a list of tools that should be kept in well-organized toolboxes, for inspecting in the field. Most shop inspection departments will have just about all of these things on hand and make them available to the buyer's inspector. However, it is prudent to include a paragraph in the inspection brief that requires the shop to provide all tools and instruments necessary to permit complete inspection and to provide whatever assistance your inspector will need to do the work.

Table 1.3 lists equipment and instruments used primarily for maintenance inspections. Because there are so many possible ways to examine and inspect equipment in the field, the list is necessarily incomplete.

Inspection Reports and Data Collection

Make Inspectors' reports and the data collected part of the permanent maintenance file. Having full knowledge of what took place during the work is invaluable for troubleshooting causes of failure later on.

A model form for the inspector's notebook is suggested below. Tell the inspector to attach to the final inspection report the following documents:

- An as-built dimension drawing showing originally approved dimensions and marked to show all deviations.
- A weld seam location sketch.
- A location sketch of tubes plugged by the manufacturer or repairer, with supporting information about the type of plug used and the

TABLE 1.2 Toolbox Inspection Equipment for Shop Inspections and In-Service Equipment Inspections

Sulfur-free marking crayon
Low-voltage-powered drop light with extension (explosionproof where required)
Feeler gage set
Flashlight (explosionproof where required)
Inspector's hammer (sparkproof where required)
Lightweight ball peen hammer (sparkproof where required)
Hook gage to measure internal welds and cavities
10-power magnifying lens; 30-power illuminated pocket microscope
Inspection mirrors (jointed, magnifying, etc.)
Pit depth gage to determine depth of pit corrosion
Scrapers: steel for scraping unmachined surfaces; brass or copper for scraping machined surfaces
Wire brushes (use nonferrous or stainless where required)

TABLE 1.3 Larger Supplementary Inspection Equipment (Not Carried in Toolboxes)

Borescope set
Borescope videorecorder, keyboard, and console
Eddy-current testing system equipment
Fluid penetrant examination kit, special lamps (when required for method)
Halogen leak sniffer and accessories
Helium leak sniffer and accessories
Hydroblasting equipment
Magnetic particle testing equipment
Pressure-testing pumps, gages, and recorders
Trepanning, hole-drilling, and hole-plugging equipment
Telescopic transit
Split-image transit
Ultrasonic inspection apparatus (IRIS and pulsed types)
X-ray or isotope camera, shielding, radiation-zone exclusions, warnings, personnel radiation recorders and badges (dosimeters)
Videocamera (handheld)

plugging procedure. This should include sketches and descriptions of how broken tubes are stabilized in a bundle.

- A tube heat map when more than one heat of tubing is used (see Fig. 1.3). In addition to showing the location of the different heats of tubing, record the torque values or hydroexpanding pressures used for expanding the tubes in each heat.

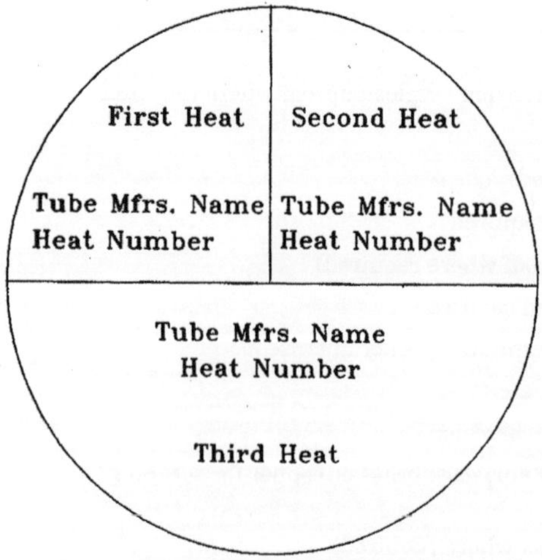

Figure 1.3 Illustration of a tube heat map.

- Certified copies of mill test reports and whatever documents the shop has used to identify where the material has been used. Your inspector should verify that the heat numbers are stamped or stenciled on all critical parts and correspond with the mill test reports. The mill test reports should be examined for conformity with applicable specifications.
- Rubbings of the nameplate, whether a manufacturer's or a repair or alteration one, or a legible print of a photograph of the nameplate(s).
- For units constructed to the ASME Code, a signed original of the U-1, U-2, A-1, and A-2 Manufacturer's Data Report forms and attached partial data reports, if any, or the NBIC form R Repair or Alteration Certificate. For exchangers constructed to other pressure safety codes, equivalent documentation.
- Records of nondestructive examinations.
- Records and reports of leak and hydrostatic tests.
- Manufacturer's Design Report (for ASME Code sections requiring such reports); copies of ASME Code calculations for exchangers constructed to Section VIII, Division 1, calculations required by any of the national pressure safety codes that apply.

Inspector's log

Require the inspector to maintain a bound, page-numbered notebook in which are recorded any deviations accepted in new construction and observations during other inspections. This is in addition to all other documents such as mill test reports, Manufacturer's Data Reports, nameplate rubbings, heat-treatment charts, and pressure-test records and charts.

Entries should be chronological, with a separate page for each item inspected on a given day. They should be indexed by item or tag number or whatever other numbers your organization uses to track or identify exchangers and parts. A model form for inspecting new construction is shown in Fig. 1.4. Instruct the inspector to photocopy each day's entries and to collate them by item. Make the collated photocopies part of the permanent inspection file.

Inspection Records as a Maintenance Tool

Inspection records are a basic maintenance tool. The collated records of new-equipment and parts inspection and the ones for inspection on-line and during outages are essential for troubleshooting the causes of failures and for planning repairs, alterations, and replacements.

SAMPLE COMPANY

Refineryville, Texas

Record of Inspection

Date_________ P.O. No._______________ Item No._______

Tag No.______ Manufacturer______________________________

Mfrs. Serial No.__________ Nat'l Bd or Reg. No.____________

Part Inspected	Accepted	Rejected	Comments
_______________	__________	__________	__________
_______________	__________	__________	__________
_______________	__________	__________	__________
_______________	__________	__________	__________

Deviations Accepted:

Figure 1.4 Suggested form for entries in inspector's notebook.

Records of new-equipment and replacement-parts inspections are basic to troubleshooting subsequent failures. For example, persistent leaks in expanded joints in one region of the tube field may sometimes be traced to the manufacturer's having failed to adjust the rolling torque from heat to heat of tubes expanded into the tubesheets. When this happens, if the torque was suitable for the yield stress of one heat, the tubes may be underrolled in a heat that has a higher yield strength.

Knowing the yield strengths of the various heats of tubing also is essential when it is necessary to reexpand hydraulically or by rolling tubes in which operation at high temperature has relieved the locked-in stress necessary for tightness and strength. Knowing the chemical composition of the tubes in each heat is also helpful when repairing welded tube-to-tubesheet joints. This information may also affect a decision about whether to attempt to seal-weld expanded tube-to-tubesheet connections that chronically leak.

The deviations record helps to find causes of failures and plan corrections. A deviation from approved drawings that was accepted because it appeared to be harmless may be the cause of trouble. Inlet nozzle placement deviations and deviations in the location of the first tube support or baffle may be the cause of tube erosion and tube vibration problems.

The record of tube-wall-thickness measurements made during outages guides maintenance in two ways: (1) it indicates when to plug tubes that have reached the minimum thickness suitable for the operating conditions; and (2) by plotting wall thicknesses against hours of operation, it permits predicting when to retube or replace the exchanger. Records of thickness measurements of channels, shells, and other parts subject to pressure can similarly be plotted against time of operation to predict when derating or replacement will be required. In recent years, several commercial organizations have developed effective, efficient systems for computerized record keeping.*

If the records show that paperlike cylinders of mostly copper chemical composition (called "snakeskins") were found on the inside of nonferrous tubes, anticipate tube failures and downstream problems. Such a finding in feedwater heaters might trigger checking on the boiler feedwater treatment system. Copper dissolved out of feedwater heater tubes may also deposit on the turbine blades, reducing turbine efficiency.

The following anecdote shows how an inspection record can be used to establish the cause of failures. A chemicals manufacturer had installed a deionized-water (DIW) feed heater. For sanitary reasons, the tubing was 18-8 austenitic, welded, and polished tubing, expanded into the tubesheets. The tube ends were flush with the face of the tubesheets, and the tube ends and surrounding ligaments filed to a gentle radius and polished to a finish acceptable to the sanitary code. To make sure that no steam leaked from the shell side into the DIW stream, at regular intervals the tube-to-tubesheet joints were bubble-tested with 345 Pa (~50 lb/in^2) nitrogen in the shell and commercial, chloride-free bubble former was spread on the tubesheet faces.

Shortly after one such test circular transverse cracks were observed in the tubes where they penetrated the rear face of the tubesheet and for a short distance beyond that point. An examination of the record revealed the comment that "soap flakes and tap water" were used as the bubble former because none of the commercial bubble former was on hand. Metallurgical analysis disclosed that the failures were the result of stress-corrosion cracking (SCC). Three conditions must be met for SCC to occur: (1) a material sensitive to SCC, (2) a high level of tensile stress in the material, and (3) an environment that promotes SCC. Austenitic 18-8 alloys are sensitive to SCC. The level of tensile stress in the region of the transition from expanded to unexpanded tube is in the order of 90 percent of the yield stress. The envi-

*Examples are Karta Technology Inc. of San Antonio, Texas and Anser, Inc. of Woburn, Massachusetts, both in the United States.

ronment was provided by the high chloride ion content of the tap water and soap. Without an examination of the record, all that could have been found by metallurgical analysis would have been the SCC and presence of high concentrations of chlorides. The source would not have been known.

References

1. *Pressure Vessel Inspection Code, Maintenance Inspection, Rating, Repair, and Alteration,* 6th ed., API 510, American Petroleum Institute, Washington, D.C., June 1989. Chapters 6 and 7 apply specifically to heat exchangers.
2. *Inspection of Pressure Vessels, API Recommended Practice 572,* 1st ed., American Petroleum Institute, Washington, D.C., Feb. 1992.
3. *The ASME Boiler and Pressure Vessel Code,* the American Society of Mechanical Engineers, New York, 1995. A new edition of the ASME Code is published every 3 years, with addenda published in intervening years. Interpretations and Code Cases address questions that the rules raise.
4. *Standards for Closed Feedwater Heaters,* 5th ed., Heat Exchange Institute, Cleveland, Ohio, 1992.
5. *Standard for Power Plant Heat Exchangers,* 2d ed., Heat Exchange Institute, Cleveland, Ohio, 1990.
6. *Standards for Steam Surface Condensers,* 8th ed., Heat Exchange Institute, Cleveland, Ohio, 1984 and 1989 Addendum.
7. *The National Board Inspection Code,* The National Board of Boiler and Pressure Vessel Inspectors, Columbus, Ohio, 1995. A new edition of the National Board Code is published every 3 years.
8. *Standards of the Tubular Exchanger Manufacturers Association,* 7th ed., the Tubular Exchanger Manufacturers Association, Tarreytown, N.Y., 1988. New editions are published at approximately 10-year intervals.

Supplement 1.1: African, Asian, European, Near Eastern, and North American Nondestructive Testing Societies and Related Organizations*

Argentina	Centro Argentino de Ensayos no Destructivos de Materiales (CAND), Santiago del Estero 250-OF. 31, 1075, Buenos Aires, Argentina
Australia	Australian Institute for Nondestructive Testing, PO Box 334, Box Hill, Victoria 3128, Australia; phone (03) 898 7437
Belgium	BANT, c/o Katholieke Universiteit Leuven, De Croylan 2, B-3030 Heverlee, Belgium, fax 016 20 79 95

*This list has been abstracted from *Materials Evaluation,* **50**(2) (Feb. 1992).

Bulgaria	Bulgarian Society for NDT, Rakovski St. 108, PO Box 431, 1000 Sofia, Bulgaria; phone 89 8379
Canada	Canadian Society for Nondestructive Testing (CSNDT), Unit 47, 2400 Lucknow Dr., Missisauga, Ontario L5S1T9, Canada; phone 416 676 0785; fax (c/o Newman Hattersly Ltd.) 416 673 9225
CEN	European Committee for Standardization, Rue de Stassart, 36-1050 Bruxelles, Belgium, phone 32 2 519.68.11; fax 32 2 519.68.19; telex 206 2210097
China, People's Republic of	Chinese Society for Nondestructive Testing, c/o Nondestructive Testing Institute of the Chinese Mechanical Engineering Society, 100 Huihe Rd., Shanghai 20043, China; phone 021 5420775, ext. 483; fax 021 5420554; telex 33275CASS CN
China, Republic of (Taiwan)	Chinese Society for Nondestructive Testing, 130 Keelun Rd., Section III, Taipei, Taiwan, 10771 ROC; phone 02 3630815
Czech Republic	Czech Society for Nondestructive Testing (CSS NDT), CS 113 12 Praha-1, Opletalova 25, Czech Republic; phone 740751, 735351-linka 447 nebo 2907
Denmark	Force Institutes, Park Alle 345, DK-2605 Broenby, Denmark; phone 45 42 96 88 00; telex 3 33 88 SVC DK
Egypt	Egyptian Society for Industrial Inspection, 3A, Moaskar Romani St., Rushdi, Alexandria, Egypt; phone 859749-964888; telex 55033 DEVCO UN
Finland	Finnish Society for NDT, Mertallimiehnkuja 10, 02150 Espoo, Finland; phone 3580-4566477
France	Confederation Française pour les Essais Non Destructifs (COFREND), 32 Boulevard de la Chapelle, 75882, Paris Cedex 18, France; phone 1 46079405; OFISOUD 210335F
Germany	Deutsche Gesellschaft für Zerstörungsfreie Prufünge (DGZfP) Berlin 45, Germany; phone 030 8114001; fax 030 8114003; telex 181 788DGZFPD
Greece	Hellenic Society for Nondestructive Testing (HSNT), Sofokleous 89 St., 166 73 Voula, Athens, Greece; phone 8950880, 8956903

Hungary	Scientific Society of Mechanical Engineers, Kossuth, Lajos Ter 6-8 Budapest V., Hungary
ISO	International Organization for Standardization (ISO), case postal 56, CH 1211, Geneva 20, Switzerland; phone 41 22 34 1240; telex 23887 ISO CH
India	Indian Society for Nondestructive Testing, E-1, 2d floor, J. P. Tower Nungambakkam High Rd., Madras 600 034, India; phone 86 59 84; fax 91 44 836644
Israel	Technion, Israel Institute of Technology, 32000 Haifa, Israel; phone 972 429430/23077; fax 972 4221581
Italy	Italian Society for Nondestructive Testing, via Foresti 5, 25126, Brescia, Italy; phone 030 391716; fax 030 392156
Japan	Japanese Society for Non-Destructive Inspection (JSND), 5-4-5 Askusabashi Taitoh-Ku, Tokyo, 111, Japan
Korea, Republic of	Korean Society for Nondestructive Testing, c/o KID 1473-10, Seocho-dong, Seocho-ku, Seoul, 137-070 ROK; phone 02 582 2743/583 7564; fax 02 582 2743
Philippines	Philippine Society for Nondestructive Testing, Inc., c/o Philippine Nuclear Research Institute, Commonwealth Ave., Diliman, Quezon City, Metro Manila, Philippines; phone 976011 12 13 14 15; fax 632 817 9870
Singapore	Singapore Institute of Standards and Industrial Research, 1 Science Park Dr., Singapore 0511, Republic of Singapore
South Africa	South African Association of Nondestructive Testing, PO Box 670, Bergvlei 2012, Republic of South Africa
Netherlands	KINT Dutch Quality Surveillance and Non-Destructive Testing, PO Box 390, 3330 AJ Zwijndrecht, Netherlands; phone 078 19 26 55
Norway	Norsk Forening for Ikke-Destruktiv Proving, c/o Robit Training, PO Box 100, N-136 Billingstadsletta, Norway; phone 02982501; fax 02982333
Romania	Romanian Association for Nondestructive Testing (ARoEND), Str. Frumoasa 26, Bucharest 78116, Romania; c/o fax 90 657896

Russia	Russian Society for Nondestructive Testing and Technical Diagnostics (SSNTTD), 35 St. Usacheva, Moscow, 119048, Russia; phone 245 5656; fax 245 5050; telex 54 RUBIN SU
Spain	Associación Españ de Ensayos No Destructivos (AEND) (Spanish Association for Nondestructive Testing), Zurbano 92 10 D 28003 Madrid, Spain; phone 441 77 33
Sweden	Swedish Society for NDT, FOP Kansli, Materialnormcentralene, Box 4094, S-102 62 Stockholm, Sweden; phone 46 8 644 93 75; fax 46 8 640 87 25
Switzerland	Swiss Society for Nondestructive Testing (SSNT) [SVMT/SGZP], c/o EMPA Dübendorf, Überlandstrasse 129, CH-8600, Dübendorf, Switzerland; phone 01 8234411; fax 01 8216244; telex 825345
Thailand	Thai NDT Society, Office of Atomic Energy for Peace, Vibgasadee-Rungsit Rd., Bankehn, Bangkok 10900, Thailand
Tunisia	Permanent National Committee for Nondestructive Testing and Control, Technical Center of Mechanical and Electrical Industries, 1 Rue, 7037 Menzah IV, Tunisia
United Kingdom	British Institute of Non-Destructive Testing, 1 Spencer Parade, Northampton NN15AA, UK; phone 0604 30123; telex 31612; fax 0604 231849
United States	American Society for Nondestructive Testing (ASNT), 1711 Arlingate Lane, PO Box 28518, Columbus, OH 43228-0518, USA; phone (614) 274 6003, toll-free in USA (800) 222-2768; fax (614) 274-6899; telex 245347

Chapter

2

Inspecting in the Heat Exchanger Factory

Introduction

We address this chapter to engineers who prepare inspection briefs for inspectors and who may themselves sometimes perform inspections. One of its objectives is to provide insight into inspectors' work and what they encounter when inspecting new tubular heat-transfer equipment, replacement parts, and factory retubing and repairs. Another is to give guidelines for preparing inspection briefs. We discuss inspection hold points, review safety requirements, and make suggestions about pre-shipment cleaning and inspection of exchangers to be factory-repaired or -retubed. We provide checklists with commentaries for inspecting during construction of new exchangers and replacement parts and for inspecting before and during factory alterations, retubing, and repairs. The chapter closes with a brief discussion of pressure testing.

Inspection Briefs

Previously, we emphasized the importance of reviewing inspection briefs at preinspection conferences. Here we discuss their purposes, the circumstances for which they should be written and how to write them.

Purposes of inspection briefs

The purposes of inspection briefs are to

- Describe what to inspect and examine
- State when to inspect and establish hold points

- Address the inspector's safety and health
- Affirm rules and procedures to be followed in performing inspections agreed on between purchasers and manufacturers and suppliers
- Provide acceptance/rejection criteria
- Establish guidelines for resolving disagreements about acceptability of work
- State how to record inspection results uniformly and in a way that facilitates subsequent maintenance
- Specify data and reports to acquire and make part of the maintenance file
- Familiarize the inspector with the tools and instruments to be used
- Prepare for examining shutdown equipment by stating what to inspect and the inspection methods and devices to use
- Provide criteria for interpreting the results of nondestructive examinations
- Implement an order for inspection

Issue briefs as much in advance of the initial inspection as possible. To make sure that the supplier is fully aware of your inspection requirements, include a copy with the purchase specifications whenever possible. Provide copies to inspectors and to the engineers responsible for resolving discrepancies or dealing with deficiencies that the inspector uncovers well before the onset of construction.

When you delete a paragraph from a manufacturer's copy to maintain confidentiality or secrecy, substitute the following statement: "This paragraph has been deleted from seller's copy of this document because of confidentiality requirements." Leave the paragraph numbering intact.

Write briefs for the following:

- Inspecting exchangers under construction
- Inspecting manufacture of replacement parts
- Inspecting mothballed or previously used equipment being considered for service
- Inspecting in-service equipment during planned outages
- Inspecting during unplanned outages

Organization and style

Do not include criteria for accepting or rejecting materials and construction in the body of the inspection brief. However, refer to the documents that have the criteria (usually the purchase specification, applicable manufacturers' standards, drawings, and construction code). This will avoid unnecessary paperwork and the possibility of creating an inadvertent conflict between the inspection brief and purchase documents.

Use the format your organization uses for specifications. Write concise, descriptive captions. The following outline, typical for exchangers under construction, can be adapted with a few modifications for replacement parts, repairs and alterations, in-service units, and used and mothballed units.

1. Intent
2. Scope
3. Specifications, codes, standards, and drawings
4. Safety
5. Tools and instruments
    5.1. Company-provided
    5.2. Manufacturer-provided
6. Manufacturer's assistance to inspector
7. Reports and documents
8. Nameplate information
9. Materials of construction
    9.1. Materials specifications
    9.2. Materials documentation and tracking
10. Bundles
    10.1. Tubesheets
        10.1.1. Tube holes
        10.1.2. Annular grooves
        10.1.3. Weld-joint preparation
        10.1.4. Packed floating-head skirts
    10.2. Baffles and tube supports
    10.3. Tie-rods and spacers
    10.4. Sealing devices, dummy tubes, and seal strips
    10.5. Shell-side pass dividers and shrouded passes
    10.6. Tubes
        10.6.1. Maximum permissible residual stress
        10.6.2. Cleanliness requirements
        10.6.3. Condition of tube ends
        10.6.4. U-bend requirements
            10.6.4.1. Postbending heat treatment
        10.6.5. Straight-tube loading
        10.6.6. U-tube loading
        10.6.7. Heat maps
    10.7. Tube-to-tubesheet joints
        10.7.1. Hole and tube-end preparation for tube loading

10.7.2. Tube expansion process
10.7.2.1. Tube expansion parameters
10.7.2.2. Expanded joint production method
10.7.2.3. Quality control of tube expansion
10.7.3. Welding procedures and weld quality
10.7.3.1. Weld-joint examination
10.7.4. Nondestructive examinations
10.7.5. Tightness requirements and testing
10.7.6. Location of shop-plugged tubes
10.8. Impingement protection
10.8.1. Size and escape area requirements
10.8.2. Attachments to bundles and shells
10.9. Bundle assembly
11. Channels and bonnets
11.1. Handling fixtures
11.2. Pass partitions
11.3. Flanges
11.4. Gaskets
11.5. Bolting
11.6. Channel and manway covers
11.6.1. Cover handling fixtures
11.6.2. Diaphragm seal plates
11.7. Channel and bonnet flanges
11.8. Channel and bonnet gaskets
11.9. Channel and bonnet bolting
12. Return covers
12.1. Turnaround space
12.2. Pass partitions
12.3. Return-cover flanges
12.4. Return-cover gaskets
12.5. Return-cover bolting
13. Shells
13.1. Roundness and fit-up
13.2. Rolled-plate shells
13.2.1. Fit of courses
13.2.2. Girth welds
13.2.3. Longitudinal welds
13.3. Pipe shells
13.4. Shell flanges
13.5. Shell expansion joints
13.5.1. Thin-walled expansion joints
13.5.2. Thick-walled expansion joints
13.6. Saddles, lugs, trunnions, skirts, supports, and appurtenances

- 13.7. Nameplate brackets
- 13.8. Warning-tag brackets

14. Shell covers
    - 14.1. Shell-cover flanges
    - 14.2. Shell-cover flange gaskets
    - 14.3. Shell-cover bolting
15. Nozzles and connections
    - 15.1. Location and position
    - 15.2. Flange finish
    - 15.3. Weld-end preparation
    - 15.4. Attachments and reinforcements
    - 15.5. Pressure and temperature taps
    - 15.6. Load capability
16. Assembly
    - 16.1. Bundle insertion and withdrawal
    - 16.2. Match marks on flanges, tubesheets, and covers
    - 16.3. Pressure testing
    - 16.4. Tightness testing
    - 16.5. Drying and cleanup
        - 16.5.1. Nitrogen blanket
    - 16.6. Preparation prior to coating and painting
    - 16.7. Preparation for shipment
17. Documentation

Where possible, incorporate illustrative photographs or sketches or reproduced segments of drawings in the text. Guard against legalistic jargon and phrases. The goal of inspection briefs is to communicate, not obscure meanings. Advance this end by using simple, straightforward, declarative statements. Here is how you might write text for paragraph 16.1, "Bundle insertion and withdrawal":

> Make sure you watch stabbing the bundle into the shell. It should go in smoothly, without too many hangups. It is acceptable for the shop to use air or hydraulically operated bundle extraction-installation machines if they do not apply so much force as to damage the baffles or tube supports. This means that if the bundle jams while going into the shell, the machine must be stopped until the cause of the jam is removed. It is also permissible to use matched pairs of come-alongs, hydraulic jacks, or homemade systems. The important thing is that the baffles and tube supports must not be damaged or hang up as the bundle moves in or out of the shell. It is allowable to apply a vibrator to the shell to make it easier to slide the bundle in.
>
> Unless there is a cleaning procedure that will positively remove any material used to lubricate the baffle or support edges (or bundle skids for bundles built with them), lubrication is not permissible. If the manufacturer intends to use lubricant, discuss it with the factory's quality con-

trol (QC) manager. If you cannot come to an amicable agreement, ask to hold off on installation while you telephone the engineering department for instructions. This will give us a chance to talk with the plant management and decide how to proceed.

Inspection briefs for new construction

In general, follow the outline suggested above. Comments on a few of the outline's paragraphs follow.

Intent. Use the intent statement to convey (1) your intentions about the kinds of inspections to be carried out before, during, and on completion of construction; (2) the numbers and times of inspections; and (3) your intentions about acceptance and rejection. A carefully written intent statement will eliminate legalistic nitpicking and cover minor omissions, misspellings, improper punctuation, etc. A typical intent statement for new construction is as follows:

> It is the intent of this inspection brief to make clear to all concerned persons that inspections will be conducted as required by Sample Company's purchase order no. XXXX and purchase specification no. YYYY. This means that Sample Company will inspect to verify that all construction conforms with its *General Specification for Pressure Vessels* no. GP-100; its *Mechanical Specification for Tubular Heat Exchangers Used in Refinery Service* no. H-101; the *TEMA Standards,* 7th ed.; Class R and API Standard no. 660, *Shell-and-Tube Heat Exchangers for General Refinery Services*, 4th ed, except that references in API 660 to the 6th edition of the TEMA Standards shall be replaced by the equivalent paragraphs in the 7th edition.
>
> We specifically intend Sample Company's inspector to inspect the following:
>
> 1. Raw materials, including witnessing of mill tests at their source and documentation and tracking of materials used in construction of our heat exchanger(s).
> 2. Parts before assembly into bundles, channels, shells, and other subassemblies, including nondestructive examinations where required by the construction code or listed on the manufacturing drawings. This includes channel, shell and shell-cover flanges, return covers, split rings, baffles and tube supports, and any other part that can affect the operation of the heat exchanger or its useful life.
> 3. All subassemblies and pressure tests of subassemblies.
> 4. Installation of subassemblies into the heat exchanger, especially insertion of the bundle into the shell.
> 5. All leak and pressure tests of the channel, bundle, shell, and assembled heat exchanger.
> 6. Draining, cleanup, surface finish, and preparation for shipment.
>
> We further intend that the Sample Company inspector and the manufacturer agree on the number of inspection points necessary for adequate mon-

itoring of fabrication progress. These are categorized as (1) routine inspections that the manufacturer's QC inspectors ordinarily perform; and (2) hold-point inspection required by the purchase specification or which may otherwise be agreed on by the manufacturer and the Sample Company inspector. Manufacturer's routine QC inspections include radiography, dimensional checking, verifying use of correct welding procedures by the workforce, and various nondestructive examinations. Typical hold points are at the time of tube-hole examinations, when tube-to-tubesheet joint tests are performed and when bundles are inserted into shells.

The manufacturer may carry out routine inspections that the purchase specification do not designate as hold points when the work is ready.

The manufacturer shall notify Sample's inspector when it is anticipated that a hold point will be reached in ample time for scheduling inspection. The manufacturer shall not proceed with work beyond the hold point until Sample Company's inspector has witnessed or performed the inspection unless Sample's inspector has provided the manufacturer with a written release to proceed. Sample Company shall agree with the manufacturer that if inspection is not made within the agreed-on notification time, the manufacturer may proceed and will not be required to halt production for the missed inspection.

Scope. Write the scope statement to set limits on what is to be inspected. Use short, declarative, unambiguous statements. Here is a typical one:

This inspection brief applies to all TEMA Class R tubular heat exchangers purchased on Sample Company's purchase order no. GGGGG-HHH. It covers raw materials, parts, subassemblies, assembly, and testing of these exchangers. It does not cover inspecting other tubular or nontubular heat-transfer equipment.

Reference documents. Paragraph 3, "Specifications, codes, standards, and drawings," of the outline is especially important for new construction and replacements. It covers the details of agreed-on configurations, sizes, materials, construction methods, pressure-vessel code requirements, stamping, and documentation criteria for acceptance or rejection. These are the documents that form the basis for the inspector's work. Include as many of these items as possible:

- Purchase specifications
- Approved setting plan
- Approved assembly drawings
- Approved detail drawings
- Manufacturers bill of materials keyed to mill test reports for all critical parts

- Qualified procedures and manufacturer's procedures not requiring qualification listed on manufacturing drawings, but not requiring user's approval

Safety. Many organization have regular safety meetings and programs to ensure the safety and health of their personnel in their own facilities. However, they may not have addressed safety requirements for inspecting in a heat exchanger factory or inspecting previously used exchangers for which formal safety requirements may be nonexistent.

Safety consciousness and requirements vary from factory to factory. Some places require wearing hard hats, safety shoes, safety goggles, gloves, hearing protection, face masks in dusty areas, long-sleeved shirts and full-length trousers, etc. Others observe no precautions. For these reasons, it is prudent to spell out the full list of safety precautions that the inspector must follow and to include references to governmental safety standards and your company's rules. Safety requirements are discussed in more detail below.

Tools and instruments. Inspectors cannot carry all the tools and instruments listed in the tables of Chap. 1. When the design of an exchanger or the specifications for its construction require testing and measurement with instruments not usually in inspectors' possession, the inspection brief should clearly state that the manufacturer shall have and make available such tools and instruments and the personnel skilled in their use.

Manufacturer's QC manual. When possible, obtain a copy of the manufacturer's QC manual before writing the inspection brief and refer to the paragraphs that affect the quality of the exchangers.

Hold points. It is essential to establish, before construction begins, the points at which the manufacturer must stop construction to enable the inspector to inspect parts and assemblies. The inspection brief should detail the hold points.

Inspector's log. Make sure to describe fully the information the inspector must include in this document.

Inspection briefs for inspecting replacement parts

Briefs for inspecting replacement parts are similar to those for new construction. However, they must include precautions to make sure that the replacement parts fit precisely into tubulars with the same clearances as the original. It is dangerous to rely on the nominal dimensions shown on the original manufacturer's drawings when buying a replacement bundle. The replacement may not fit into the shell, or the clearances between the bundle baffles and shell inside diameter (I.D.) may be too large for efficient shell-side sensible-heat transfer. Oversized baffles of replacement bundles may be ground down so the bundle fits into the shell, provided care is taken not to damage the peripheral tubes and to make the fit uniform. But nothing can be done to correct an excessive bundle-to-shell clearance too large to achieve efficient shell-side heat transfer short of building a replacement shell to fit the bundle. Therefore, make sure to provide the manufacturer and inspector with as-built or measured dimensions of the parts being replaced or the parts into which they are to fit. When there are no as-built dimension drawings, plan ahead to allow enough time to take the necessary measurements. The time required may be as long as a year!

Purchase specifications for stack-drilled baffles or stack-drilled tubesheet-baffle assemblies should require the manufacturer to number the baffles from topmost to bottommost in the stack and to label the tubesheets' front and rear. The purpose is to minimize the effect of drill drift. The brief should instruct the inspector to make sure that the works follows this practice.

Purchase specifications should require the factory to demonstrate roundness of replacement shells by means of templates passed through the shell.* The inspection brief should restate this requirement.

The brief should require checking to make sure that serial numbers of replacement pressure parts are permanently stamped onto the parts and match the ones shown on the manufacturer's documentation.

Inspection briefs for inspecting in-service units

Use an in-service inspection brief to focus the inspector's attention on the following: (1) evidence of deterioration with use, such as tube-end thinning, ligament cracking, or tube-hole out-of-roundness; (2) widespread severe tube metal loss in feedwater heaters; (3) prospective requirements for maintenance, such as a group of failures in a discrete location in a tube nest or evidence of pass-partition bypassing; and (4) indications of conditions that can affect performance, such as excessive scale or fouling.

---

*A suitable template is one that has at least two 12.5-mm (~0.5-in)-thick full circles of the same diameter of the baffles set 305 mm (~12 in) apart, welded concentrically on an approximately 60-mm-diameter pipe shaft [2 in iron pipe size (IPS)].

Inspection briefs for inspecting repairs and alterations

Almost always, repairs or alterations to pressure parts will be made to either the original pressure safety code rules or those of an inspection code. Both the NBIC and API-510 apply by reference the most suitable edition of ASME Code rules. The authority responsible for on-site inspection and the regularly assigned Authorized Inspector or Authorized Nuclear Inspector (ANI) who inspects on site must be fully apprised of the proposed work. The AI or ANI will set the kind and number of inspections to be performed to satisfy code requirements. However, in all likelihood they will not concern themselves with dimensions and tolerances that do not affect the pressure-bearing capability or safety of the repaired or altered structure. With this in mind, write the inspection brief to cover verifying dimensional accuracy and documentation of the work.

If the work is to be done on site, your organization's safety engineer should provide instructions concerning safety for incorporation in the brief.

Inspection briefs for inspecting used and mothballed units

Full information is rarely available about hazardous, noxious, or lethal contaminants in previously used exchangers. Therefore, discuss with the safety department which precautions to include. The brief should instruct the inspector not to inspect unless chemical, sand- or water-blast precleaning has been done and Materials Safety Data Sheets (MSDSs) have been issued.

Briefs for inspecting used and mothballed exchangers should concentrate on serviceability and should include

- Examining for cleanliness and estimating the degree of tube interior and exterior fouling.
- Measuring the degree of metal wall loss and corrosion of such parts as the shell, channels, tubes, and tubesheets.
- Checking for gross damage such as flange and tubesheet deformations; fatigued, squirmed, dented, or blown-out bellows expansion joints; and fatigued thick-walled joints.
- Inspecting for previously performed alterations and repairs that could affect usefulness—this includes comparing repairs or alterations with nameplate and repair and/or alteration certificate data.
- Examining conditions of tubesheet ligaments.
- Testing tube-to-tubesheet joint integrity.
- Pressure testing for integrity of the pressure envelope.

Safety Precautions for Entering Closed Spaces

Some hazards of entering closed spaces are obvious; others are not. Do not leave the determination to inspectors' judgment. Apply the more restrictive of prevailing governmental safety requirements for entering closed spaces or your company's safety regulations. *Caution:* No one should enter an enclosed space unless (1) the atmosphere is sampled to determine that there is sufficient breathable air free from harmful substances, or the person entering is provided with a self-contained source of breathable air fully isolated from the atmosphere in the closed space and is clad in protective clothing; (2) the temperature of the closed space and its containment does not exceed those listed in the applicable set of safety standards; (3) one is provided with and wears a safety harness that will permit one to be pulled out quickly in the event of a problem; (4) a safety watch of personnel, trained to effect rescue and apply first aid, is constantly present, and is in communication with the one in the closed space and equipped to render assistance if required; (5) first aid and emergency procedures and practices have been established and drilled into the concerned people; and (6) an authorization and record-keeping procedure for monitoring safety is in effect.

Safety When Inspecting in the Factory

One cannot rely on all manufacturers to provide safe conditions for inspecting. Busy works may be so crowded that there are either no well-defined aisles or work encroaches on safety aisles. They may not provide curtains to prevent welding-flash eye injury. Manufacturers may pressure-test in the open without protective shielding in the event of a blowout or fracture. Following are recommendations for securing the safety of the inspector.

Signing in. Require logging into and out of the inspection sites to permit tracking the inspector's whereabouts. Compelling an inspector to sign in on visiting the works is a safety measure, not a courtesy. It tells the management that a caller is on the premises and provides some assurance that if danger or a hazard is anticipated, the visitor will be alerted. If the facility does not have a formal log-in procedure, require your inspector to telephone on arriving at the site and just prior to leaving it.

Do not take for granted that the inspector will follow safe procedures. Include appropriate requirements in the inspection brief to address the hazards of witnessing pressure tests, being present during radiography and when entering closed spaces such as channels, long shells, or reboiler bodies through manways.

Wearing protective clothing. Many works do not require wearing some or all of the following: hard hats; hairnets for long hair; safety shoes; noise protection; long-sleeved shirts; and safety glasses with side shields or protective goggles. Because some factories do not supply safety clothing, require inspectors to have and use their own personal safety gear.

Following the works' safety rules. Briefs should require inspectors always to follow the factory's safety rules. If a formal set of rules is not posted or made available in writing, the brief should instruct the inspector to inquire about the exchanger shop's safety policies and rules.

Personally judging risks. Most veteran inspectors have learned from experience when their health or safety is threatened by conditions in the factory. Inspectors must have full freedom to exercise and act on their personal judgment of risks. Briefs should include a provision that permits an inspector who feels endangered by unsafe practices to withdraw until the hazardous procedure is corrected. Require any inspectors who believe the conditions to be unsafe to notify the manufacturer and the individuals to whom they report.

Safely observing pressure tests. Although witnessing pressure tests falls within the Authorized Inspector's purview, many inspection briefs require the purchaser's inspector to be present. A basic concern is that the test not be conducted at temperatures where the ductility of the parts is so low that brittle fractures can occur. Inspection briefs should require inspectors to make sure that the pressure-vessel code's rules for raising the pressure are exactly followed and its limits precisely observed when tests with inert compressed gases or compressed air are performed. The brief should clearly state that under no circumstances should an inspector remain present if an air or inert-gas pressure test is performed at a pressure higher than the maximum allowable working pressure stamped on the nameplate.

Inspecting Factory Repairs, Retubing, or Alterations

The kind and degree of inspection for alterations, retubing, and repairs depends on the type of heat exchanger and the extent of the work to be done. The condition of an exchanger can be assessed more fully in the factory than can conveniently be done elsewhere. It can be determined if repairs, refurbishing or replacements not initially planned for are required. During factory retubing, it may be found that parts intended for reuse require repairs or replacement to restore the exchanger to good condition. Inspecting as the work is done and on completion of repairs, alterations, or retubing helps ensure that the required quality is achieved. Following are suggestions for inspecting factory alterations, retubing, and repairs.

Preparations before shipment

Factories usually demand owners or users to make some preparations before they ship the exchanger. Here, the owner's interests coincide with the factory's. For both, it is helpful to have at hand as much information about the exchanger's construction and maintenance history as possible. Such information is usually necessary to permit doing the work: (1) to satisfy the requirements of the prevailing inspection code; (2) to satisfy the governing political jurisdiction; (3) to facilitate the work by providing details of construction, alterations that may have been made, and previously performed repairs and retubing; and (4) to facilitate inspection. Assemble the following:

- Nameplate data
- Heat exchanger, feedwater heater, or condenser specification sheets
- Manufacturer-supplied drawings, procedures, and records; also reports and code documentation
- Records of alterations and repairs
- Records of previously performed maintenance

A complete set of manufacturer-supplied documents should include

- Setting plan
- Assembly drawings with as-built dimensions
- Detail drawings of the components as built
- Procedure specifications
- Records of nondestructive test results and interpretations
- Mill test reports
- Manufacturer's data report form

Preshipment cleaning. Preshipment cleaning is necessary for the safety of the exchanger shop's workforce and to permit a complete inspection of the exchanger to determine its condition and requirements for repair of its components. Before cleaning, completely remove all plugs (whether welded, driven, expanded, etc.) from the tubes. This is especially important when the tubes may be sawed or burned. Make sure that the tubes are not blocked, either by sighting down straight tubes or blowing air through U-tubes. This is because—despite the best efforts to clean the exchanger—some tubes may remain plugged with foreign matter. Some process materials partially polymerize into plugs of solid matter separated by regions of flammable monomer. Heat developed by cutting saws, or applied when using burning torches to cut apart the unit, may ignite the monomer, causing jets of flame to shoot out. Taper plugs driven into the ends of tubes in which monomer is entrapped may be propelled with considerable velocity when the fluid is ignited.

Fully clean both sides by chemical cleaning, solvent washing, sand or grit blasting, hydrocleaning, or a combination of methods. The purposes of preinspection and preshipment cleaning are to (1) protect workers from exposure to hazardous, noxious, and irritating substances resident in the equipment or on its surfaces; (2) expose the bare surfaces to view; and (3) prevent uncontrolled dispersing of undesirable materials into the environment.

Be aware of and take appropriate measures to deal with the hazards that each cleaning method presents. If on-site cleaning facilities are not available, employ the services of a professional cleaning service. When possible, disassemble and inspect the exchanger before it leaves your facility.

Even carefully cleaned exchangers may not be free of residual hazardous substances. Therefore, provide the factory doing alterations, repairs, or retubing with the information required for protecting its workforce from exposure. Manufacturers doing retubing, repairs, and alterations are well advised to require owners to provide certificates stating that the exchangers to be worked on are free of hazardous substances.

Preinspect removable-bundle exchangers to determine whether the factory must manufacture and install a new set of replacement baffles and supports, tie-rods and spacers, and other components. Similarly, evaluate whether the tubesheet(s) can and should be replaced or reused. If it is decided to reuse the tubesheet(s), assess the extent of refurbishment required. Examine the condition of the other components to ascertain if repairs are feasible and worthwhile. Following are recommendations for inspecting various types of exchangers before shipment for factory rework.

Floating-head exchangers. The principal parts of floating-head equipment to look at are

- Channel (or bonnet)
- Channel cover

- Shell
- Return cover
- Floating head
- Tubes
- Tubesheets
- Tube-to-tubesheet joints
- Baffles and/or tube supports and end support plate

U-tube exchangers. Where possible, examine these parts of U-tube units:

- Channel (or bonnet)
- Tubesheet
- Shell
- Tubes
- Tube-to-tubesheet joints
- Baffles and/or tube supports and end support plate(s)

Fixed-tubesheet exchangers. Preshipment inspection of the shell sides of fixed-tubesheet exchangers is limited to what can be seen externally. The tube side can be fully inspected by removing the bonnets or channels and channel covers. Use API 520, *Pressure Vessel Inspection Code* and API 572, *Inspection of Pressure Vessels* as guides to inspecting the various parts.

If time does not permit a preshipment inspection, try to inspect after the unit is disassembled at the heat exchanger works but before any further work is done. The factory will make its own inspection to determine what is reusable, what must be replaced, and what can be restored.

Inspection Checklists

Following are two inspection checklists with guiding comments. The first is for inspecting during construction of new exchangers and replacement parts; the second is for inspecting before and during factory alterations, retubing, and repairs. Checklists without comments are provided in Supplement 2.1 for inspecting during construction of new exchangers and replacement parts and Supplement 2.2 for inspecting before factory alterations, retubing, and repairs.

After completing all inspections listed, the inspector should verify that all parts have been fitted and are assembled without undue distortion or damage during assembly. Except for bundles built in the shell, the joining of the bundle to the shell is another critical manufacturing operation that should be a required hold point to permit the purchaser's inspector to witness. This is so regardless of whether the bundle is inserted into the shell, as is customary for the vast majority of process heat exchangers; or the shell is pulled over the bundle, as is common in feedwater heater construction. In either case the shell must be round.

Inspection checklist for new construction and replacement parts

☐ CHECK RAW MATERIALS FOR DIMENSIONAL CONFORMITY

Comment. Check raw materials in the shop in the following ways:

- ☐ Measure to verify that minimum thickness requirements have been met.
- ☐ Measure the out-of-roundness of pipes and subcontractor-supplied rolled-and-welded cylinders. *Out-of-roundness* is expressed as $100(D_{max} - D_{min})/D_{nom}$, where D_{max} is the largest inside diametral measurement, D_{min} is the smallest inside diametral measurement, and D_{nom} is the nominal shell inside diameter. *Note:* The ASME Code's permitted 1 percent of diameter out-of-roundness for cylindrical shells is not acceptable for tubular heat exchangers that must conform to the requirements of the HEI *Standard for Power Plant Heat Exchangers* or the TEMA Standards.

☐ INSPECT RAW-MATERIALS SURFACES

Comment. Check raw-materials surfaces for the following:

- ☐ Blemishes
- ☐ Mill scale removal (when specified in purchase documents)
- ☐ Scabs and spalls
- ☐ Edge laminations
- ☐ Surface profile of sand- or grit-blasted parts as specified
- ☐ Surface smoothness and finish as specified

Comment. Comparison samples may be adequate. Here, the shop finishes a sample of material to a surface acceptable to the user. The specimen is cut in half. The inspector uses one-half for tactile comparison; the shop uses the other. Alternatively, commercially available plastic surface finish comparison gages may be used:

- ☐ Excessive internal bead crown height of welds in welded tubing
- ☐ All tubes clean and free of longitudinal and spiral scores, dents, and crimps
- ☐ Holes in tubesheets meeting specified finish and free of axial or spiral scores
- ☐ Tube-hole groove edges free of upset metal
- ☐ All parts free of gouges, weld drags, and remnants of temporary handling attachments
- ☐ Surfaces damaged in removing temporary attachments restored to acceptable condition by welding and grinding

☐ VERIFY CERTIFICATION OF SUB-VENDOR-SUPPLIED PARTS

Comment. Authorized Inspectors must verify that sub-vendor-supplied pressure parts conform with code requirements for stamping and documentation. However, if they don't do so until just before or after the part is installed, serious delays can result. This is especially frustrating if it is discovered that uncertified parts have been used. A double check is important.

Always check the documentation of shell expansion joints for fixed-tubesheet units before they are installed. It is especially important to verify that specified hydrostatic testing has been performed by the expansion joint manufacturer. *Note that AIs are concerned only that the expansion joint meets the final hydrostatic test. A failure at that time could result in a severe delay in when the user receives the unit.*

☐ CHECK PARTS AND MATERIALS PROTECTION DURING MANUFACTURE

Comment. Parts and materials not protected during storage and construction may fail after the unit has been installed because of damage or abuse. These are things inspectors should look at:

- ☐ Storage

Comment. Tubes, expansion joints, etc. should be stored in their original containers until installation. Good storage practices are implicit in the checklist items:

- ☐ Exclusion of dirt and foreign matter

Comment. It is especially important to enclose stainless-steel and nonferrous bundles in sealed plastic wrapping to prevent ferrous particles from grinding and welding settling on their surfaces.

- ☐ Storage of tubes and corrosion-sensitive parts in plastic wrapping.
- ☐ Austenitic tubes not stored in direct contact with carbon-steel storage racks.
- ☐ Carbon-steel tubes stored in moisture-controlled containers or carbon-steel tubes stored with rust preventatives. (Oily or greasy rust preventative coatings must be removed before start-ups.)
- ☐ Tube ends covered or plugged with plastic tube-end protectors.
- ☐ Machined parts temporarily coated with rust-preventive agent and machined surfaces otherwise protected.
- ☐ Thin-walled expansion joints protected from damage and dirt by enclosing in appropriate containers.
- ☐ Handling

Comment. These are recommended practices for handling materials and parts during construction:

- ☐ Nylon handling slings used to prevent scoring, gouging, or denting.
- ☐ Tubes placed where they will not be walked on or mechanically damaged during installation.
- ☐ Tube guides ("bullets") shown in Fig. 2.1 used to protect tube ends during loading tubes into the bundle. (See section on fitting and assembly without damage and distortion, below.)

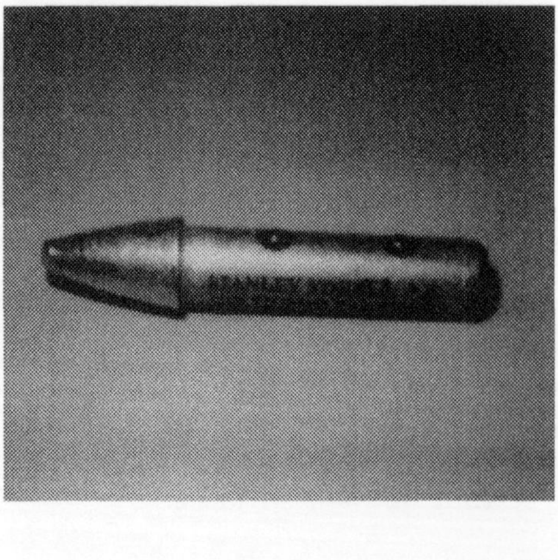

Figure 2.1 Tube guide. (*Photograph by MGT Inc., Boulder, Colo.*)

- ☐ Leather- or rubber-headed mallets used for tapping tubes into position. (See section on fitting and assembly without damage and distortion, below.)

☐ CHECK FOR PROPER CONSTRUCTION

Comment. Inspectors should make sure that what the manufacturer does during construction will not cause premature failure. These are some things of concern:

- ☐ Use of appropriate weld metals for welding fitting lugs or clips to the unit.

Comment. If the shop welds temporary steel construction fittings directly to alloy parts, and uses improper weld rod metal, dilution will occur and impair the alloy part's corrosion resistance.

- ☐ Removal of fitting lugs and clips by burning and grinding; repair damage to surfaces after removal.

Comment. It is not acceptable for the manufacturer to bend a lug or clip back and forth by hammering until it tears out of the part to which it was welded.

- ☐ Proper installation of thin-walled expansion joints.

Comment. Protective sheetmetal covers furnished by the joint manufacturer should remain in place throughout fabrication. They should be removed for the final hydrostatic test, then reinstalled. The expansion joint manufacturer's temporary shipping bridges should remain in place throughout fabrication and be removed only after the unit is installed. This is to avoid deflecting the expansion joint to accommodate fitting misalignments or changing the amount of expansion joint manufacturer's precompression or preextension. Any variation from the joint's designed installation length will shorten its cycle life.

- ☐ Use of appropriate forming and parting methods

Comment. Inappropriate methods produce poor quality that may lead to early failure. The following unacceptable practices should be covered in the checklist and rejected:

- ☐ Press breaking that leaves unacceptable marks on the surfaces
- ☐ Carbon-arcing austenitic materials without subsequent grinding back approximately 6 mm (~¼ in)
- ☐ Cold forming carbon-steel by blows, which is impermissible.
- ☐ Improper or poorly done cutting, shearing, or beveling.

Comment. Poor technique for parting, trimming, or edge preparation of metals may predispose them, and the welds joining them, to failure.

☐ Improper or incorrectly followed weld procedures.

Comment. Either may cause excessive distortions and intensify the effects of discontinuities. Longitudinal seams of adjacent courses must be offset at least 15° or the junctions of girth and long seams radiographed.

☐ Lifting shells, bundles, and parts properly. (See section on fitting and assembly without damage and distortion, below.)

Comment. As a general rule, the shop should lift parts in such a way that they do not distort under their own weight. When bundles are not lifted or handled properly, the tubes may crimp and flatten. The preferred way to handle them is with nylon slings (Fig. 2.2). Second best is to use cable slings. Chain slings wrapped about a bundle without wooden dunnage spanning the baffles or supports can crush the tubes. When slings are not placed at the appropriate weight centers or positioned to bear on tube supports or baffles, they may damage the tubes. For long, slim structures, appropriate spreader beams and handling beams should be used.

Figure 2.2 Nylon slings used to lift a bundle and load it into the shell. (*Engineers and Fabricators, Houston, Tex.*)

Using nozzles and nozzle flanges for lifting should be avoided. Even large nozzles are easily bent under lateral loads. This may cause distortions in the shell or channel which they penetrate. Shell expansion joints in fixed-tubesheet exchangers can be sprung if the unit is not properly lifted and the shell supported during lifting.

☐ Making layout and construction marks with materials that will not initiate subsequent corrosion

Comment. Sulfur-bearing crayons must not be used on nickel surfaces.

☐ VERIFY USE OF QUALIFIED PROCEDURES AND PERSONNEL

Comment. The Authorized Inspector is responsible for making sure that all procedures required to be qualified by the code are on hand and that the qualifications of the welders and nondestructive examiners who use them are current. The user's inspector should examine documentation of the qualifications. This includes checking that welders identifying symbols or numbers have been stamped adjacent to the welds they produced. Documents should be viewed that show the organization that qualified radiographers, eddy-current testers, ultrasonic testers, and others and the level to which they are qualified. *Note:* Workers who perform nondestructive examinations are examiners and do not have authority to accept and reject. Following are things to check.

☐ Use of specified procedures

Comment. Inspectors should verify that procedures specified in purchase documents, heat exchanger specifications, and drawings are followed. They should note whether copies of these procedures are readily available to the people doing the work. Verify the use of these procedures:

☐ Welding procedures

☐ Tube-to-tubesheet joining procedures

☐ Cleaning procedures

☐ Fluid penetrant examination procedures

☐ Bubble-test procedures

☐ Pressure-test procedures

☐ Performance of work by qualified personnel

Comment. Inspectors should verify the following:

- ☐ Welders' qualifications
- ☐ Welding operators' qualifications
- ☐ Nondestructive examiners' qualifications

☐ VERIFY MANUFACTURER'S ADHERENCE TO QUALITY CONTROL MANUAL

Comment. Buyers rely on the works to follow its quality control manual. Hence, it is very important for the buyer's inspector to become familiar with it and make sure that its provisions are followed.

☐ VERIFY MANUFACTURE TO SPECIFIED DIMENSIONS, TOLERANCES, AND STANDARDS

Comment. Authorized Inspectors are not responsible for verifying dimensional accuracy beyond minimum thicknesses and dimensions of parts that contain pressure. The inspector should verify that non-pressure-bearing essential parts are manufactured to the dimensions and tolerances on the approved drawings. Following is a list of parts to inspect and things to look for:

- ☐ Tubesheets

Comment. Tubesheets should be inspected before assembly to the bundle or shell for the following:

- ☐ Flatness over the machined surfaces and at the gasket surfaces after drilling and machining
- ☐ Correct drilling template
- ☐ Hole size, ligament width, and drill drift within specified tolerances
- ☐ Tappings, vents, and drains installed and properly located
- ☐ Specified surface finishes

Comment. Tube holes must meet the smoothness specified (if any) in the approved drawings or applicable standards and be free from spiral or axial scores.

- ☐ Position, dimensions, and finish of annular grooves.

Comment. Rectangular grooves must be checked for squareness. There must be no upset metal at the groove edges.

- ☐ Weld-joint preparation for tube-to-tubesheet welds.
- ☐ Relief of backsides of holes.

☐ Baffles and tube supports

Comment. The following should be checked:

- ☐ Correctness of drilling template
- ☐ Hole size and finish
- ☐ Position of cutoff lines
- ☐ Removal of upset metal burrs at backfaces

☐ Tie-rods, spacers, seal strips, and dummy tubes

Comment. Tie-rod threads should be clean. Spacers should be machine cut to the proper lengths and have square ends. Seal strip attachments should leave no crevices that could promote corrosion. Dummy tube installation should not lead to entrapping shell liquids.

☐ Tube loading and joining

Comment. Loading the tubes into the exchanger cage and joining them to the tubesheet are the two of the most crucial operations in heat exchanger construction. Therefore, make tube loading a hold point in the inspection brief. In addition, the inspector should be present during the tube joining to make sure that the factory follows the approved joining procedure with special regard to cleaning procedures, joining sequence, and recalibration of torque controls and hydraulic pressure settings.

☐ Shells

Comment. The following should be measured and checked:

- ☐ Diameter and thickness
- ☐ Roundness

Comment. The roundness tolerance required to meet the limits of diametral clearance between the baffles and shell prescribed by TEMA Standards is more restrictive than that required by most pressure-vessel codes. For example, the out-of-roundness permitted by the ASME Code is 1 percent of the diameter as shown in Fig. 2.3. Table 2.1 shows that for some shell sizes, there would be interference of the baffles and shell if the shell were manufactured to the limits of ASME Code–permitted out-of-roundness. The figures in bold italicized print

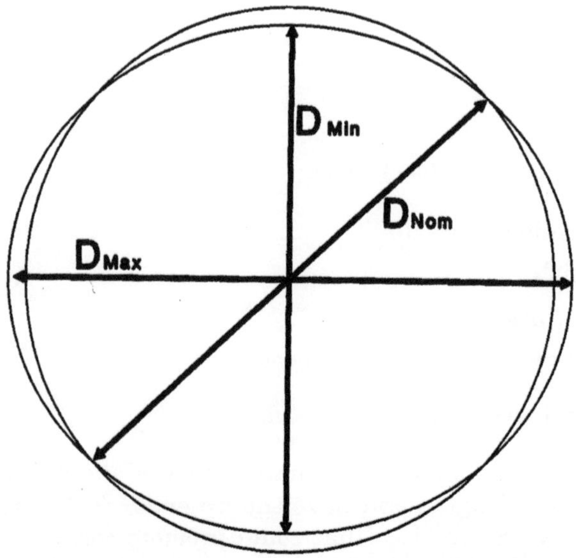

Figure 2.3 ASME Code permissible out-of-roundness for pressure-retaining cylinders.

TABLE 2.1 Comparison of ASME Code Allowed Out-of-Roundness for Plate Formed into Larger-than-Pipe-Size Cylinders with TEMA Maximum Permissible Diametral Clearance*

English units				Metric units			
Nominal O.D., in	Code maximum minus minimum O.D., in	Code undertolerance from mean I.D., in	TEMA maximum diametral baffle to shell clearance, in	Nominal O.D., mm	Code maximum minus minimum O.D., mm	Code undertolerance from mean I.D., mm	TEMA maximum diametral baffle to shell clearance mm
25¾	0.2575	0.1238	0.175	654.0	6.540	3.370	4.445
27¾	0.2738	0.1369	0.175	704.8	7.048	3.524	4.445
29¾	0.2975	0.1488	0.175	755.6	7.556	3.778	4.445
31¾	0.3175	0.1588	0.175	806.4	8.064	4.032	4.445
33¾	0.3375	0.1688	0.175	857.2	8.572	4.286	4.445
35¾	0.3575	***0.1788***	***0.175***	908.0	9.080	***4.540***	***4.445***
38	0.3800	***0.1900***	***0.175***	965.2	9.652	***4.876***	***4.445***
40	0.4000	***0.2000***	***0.175***	1016.0	10.160	***5.080***	***4.445***
42	0.4200	0.2100	0.225	1066.8	10.668	5.334	5.715
44	0.4400	0.2200	0.225	1117.6	11.176	5.588	5.715
46⅛	0.4613	***0.2306***	0.225	1171.6	11.716	***5.858***	***5.715***
48⅛	0.4813	***0.2406***	***0.225***	1222.4	12.224	***6.112***	***5.715***
52¼	0.5225	***0.2612***	***0.225***	1327.2	13.272	***6.636***	***5.715***
56¼	0.5625	0.2812	0.300	1428.8	14.288	7.144	7.620
61½	0.615	***0.3075***	***0.300***	1562.1	15.621	***7.816***	***7.620***

*Consider requiring the manufacturer to pass a template through the shell as described in API 660.[1]

highlight those sizes in which the permissible ASME Code under tolerance is greater than the permissible TEMA clearance.

- ☐ Internal smoothness

Comment. All shell internal welds should be ground smooth and flush.

- ☐ Positions, projections, and angularity of nozzles relative to specified working points or those shown on the approved drawings

☐ Shell internals

Comment. These internals should be inspected for conformity with approved drawings.

- ☐ Zone separators and shrouds
- ☐ Rails and guides
- ☐ Weir and dam overflow heights or underflow clearances
- ☐ Standpipe insertion depth
- ☐ Snorkel position
- ☐ Collection devices for noncondensible venting
- ☐ Distribution devices for cascading drains
- ☐ Demister mesh placement

☐ Main supports and supports for stacking

Comment. It is costly to relocate, in the field, the position of a unit's supports or holes and slots for hold-down bolts. Improperly located supports, or those in which boltholes and slots are misplaced, may cause loads to be imposed on connecting piping and nozzles not foreseen in design. These should be checked for conformity with the drawings:

- ☐ Positions of supports relative to specified working points
- ☐ Support bolthole and bolt slot templates
- ☐ Welds used for joining supports to the unit

☐ Channels and covers

Comment. These are items to check in channels and covers:

- ☐ Diameter and thickness
- ☐ Positions, projections, and angularity of nozzles relative to specified working points or those shown on the approved drawings

- ☐ Pass-partition positions, installation, and finish; attachment welds
- ☐ Machining
- ☐ Dimensions, fit, and operation of special high-pressure and/or quick-opening closures

Comment. The fabrication shop should be required to open and close and retest high-pressure, quick-opening, and other special types of closures in the presence of the inspector.

- ☐ Dimensions of manway and cover of manway-access feedwater heater channels
- ☐ Diaphragm attachment for diaphragm seal type
- ☐ Gasket and cover for bolted-cover, gasket sealed type
- ☐ Dimensions and attachment for internal pressure-sealed type

☐ FITTING AND ASSEMBLY WITHOUT DAMAGE AND DISTORTION

Comment. Ideally, the shell-and-bundle assembly position would be vertical. In this position, if the bundle and shell centerlines are aligned, vertical assembly allows gravity to provide the force required to join them. However, very few factories have the headroom to permit sliding the bundle into a vertical shell or a vertical shell over the bundle. Therefore, bundle-to-shell assembly is almost always done in the horizontal position. This means that for designs without skid bars or rails, the baffles or tube supports drag along the bottom of the shell. It also means that if the bundle is not adequately supported along its length, it will sag, possibly causing the baffles or supports to cut into the tubes.

The use of paired ratcheted chain pulls (come-alongs), hydraulic, air-operated, or mechanical bundle-loading machines is acceptable providing the bundle is not forced into the shell or the shell pulled over the bundle in such a way that the baffles and tubes are damaged. Sledge-hammer blows to the shell during bundle loading are acceptable only if stakes are imposed between the sledge hammer and the shell to protect the shell. Vibrators attached to the shell are also acceptable. However, lubricants other than water may not be used unless there is an assured way to fully remove them during final cleaning.

Inspectors should make sure that the welds and members of temporary shipping bridges on thin-walled (bellows) expansion joints remain intact during construction and shipment. Stretching or compressing the joint to accommodate poor fit-up will materially shorten the joint's cycle life. Protective covers should be slid back only during pressure testing, and restored to their protective position on completion of testing. The shipping bridges must be marked with the legend "REMOVE AFTER FIELD INSTALLATION."

☐ WITNESS NONDESTRUCTIVE EXAMINATIONS AND TESTS

Comment. The inspector should witness nondestructive examinations and tests that the manufacturer performs. Individuals who perform and interpret the tests, whether employed by the manufacturer or a contract nondestructive examination firm, should be qualified in accordance with the requirements of the prevailing pressure-vessel code and the standards of the nondestructive examination society in the country in which the exchanger is manufactured. (See Supplement 1.1 in Chap. 1 for a list of international nondestructive testing societies and related organizations.)

Inspection checklist for retubing, repairs, and alterations

☐ INITIAL INSPECTION

Comment. Perform an initial inspection after cleaning and before any disassembly. If a preshipment inspection is to be done, it may be cost-effective to have the manufacturer's inspector present to conduct a joint inspection. Similarly, if the manufacturer is to perform the initial inspection, it is cost-effective and saves time for the owner's inspector to be present. In both circumstances it may be possible to shorten the time required to make repair or replace decisions. The checklist for initial inspection before the unit is shipped or on its arrival at the factory is necessarily incomplete. However, it can be used as a guide. Following are checklist items for the initial inspection.

☐ Inspect channel(s) [or bonnet(s)]

Comment. The purposes of this examination are (1) To determine the degree of internal and external corrosion;* (2) to establish the remaining, corroded channel, or bonnet thickness; (3) to ascertain whether specific regions of the channel or bonnet have been corroded or eroded more than others; (4) to assess the condition of welds, especially interior welds at nozzle penetrations and the welds of the pass partition to

*See Figs. 14 through 23 and 25, 30, 32, and 33 of API 572[2] for illustrations of various types of corrosion in heat exchangers.

the channel or bonnet, and for those configurations where the channel is integral with the tubesheet, the welds of the pass partitions to the tubesheet, and the transitions or corner welds of the tubesheets to the channel barrel; (5) to determine whether the pass-partition plates have been deformed by excessive pressure differential and eroded at gasket contact surfaces; (6) for pass-partition cover designs, to determine the condition of pass-partition cover flanges and their attachment to the channel barrel, fastening devices, gaskets, and pass-partition covers; and (7) to estimate the feasibility of making repairs and restorations to the channel. Look for the following:

☐ General pit corrosion

Comment. When there is pit corrosion, measure the depth of the pits using a pit gage to assess whether the channel should be reused or its pressure-bearing capability reduced.

☐ Erosion-corrosion

Comment. Erosion-corrosion of channel barrels, pass-partition plates, and welds may occur in the vicinity of the inlet and outlet nozzles.

☐ Condition of inside of welds

Comment. Frequently the cover pass and underlying weld passes are washed away or corrode. This is especially prevalent in inlet and outlet nozzles because of entrance and exit turbulence. Explore the depth and length of the erosion and/or corrosion to determine the kind of repair that must be made.

☐ Condition of channel flanges

Comment. Inspect the channel flange and pass-partition gasket surfaces for pitting, erosion, crevice corrosion, and other forms of corrosion. Measure the flange flatness and the flatness across the pass partitions with a straight-edge. Take notice of the pattern of marks made on the gaskets. Uniform marks indicate flatness, adequate bolting, and proper bolting and assembly. Deep impressions at the inside or outside of the gasket ring indicate flange distortions. Uncompressed gasket ribs indicate inadequate bolting, out-of-flatness across the pass partitions, or both. Note that if the gasket ribs have not been compressed, bypassing with consequent erosion of the pass-partition edges or tubesheet may have occurred. Examine the flanges for distortion that may have been induced by excessive bolt loads applied in efforts to seal off leaks. Check the bolt holes. They should be round and their edges should be sharp.

☐ Condition of pass-partition plates

Comment. Examine pass-partition plates for deformation and erosion or corrosion where they bear against gaskets.

C and N channels

Comment. These are configurations in which the channel is integral with the tubesheet. Check the following:

☐ Welds of pass partitions to tubesheet and barrel

☐ Transitions and/or corner welds of the tubesheets to the channel barrel

Comment. Older, high-pressure units may have been built with small corner radii that promote high stress levels at the discontinuity. Often fluid penetrant examination will reveal cracks in this region. Most of these can be repaired and the unit restored to operation. However, analysis of the effects of continued operation without repairs and the effects of the proposed repairs is required.

☐ Inspect the channel cover.

Comment. Inspect the channel cover for bowing, corrosion, and erosion. When pass-partition gasket retaining grooves are milled into covers, the groove depth is usually allowed for corrosion. However, corrosion and/or erosion of the grooves can either cause bypassing the pass partitions or be caused by bypassing resulting from inadequate bolting. Similarly, examine ungrooved covers and those of single-pass units for distortions and corrosion.

C and N channels with internal pass-partition covers with bolted removable covers

☐ Condition of pass-partition cover flanges.

Comment. Pass-partition cover flanges are not pressure-resisting parts. However, they are welded to the channel which is. Lack of flexibility in the attachment produces thermal stress. The discontinuities at the corners of the channel flanges may lead to the kind of cracking illustrated in Fig. 2.4, copied from a report of inspection and repairs to feedwater heaters.

In addition to such cracks, pass-partition flanges may warp and pull away from the channel barrel. In addition, the tube-side stream may erode the pass-partition cover-to-flange bolting.

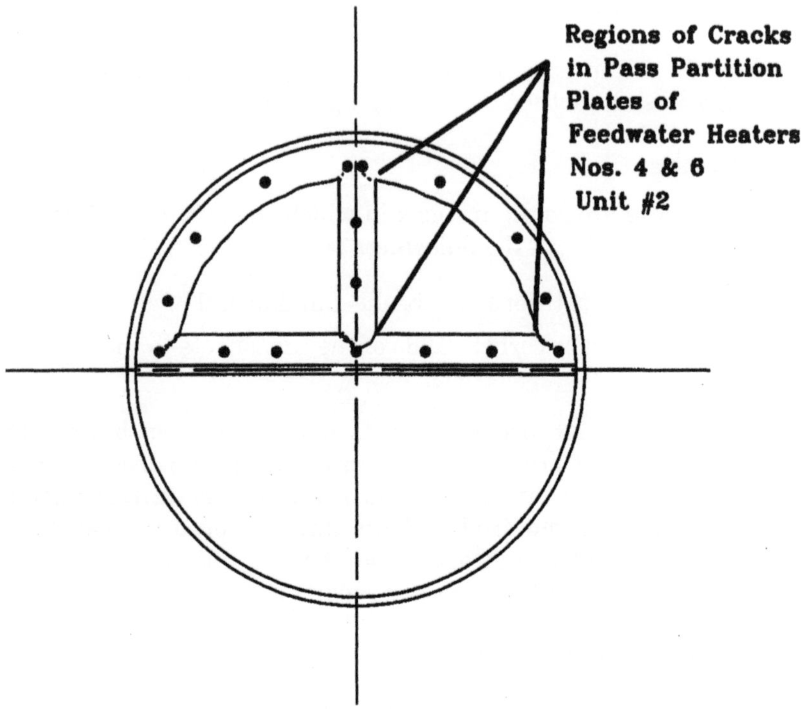

Figure 2.4 Cracks in welds of pass-partition flanges to channel.

☐ Inspect tubes.

Comment. Inspection of tubes is required only when the repair or alteration does not consist of a complete retubing. When repairs or alterations to be performed at the factory do not involve the tubes, it is only prudent to make sure that there is sufficient remaining tube life to make the repair or alteration worthwhile. For this purpose eddy-current or ultrasonic estimates of the wall thicknesses and tube condition should be made.

☐ Inspect tubesheets.

- ☐ Make an initial visual inspection under high-intensity light. Measure the ligament widths. If a large number of ligament widths are narrower than those permitted for new construction to the TEMA standards, the tubesheet(s) should be replaced. Also look for readily apparent transverse cracks and ligament erosion. Examine the condition of machined surfaces and gasket retaining grooves.
- ☐ Fluid penetrant examining of the tubesheet(s).

Comment. Fluid penetrant examination may reveal cracks in welds and ligaments and porosity in tube-to-tubesheet welds. Where porosity or weld cracks are encountered, inform the manufacturer to investigate the possibility of wormholing.

☐ Measure the tubesheet O.D. and flatness.

Comment. A substantial increase in tubesheet diameter indicates that expanding and reexpanding the tubes has caused radial growth. This is usually accompanied by tubesheet bowing. If the out-of flatness prevents gasket sealing without excessive bolting force or the larger-than-designed tubesheet diameters interfere with disassembly and reassembly during maintenance, the tubesheet(s) may have to be replaced.

☐ Examine tube-to-tubesheet joints

☐ In a representative number of tube I.D.s in each quadrant of the tubesheet, insert and rotate a tube gage (Fig. 2.5) in 25-mm (~1-in) steps through the tubesheet thickness and in the tube beyond the tubesheet.

Comment. This will indicate (1) the difference between the unexpanded tube I.D. from which an estimate of the wall reduction may be

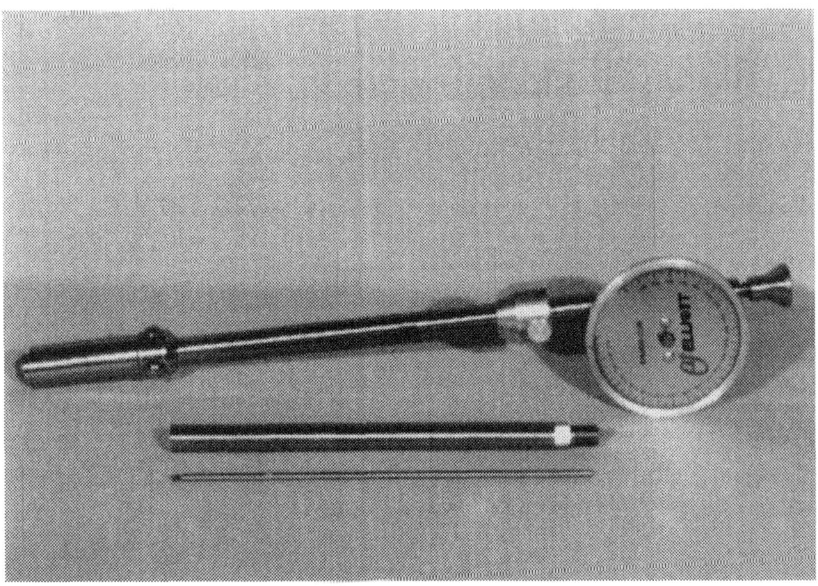

Figure 2.5 Tube gage. (*Elliott Tool Technologies, Dayton, Ohio.*)

made and (2) ovality through the tubesheet depth. Estimated wall reductions in the range of 12 percent or more indicate that the tubes should be replaced. If the excessive wall reduction is accompanied by considerable ovality, it is likely that the ligaments have been deformed to the extent that the tubesheet should be replaced.

Removable-bundle exchangers

☐ Examine the shell and return cover.

Comment. The purposes of this examination are to (1) determine the degree of internal and external corrosion; (2) establish the remaining corroded shell thickness; (3) ascertain whether specific regions of the shell have been corroded or eroded more than others; (4) assess the condition of welds, especially interior welds at nozzle penetrations; and (5) estimate the feasibility of making repairs and restorations to the shell.[2] Look for the following:

- ☐ General pit corrosion

Comment. When there is pit corrosion, measure the depth of the pits using a pit gage to assess whether the shell should be reused or its pressure-bearing capability reduced.

- ☐ Erosion-corrosion in the vicinity of cross-flow baffles

Comment. A common form of deterioration of the shell is the corroding and eroding away of rings of metal at the positions of semisupports and segmental cross-flow baffles. Such erosion-corrosion is often paralleled by erosion-corrosion of the supports or baffles.

- ☐ Condition of inside of nozzle welds

Comment. Frequently the cover pass and underlying weld passes are washed away or corrode. This is especially prevalent in inlet and outlet nozzles because of entrance and exit turbulence. Explore the depth and length of the erosion and/or corrosion to determine the kind of repair that must be made.

- ☐ Impingement protection attached to the shell or contained in nozzles.

Comment. Impingement protection attached to shells or nozzles bears the brunt of the turbulent effects of the incoming stream. Examine the attachments and condition of the parts.

- ☐ Condition of the shell and cover flanges.

Comment. Inspect the shell and cover flange gasket surfaces for pitting, erosion, crevice corrosion, and other forms of corrosion. Measure the flange flatness with a straight-edge. Note the pattern of marks made by the gaskets. Uniform marks indicate proper bolting and assembly. Deep impressions at the inside or outside of the gasket ring indicate flange distortions. Examine the flange for distortion that may have been induced by excessive bolt loads applied in efforts to seal off leaks. Check the boltholes. They should be round, and their edges should be sharp.

☐ Examine the shell for any attachments, alterations, and/or repairs made subsequent to installing the exchanger.

Comment. Verify that attachments, alterations, and/or repairs were made in accordance with the prevailing inspection code. Attachments, repairs, and alterations not made in conformity with the inspection code should be removed and replaced with properly authorized and inspected ones.

☐ Examine the bundle.

Comment. Look particularly at the baffles and/or tube supports and end support plate and tie-rods and spacers. If there is substantial erosion or corrosion of the baffles and/or supports, evidenced by thinner-than-specified thickness or enlarged tube holes, a new set will have to be manufactured. Even modest corrosion or deformation of tie-rods or spacers should be cause for replacing these parts.

U-tube exchangers

☐ Examine the condition of the U-bends.

Comment. Look particularly for abrasion where the ends may have rubbed against each other. Shiny surfaces or surfaces rubbed flat are cause to require retubing.

Floating-head exchangers

☐ Inspect the floating head

Comment. Look for the following in all floating heads:

☐ General pit corrosion

☐ Erosion-corrosion

☐ Condition of inside of welds

- ☐ Condition of the closure flanges
- ☐ Condition of the pass-partition plates

TEMA type T floating heads

Comment. Inspecting pull-through (TEMA type T) floating heads is similar to inspecting bonnets.

TEMA type S floating heads

Comment. To inspect floating heads with backing devices (TEMA type S), examine, in addition:

- ☐ Split backing ring or snap ring condition
- ☐ Conditions of the peripheral groove in the tubesheet

Packed floating-head types:

- ☐ Check the finish of the floating-tubesheet barrel.

Comment. To properly seal the packed floating end, the finish of the outside of the tubesheet-barrel-cylinder structure should be no coarser than 63 rms. Any required refinishing must not remove so much material that the floating-tubesheet–packing-gland clearance is excessive.

Pressure Testing

Liquid pressure tests

Most pressure tests are static liquid pressure tests. Rarely, a pulsed liquid pressure test is performed. Static testing is routinely hydrostatic. However, if residual moisture would cause rapid corrosion or a hazard in contact with the process stream, another medium may be used. The authors have had experience with static testing with mineral oil and dowtherm.

It is important to consider that Authorized Inspectors are not concerned about the test water chloride ion content or dissolved or suspended foreign matter that does not interfere with the test. Similarly, the AI may not care whether a liquid medium other than water deviates from the specified chemical content.

Authorized Inspectors have no further responsibility after the hydrostatic test has been successfully performed and the unit accepted. They need not witness draining and drying and may not be present to witness subsequent tightness testing not required by the code.

Purchasers' inspectors must make sure that the test water or other medium meets the agreed-on cleanliness, chloride ion content, etc. When the test is completed, it is desirable for them to make sure that all liquid is drained, and that the exchanger, bundle, or other part be dried with filtered, dry, warm air. If the specification requires dehydrating and padding the exchanger or bundle with nitrogen, the inspector should be present when the work is done. Authorized Inspectors will most probably not verify fulfillment of this requirement.

Pneumatic testing

Pressure-vessel codes are very specific in setting requirements for safe pneumatic testing. Your inspector should be aware of the danger created by the stored energy and should leave the area of the test if the workforce deviates from the code requirements for safe testing.

References

1. *Shell-and-Tube Heat Exchangers for General Refinery Services, API Standard 660,* 4th ed., American Petroleum Institute, Washington, D.C., Sept. 1982, reaffirmed Dec. 1987.
2. *Inspection of Pressure Vessels, API Recommended Practice 572,* 1st ed., American Petroleum Institute, Washington, D.C., Feb. 1992.

Supplement 2.1: Inspection Checklist for Factory Inspection of Heat Exchangers*

Date: ______________

Client ______________________________

____

____

Address ______________________________

____

____

____

Purchase order number ______________ Item/tag number______________

Manufacturer ______________________________

Address ______________________________

____

____

____

Factory contact ______________________________

Telephone ______________ Ext ______________ Fax ______________

Inspector ______________________________

☐ CHECK MATERIALS FOR DIMENSIONAL CONFORMITY

Measured thicknesses of:

- ☐ Channel Cover or Bonnet______________________
- ☐ Channel or Bonnet Barrel______________________
- ☐ Stationary or Front Tubesheet(s) / at Depth of Grooves
  ______________________ / ______________________
- ☐ Floating or Rear Tubesheet(s) / at Depth of Grooves
  ______________________ / ______________________
- ☐ Flanges at depths of recesses
  - Channel Cover Flange ________O.D. x ________Thick
  - Channel Shell Flange ________O.D. x ________Thick
  - Shell Flange at Channel ________O.D. x ________Thick

*MGT Inc., Boulder, CO 80303-3607.

Return Cover Flange ________O.D. x ________Thick
or
Channel Shell Flange ________O.D. x ________Thick
Channel Cover Flange ________O.D. x ________Thick

Tubes:

No. Measured ____ Largest Dia. _____Smallest Dia. _____

Thinnest Wall _____Thickest Wall _____ Maximum Ovality _____

Maximum Weld Bead Crown Height _____

Shell:

Pipe _____ or Rolled Plate _____ or Forged _____

No. of Courses _____ Max. I.D. _____ Min. I.D. _____

Alignment of Courses ______________________________

☐ SURFACES OF RAW MATERIALS

- ☐ Shell and Channel Barrel Plates______________________________
- ☐ Tubesheets______________________________
- ☐ Bonnet Heads and Channel Covers______________________________
- ☐ Tubes______________________________
- ☐ Tube Holes in Tubesheets______________________________
- ☐ Tube-hole grooves______________________________
- ☐ Restoration of Surfaces______________________________

☐ VERIFICATION OF CERTIFICATION OF SUBVENDOR SUPPLIED PARTS

- ☐ Shell Expansion Joint(s)

  Mill Test Reports Acceptable? _____ Hydrostatically Tested? _____

☐ CHECK OF PARTS AND MATERIALS PROTECTION

Materials Storage and Handling:

- ☐ Tubes______________________________
- ☐ Expansion Joints______________________________
- ☐ Machined and Drilled Parts

  Are machined surfaces protected? ________
- ☐ Plate______________________________

Handling of Parts:

- ☐ Use nylon handling slings? _____ or Chain Slings? _____
- ☐ How are tubes stored and handled ______________________________
- ☐ Tube Loading

  Guides used? _____ Rubber/Leather-headed mallets used? _____

☐ **CONSTRUCTION**

☐ Use appropriate weld metals for welding fitting lugs or clips? ________

☐ How are fitting lugs and clips removed? ________________

<u>Installation of thin-walled expansion joints:</u>

Protective sheet metal covers kept in place until hydro? ________

Removed for the final hydrostatic test, then reinstalled? ________

Temporary shipping bridges intact throughout fabrication? ________

<u>Fabrication methods:</u>

Brake marks acceptable? ______ Cold forming of carbon steel? ________

How are high-alloy materials cut? ________________

If carbon arced, is an adequate amount of edge ground off? ________

Is edge condition of cut, sheared or bevelled parts acceptable? ________

<u>Weld procedures:</u>

Are the approved weld procedures available to the welders? ________

Are the approved procedures followed? ______ Do the welders apply their stamps? ________

<u>Lifting shells, bundles and parts:</u>

Nylon slings used for bundles? ______ Spreaders between slings? ______

Cradle large bundles? ______ or Chain Slings wrapped around without wooden dunnage spanning the baffles or supports? ______

Lift on nozzles? ________ Temporary lift lugs? ________ Slings? ______

☐ Are marking materials for layout and construction sulfur-free? ________

Which are used? Soapstone ______ Prussian blue ______ Other ______

☐ **VERIFY USE OF APPROVED QUALIFIED PROCEDURES AND QUALIFIED PERSONNEL**

<u>Welding procedures:</u>

Are all welding procedures being used qualified and current? ________

Are they: All as approved? ______ Available to the welders? ________

☐ Tube-to-tubesheet joining procedures:

Are the tube holes and tube ends cleaned before joining? ______

How? ________________

____

Is the approved expanding procedure being followed? ________

Rolled Tube-to-Tubesheet Joints:

Condition of rolling tool ______________________

Lubricant/coolant used (must be approved chloride free)__________

Torque setting ______ Frequency of calibrating torque control ______

Calculated wall reduction ______ Measured wall reduction __________

Maximum torque setting used for sealing leakers __________

Check expanded depth (must not be beyond specified depth) ______

Condition of rolled end acceptable (free of scores and scratches)? ______

Hydroexpanded Tube-to-Tubesheet Joints:

Tubesheet #1 yield stress ______ Tubesheet #2 yield stress ______

Heat #	Tube yield	Expanding pressure	Heat #	Tube yield	Exp. pressure
______	______	______	______	______	______

Heat #	Tube yield	Expanding pressure	Heat #	Tube yield	Exp. pressure
______	______	______	______	______	______

Welded Tube-to-Tubesheet Joints:

Is the approved procedure qualified? ______ Current? __________

Welders' qualified in the approved procedure? __________

Welded-and Expanded Tube-to-Tubesheet Joints:

Minimum measured set-back of start of expansion __________

Welds bubble tested before onset of expansion? ______ Fluid penetrant examined before and after expansion? ______ No. of repair welds ______

Sketch location of repair welds or identify position in tube field.

- ☐ Are specified cleaning procedures on hand and followed? ______
- ☐ Approved fluid penetrant examination procedures followed? ______
- ☐ Approved bubble test procedures followed? ______

  ubble test witnessed? ______ Describe repairs required and made ______

- ☐ Pressure tests comply with code and safety requirements ______

Personnel qualifications

Welders' and welding operators' qualifications

Are all welders and welding operators qualified? ______ Current ______

Symbols adjacent to welds? ______ Satisfactory to AI? ______

Nondestructive Examiners' qualifications:

Fluid-penetrant examiners' qualification organization and level ______

Radiographers' qualifying organization and level ______

Magnetic flux examiners qualification organization and level ______

Other ______

☐ **VERIFY ADHERENCE TO QUALITY CONTROL MANUAL**

Are Code and QC manual tracking procedures followed? ______

Has the manufacturer's Inspector signed off where required? ______

Has hold point notification been in accordance with the specification? ______

Does manufacturer cease manufacture pending release of hold points? ______

☐ **VERIFY SPECIFIED DIMENSIONS, TOLERANCES AND STANDARDS**

Mark up two copies of the applicable drawings with "as built" dimensions. Include any deviations that have been accepted. Where repairs or alterations have been made because of construction errors, add a note describing same to drawing. Retain one copy for the maintenance file and provide one copy to the manufacturer. Manufacturer will be required to issue an "as-built" final certified drawing, including the note regarding repairs and alterations.

TUBESHEETS:

- ☐ Flatness (after drilling and machining) at all gasket surfaces ______
- ☐ Correct drilling template? ______ Correction of errors ______
- ☐ Hole diameter ______ Maximum ______ Minimum Drill Drift ______

  Ligament width tolerance (show worst-case measured) ______

- ☐ Tappings, vents and drains per drawings? ______ If not, describe deviations. ______

- ☐ Gasket surfaces acceptable?______ rms of finish ________________
  Hole finishes per drawing and standard? ______ approximate rms ______
- ☐ Annular groove position, dimensions and finish OK? ______ Describe deviations. ____

  ____
- ☐ Weld-joint preparation for tube-to-tubesheet welds acceptable? ________
- ☐ Are back sides of holes relieved per drawing and specification? ________

<u>BAFFLES AND TUBE SUPPORTS:</u>

- ☐ Does drilling template match tubesheet? Accept ______ Reject ________
- ☐ Hole size meet drawing and standard? Accept ______ Reject ________
  Back face deburred? ______ Cut-off lines per drawing? ______ If not, describe deviations. ____

  ____

<u>TIE RODS, SPACERS, SEAL STRIPS AND DUMMY TUBES:</u>

Tie rod threads ________ Spacer cuts ________ Seal strip attachment

____

<u>SHELLS:</u>

- ☐ Measured I.D. ________ Maximum ________ Minimum ________
  Thickness ________
- ☐ Witnessed insertion of roundness test template ________ (when specified)
- ☐ Internal welds ground smooth and flush? Accept ________ Reject ________
- ☐ Fit of multiple courses: Accept ________ Reject ________ If reject, describe proposed "fix".

  ____

  ____
- ☐ Nozzle positions, projections and angularity: Accept ________ Reject ________ Describe all deviations even when acceptable.

  ____

U-TUBES:

☐ Does bending comply with the specification, drawings and standards? Accept ________ Reject ________ Describe any deviations that were accepted.

____

CHANNELS AND COVERS:

☐ Measured front channel I.D. ________________ thickness ________

☐ Nozzle positions, projections and angularity: Accept ________ Reject ________ Describe all deviations even when acceptable.

____

☐ Pass-partition positions: Accept _______ Reject ________ Describe all deviations even when acceptable.

____

____

Attachment welds: Accept ________ Reject ________ Describe repairs if reject.

____

☐ Does machining meet drawing and specification? ________ If rejected, describe rework.

____

SHELL INTERNALS:

☐ Do separator and shroud welds meet drawing and specifications? If rejected, describe rework. ________________________________

____

☐ Rails and guides____

☐ Measured weir and dam overflow heights or underflow clearance

____

☐ Measured standpipe insertion depth ________ Snorkel dimensions ________ and measured position ______________________________

☐ Collection device for noncondensible venting per drawing? ________ Describe any deviation.

____

☐ Distribution device for cascading drains per drawing? ________ Describe any deviation.

____

☐ Demister mesh placement per drawing ________ Describe any deviations.

____

MAIN SUPPORTS AND SUPPORTS FOR STACKING:

☐ Positions of supports: Accept ________ Reject ________ If rejected, describe repairs used to make locations acceptable.

____

☐ Positions of support bolt holes slots: Accept ________ Reject ______ If rejected, describe repairs to make locations acceptable.

____

☐ Are continuous welds used for joining supports to the unit? ________

FITTING AND ASSEMBLY:

☐ Were tubes loaded into the cage without excessive force or blows on tube ends or U-bend ends and the loading acceptable? ________________ If not acceptable, describe the corrections that were made to permit accepting the exchanger.

____

____

☐ Has the fit-up and assembly been done without distortion, and excessive force for loading the bundle into the shell? ________ Note any comments below.

____

☐ Were stacked units fitted together to check the fit of supports and nozzles? ________ Did they fit? ________ If not describe what was done to make the stack acceptable. ________________________________

____

If stacked units were not fitted together, show the actual measurements on the drawings and note below that a factory fit was not made.

____

Additional Comments

____

____

____

____

____

____

____

____

____

____

____

____

____

____

____

____

____

____

____

____

____

____

Supplement 2.2: Checklist for Initial Inspection of Exchangers to be Factory-Altered, Repaired, or Retubed*

Date:______

Client ____

____

____

Address ____

____

____

____

Purchase order number ______________ Item/tag number ____________

Manufacturer ____

Address ____

____

____

____

Factory contact ____

Telephone ______________ Ext ______________ Fax ______________

Inspector ____

All Exchangers:

☐ Inspect channel(s) [or bonnet(s)].

☐ Pit corrosion present? ________ Depth and area measurements

____

Reuse ________ Reduce pressure ________ Replace________

☐ Erosion-corrosion present? ________ Describe ________________

____

☐ Channel barrels: Replace ________ Reuse ________ Describe required repairs

____

____

*MGT Inc., Boulder, CO 80303-36-07.

☐ Condition of pass-partition plates and welds: OK ________

Replace pass-partitions ________ Repair welds ________

Describe required repairs ____

____

☐ Condition of inside nozzle welds: OK ________ Require work ____________

Describe required work ____

____

☐ Condition of channel flanges: OK ________ Replace ________

Repair ________ Describe required work ____________________________

____

Exchangers with C and N Channels with Internal Pass-Partitions with Bolted Removable Covers:

☐ Condition of welds of pass-partitions to tubesheet and barrel:

OK ________ Repair ________ Describe required repairs ____________

____

☐ Condition of transitions and/or corner welds of the tubesheets to the channel barrel:

OK ________ Corner cracks ________ Repairable? ________

Describe required repairs ____

____

☐ Condition of channel covers: OK ________ Replace ________

Repair ________ Describe required repairs __________________________

____

C and N Channels with Pass-Partition Covers:

☐ Condition of pass-partition cover flanges: OK ________

Replace ________ Repair ________ Describe required repairs ____________

____

☐ **Condition of tubes in quadrant:**

	1	2	3	4
Number examined	____	____	____	____
Wall thickness	____	____	____	____
Accept/reject				
Corrosion:	____	____	____	____

Describe the kind and amount of corrosion observed ______________________

____

☐ **Condition of tubesheets:**

☐ Examination methods: ______________________________________

☐ Ligaments: OK ________ Reject ________ Repair ________

Give reason for rejection or describe recommended repair __________________

____

____

☐ Machined surfaces and gasket retaining grooves: OK ________

Reject ________ Repair ________ Give reason for rejection or describe recommended repair ____________________________________

____

Measured O.D. ________ Measured flatness ______________________

Accept ________ Reject ________ Repair possible? ________

☐ **Condition of tube-to-tubesheet joints:**

Quadrant	1	2	3	4
Number examined	____	____	____	____
Wall - unexpanded	____	____	____	____
expanded	____	____	____	____
Est. % reduction	____	____	____	____
Accept/reject	____	____	____	____
Accept/reject welds	____	____	____	____

Removable Bundle Exchangers:

☐ **Condition of shell and return cover:**

☐ General pit corrosion acceptable? ________ Describe observed pitting

____

☐ Erosion-corrosion in the vicinity of cross-flow baffles?

acceptable? ________ reject ________ repair ________

Give reason for rejection or describe required repair

____

☐ Condition of inside of nozzle welds: OK ________ Require work ________

________ Describe required work ________________________

____

☐ Condition of impingement protection attached to the shell or contained in nozzles:

Thickness: OK ________ Reject ________ Repair ________

Welds: OK ________ Repair ________

☐ Condition of the shell and cover flanges: OK ________

Replace ________ Repair ________ Describe required work

____

☐ Attachments, alterations and repairs: OK ________

Repair ________ Replace _______ Describe required repair or reason for replacing

____

☐ **Condition of bundle:**

☐ Baffles and/or tube supports and end support plate OK?

________ Replace ________ Describe deterioration ________________

____

☐ Tie-rods and spacers: Reuse ________ Replace ________

U-tube exchangers:

☐ Condition of the U-bends: OK ________ Reject ________

Describe Deterioration ________________________________

____

Floating head exchangers:

☐ Condition of floating head: OK ________ Replace ________

Repair ________ Describe required repair ________________________________

____

☐ General pit corrosion: Accept ________ Repair ________

☐ Erosion-corrosion: Accept ________ Repair ________

☐ Inside welds: Accept ________ Repair ________

☐ Closure flanges: Accept ________ Repair ________

Replace ________

☐ Pass-partition plates: Accept ________ Repair ________

Replace ________

TEMA type T floating heads:

☐ Corrosion: OK ______ Repair ________ Replace ________

Describe required repair ________________________________

____

TEMA type S floating heads:

☐ Split backing ring: Reuse ________ Replace ________

☐ Peripheral groove in the tubesheet: OK ________

Remachine ________ Replace tubesheet ________

Packed floating head:

☐ Finish of floating tubesheet barrel: OK ________

Remachine ________ Replace barrel ________

Replace whole floating head ________

☐ Packing gland(s) OK ________ Repair ________ Replace ____________

Additional Comments

Chapter

3

Evaluating Heat Exchanger Factories and Repair Shops

Introduction

Most engineers concerned with tubular heat exchangers can assess manufacturers' thermal and mechanical design expertise. Few have the background and training to appraise exchanger factories or evaluate the competence of repair-alteration organizations. Although some inspectors are knowledgeable about factory layouts, capacities of equipment, workforce skills, and quality control practices, hardly any can assess engineering and design capabilities. Yet, evaluating manufacturers and repair-alteration organizations requires being able to assess both engineering proficiency and production capabilities.

Although the fundamental thermodynamics and mechanics of heat exchanger design are universal, how manufacturers employ them varies with the industries they serve. Those who cater for the dairy and commercial and industrial air-conditioning industries build exchangers on production lines and sell them from catalogs. They pick standard units from their catalogs that can most closely meet the conditions the purchaser specifies. In contrast, manufacturers who serve the process and power-generation industries use complex thermal and mechanical computer-design optimization programs to custom-design and manufacture exchangers for the specified pressures, temperatures, and flow conditions.

Exchanger service governs when to select a fixed-tubesheet or removable-bundle. Leaktightness and maintenance requirements determine when removable-bundle units should be outside packed floating-head, externally sealed floating-tubesheet, pull-through floating-head, or floating-head, with backing device. The application also is the deciding factor in whether the unit will be selected from a catalog or individually designed and constructed for a specific use.

Process and power-generation tubulars nearly always accommodate one or a small number of user-specified services. Usually, the purchaser specifies performance requirements, materials, mechanical design and construction criteria, and standards and pressure-vessel codes that apply to such exchangers.

Worldwide, most process exchangers and some power-generation ones are built to conform to TEMA standards.[1] In North America, feedwater heaters, steam surface condensers, and auxiliary power-plant heat exchangers conform to the various standards of the Heat Exchange Institute.[2–4] With the exception of main steam surface condensers, large field-erected and production-line exchangers are rarely built to TEMA or HEI standards. Instead, they meet industry and manufacturers' quality standards and practices. This does not imply low quality but simply that the TEMA and HEI standards do not apply.

Exchangers and components constructed for the chemical, oil-refining, petrochemical, pharmaceutical, and power-generating industries almost always comply with the requirements of the prevailing pressure-vessel code or codes in the countries of origin and destination. Repair-alteration organizations, whether engaged in factory or on-site repairs and alterations, work to the requirements of either the National Board Inspection Code (NBIC) or the API Inspection Code. Companies whose primary business is repair and alterations may also manufacture exchangers in their works and field-erect ones too large to ship economically. Exchangers ordered from catalogs and very large, low-pressure, partially shop-fabricated and field-assembled units may not fall within the scope of a pressure-vessel code. However, they can usually be code-constructed and -stamped when a user or governing authority requires it.

Plants that manufacture on a production line are different from those that fabricate and field-erect large exchangers. Factories that custom produce tubulars one or a few at a time are different from both. Within the latter class, different considerations apply for surveying small, intermediate-sized, and large factories, and ones that manufacture feedwater heaters and steam surface condensers. Some factors that pertain to evaluating heat exchanger field-erection companies and custom exchanger fabricators also apply to repair-alteration organizations as defined in the National Board Inspection Code.

The purposes of this chapter are to (1) describe what heat exchanger factories are like and how they function and (2) guide engineers and inspectors in grading heat exchanger manufacturing works and organizations that repair and alter tubulars. To accomplish these aims, we first describe establishments that custom-design and fabricate tubulars; produce replacement units, bundles, and parts; and do factory repairs and alterations. Next we discuss constructors and erectors of exchangers too large to ship assembled and organizations that do field repairs and alterations. We then briefly describe production-line exchanger factories. We follow with a guide to surveying the various kinds of factories.

Manufacture-to-Order Factories (Job Shops)

Factories that manufacture heat exchangers to order vary from shops as small as a two-car garage to ones occupying about 20 hectares (ha) (~50 acres) of property. Each has advantages and disadvantages for the exchanger buyer.

Small workshops

We define "small" workshops arbitrarily as those with an annual U.S. dollar volume below $2,000,000 (1997 dollars). Their limited space and lifting capacity and modest financial resources determine the maximum size and value of what they can build. Ordinarily, they work from purchaser-supplied manufacturing drawings and rarely undertake thermal or mechanical design. In North America, ones authorized to apply code symbols rely mostly on contracts with registered Professional Engineers (PEs) for engineering and preparation of quality control manuals, required procedures, code calculations, and manufacturer's data reports.

Common practice of small-shop proprietors is to personally purchase materials, track mill test reports, qualify welding and other procedures, and monitor quality. Often they operate drills, lathes, and small boring mills themselves, and subcontract drilling and machining operations that are beyond the capacity of their equipment. Customarily, they qualify themselves as welders and perform all laying out, fitting up, and welding.

Most often, small shop owners are intelligent, imaginative, skilled mechanics who understand construction requirements, know what is required to create high levels of quality, and work well with engineers. Motivated by the desire to make their own decisions and be their own bosses, most would rather work long hours for limited reward than accept the security, benefits, and supervision that go with employment for wages in someone else's factory. These characteristics make such workshops ideal for producing experimental designs and prototypes from your drawings.

Purchasing from small workshops may be very cost-effective because their capital investment is relatively meager, their overhead costs are low, and the owners frequently undervalue their time. Disadvantages are their lack of financial resources, limited ability to accept liability, and inability to procure product liability insurance. Furthermore, production may be stopped if an owner is disabled.

Intermediate-size factories

Most tubulars for the process industries, nearly all power-generation auxiliary exchangers, and some feedwater heaters are built in intermediate-sized factories. For the purpose of this book, we define them as having annual U.S. dollar volume in the approximate range of $2,000,000 to $10,000,000 (1997 dollars). There are so many exchanger factories in this category that they face an extremely competitive marketplace. To a great extent, larger works are expanded versions of such factories.

To illustrate this kind of corporation, we describe the organization, facilities, production scheme, and products of an intermediate-size heat exchanger company in some detail. For this purpose, we use the fictitious *Wonderful Heat Exchanger Manufacturing Company,* referred to in previous writings.[5] Any resemblance of the fictional characters to actual persons, living or dead, is coincidental and unintended. However, they are much like many people in the business. We suggest that readers use the following description as a guide to the survey form provided in Supplement 3.1 to this chapter.

The Wonderful Heat Exchanger Company, Inc. (WHEMCO): An Imaginary Intermediate-Sized Factory

General information

Wonderful Heat Exchanger Manufacturing Company, Inc. (WHEMCO) has its factory and offices at (nonexistent) 1000 Tubular Exchanger Street, Oilfield City, OK, 74101 (USA). Its (fictitious) telephone number is (918) 246-0000 and (fictitious) fax number (918) 245-0000. Its (fictitious) e-mail address is whemco@ix.provider.com, and it has a page on the World Wide Web.

WHEMCO's approximate 1996 dollar volume of business was just less than $8,000,000. It is a privately held, 19-year old company with manageable outstanding debt, a substantial credit line with the (fictional) Tulsa State Bank, and a good credit history and rating. It has a well-established customer base in the oil-refining, petrochemical chemical, and other process industries and has partially succeeded in penetrating the power-generation industry. The company has generally operated profitably, with gross margins varying from 25 to 33 percent. Most cash generated by depreciation allowances and a substantial amount of profits are reinvested in maintaining its plant in modern status.

Memberships

WHEMCO is a member of the Tubular Exchanger Manufacturers Association (TEMA), the Heat Transfer Research Institute (HTRI), the American Welding Society (AWS), and the American Society for Nondestructive Testing (ASNT).

Certificates of authorization and inspection agency

Wonderful holds certificates of authorization from the ASME to apply U-1, UM, and S stamps, authorization to stamp and register ASME Code–stamped exchangers with the National Board of Boiler and Pressure Vessel Inspectors, and a certificate of authorization from the National Board to apply the R symbol. Their Authorized Inspection Agency is Hartford Steam Boiler. They are currently investigating the economics of qualifying for and obtaining certificates of authorization to apply U-2 and N stamps. WHEMCO's president is investigating qualifying the company to meet ISO 9000. Wonderful has exported exchangers to countries in the European Union (EU) and Asia. Each time it met the code requirements for construction, inspection, and stamping of the country of destination. But it does not find it economical to maintain qualified status on a continuing basis. However, they have an arrangement with TUV to inspect exchangers built to European codes for shipment to EU countries.

Personnel

Wonderful currently employs approximately 85 people, with 8 in engineering, 4 in sales, and 8 in management and support staff. The factory is run by a shop superintendent to whom report a machine shop forewoman and fabrication shop foreman. Three inspectors, two of whom are qualified in various nondestructive testing methods, monitor quality control. The company is very conscious of the ratio of the costs of production to nonproduction workers. They give weight to this ratio in making decisions about whether to hire additional factory workers or to assign overtime.

Administrative. WHEMCO's president, B. B. J. (Billy Bubba Joe) Steelbender, is a mechanical engineer with about 25 years' experience in pressure-vessel and heat exchanger design and fabrication. He and the vice president of engineering, Abe Numbercruncher, cofounded the company in 1978 and are its major stockholders. Both are Professional Engineers, registered in Texas and Oklahoma. Before then they were respectively the chief mechanical engineer and chief thermal designer of a current competitor. The secretary-controller, Jay Beancounter, reports to Steelbender. He supervises the purchasing and accounting departments and an overworked support staff consisting of a bookkeeper, three clerks, and a receptionist.

Marketing and sales. Sales manager, J. B. (John Boy) Selzam, chief estimator, Hy Guess, and his assistant, Lola Lowball, report to the vice president for marketing, Jack Plansales. Jack and John are professional engineers, registered in several states and Canadian provinces. The company sells directly from the home office but has arrangements with sales representative organizations in various geographic locations. Plansales is authorized to sign proposals and accept orders for domestic shipment to a maximum value of $250,000 in accordance with standard terms and conditions established by Steelbender and Numbercruncher in consultation with Wonderful's law firm Sue'm, Litigate, and Settle. Selzam is authorized to sign such proposals and accept orders up to $100,000. Steelbender approves proposals and order acceptances for larger dollar values and for export.

Plansales establishes the priority of responding to inquiries. Although he and Selzam would like to respond promptly to all, they reject some and respond to others in a rough order of desirability to Wonderful. When they receive an inquiry, they try to find out if customers' inquiries are for estimating, budgeting, appropriation, or purchasing.

Plansales and Selzam also act as job or project managers. They maintain computerized inquiry, proposal, and job files. They write proposals and accept orders using forms with preprinted terms and conditions generated by the corporate attorney. Steelbender resolves conflicts between WHEMCO's and customers' terms and conditions.

On receiving a purchase order, Selzam assigns a job number and issues a multipart factory order. The parts of the factory order are a priced acknowledgment and order acceptance sent to the customer, a priced construction order to the control and sales departments, and unpriced construction orders to engineering, purchasing, inspection, and the factory. The description of what is to be built, materials of construction, pressures and temperatures, and other variables are identical on all the parts.

The controller's department bills customers on receiving shipping documents from manufacturing. During construction, they record engineering and manufacturing labor-hours used, and track costs of materials purchased and withdrawn from inventory using a computer program. With this program they can compare the costs incurred with the ones estimated. The program's output is restricted to the president, vice president for engineering, controller, vice president for marketing, and factory superintendent, Able Fabricant, who jointly review the data on all active jobs each week.

Guess and Lowball estimate costs for proposals and calculate charges for change orders. They use a commercially available, computer-based pricing program modified to suit Wonderful's operation. Hy has loaded WHEMCO's labor standards into the program's database. These have been derived from previous experience in manual estimating, informal time studies, return-labor-use figures, and lengthy discussions with the factory superintendent. When a customer wants an estimate or budget, Guess or Lowball use the most appropriate of the program's approximate prices scheme, an updated historical price or the cruder dollars per unit surface area.

The estimating and purchasing departments maintain updated materials costs files in the pricing program. The estimators do not rely on the database listing for major items of cost. Instead, they solicit suppliers' prices for such items as tubes, tubesheets, forgings, and bellows expansion joints. They have good relationships with the suppliers with whom the purchasing department deals.

Plansales and Selzam assist Steelbender and Numbercruncher in establishing and maintaining a profitable product mix. Steelbender has made a study of profit margins compared with the ratio cost of labor/cost of materials. Numbercruncher has established a computer program for maximizing profits based on this study, the overhead carried per worker-hour, the estimated time between incurring materials cost and shipment of the completed exchanger, and several other factors.

Wonderful's bookings and sales are market-driven, but when it has a substantial baseload of current work and a comfortable backlog, Numbercruncher's program is useful for maximizing the profits on the available production. Manufacturing heat exchangers is a cyclical business. Therefore, as much as they can, Steelbender and Plansales adjust the product mix and pricing policy to enable Wonderful to book as many hours of production as possible when they sense a dry spell in the offing.

Engineering. Numbercruncher supervises Chief Thermal Engineer Flo Calorie and Chief Mechanical Engineer Joe Stresslevel. Thermal designer Ted Heatflow reports to Flo Calorie. Mechanical designer, Pamela Modelmaker, who supervises the CAD/CAM (computer-assisted design/manufacture) work of four drafters, reports to Joe Stresslevel. All engineering and drafting personnel are computer-literate and thoroughly familiar with the CAD/CAM drafting program WHEMCO has purchased. Stresslevel, Calorie, Heatflow, and Modelmaker have taken intensive courses in computer technology. Modelmaker, who has a flair for this subject, is responsible for maintaining, upgrading, and arranging for servicing the computer network. She is currently engaged in hiring an assistant for this purpose.

WHEMCO contracts programming for new proprietary computer programs and fixes and upgrades of existing ones to Izzy Whizbang. He is the computer professor who presented the computer technology courses referred to above. Whizbang's work is relatively inexpensive since his students do a large part of it.

Stresslevel and Calorie maintain an engineering library with complete, current editions of the ASME Code, Addenda, Code Cases and Interpretations, the National Board Inspection Code, API 510, and the codes of countries to which they export. The library has a substantial number of standards to which the codes refer. It has engineering works on mechanical and thermal design, the latest edition of the *American Welding Society's Welding Handbook,* and a subscription to its *Welding Journal* and various works on welding published or sponsored by the Lincoln Welding Company and Hobart Brothers.[6] Joe keeps a private file of papers on tube-to-tubesheet joining. Flo has a similar file on thermal design methods and data. Stresslevel serves on the ASME Boiler & Pressure Vessel Code Committee's Special Working Group on Heat Transfer Equipment (SWGHTE) and the TEMA Technical Committee. From time to time Calorie has served on heat-transfer committees of the American Institute of Chemical Engineers (AIChE) and the ASME.

The company does its own thermal and mechanical designing and verifies and guarantees thermal designs provided by customers. On rare occasions when a customer supplies a complete, detailed mechanical design, Wonderful confirms its adequacy by verifying the customer-supplied calculations or by back-calculating the permissible pressure, thermal, and externally imposed loads from the specified thicknesses.

In addition to using proprietary programs, Wonderful subscribes to commercially available thermal and mechanical design programs. It uses a well-known program for finite element analysis when necessary. WHEMCO, which has the current version of TEMA software, uses it for most vibration analyses. However, when a customer requires a guarantee against vibration failure of exchangers to be used in any but the most simple services or those with unusual configurations, WHEMCO's vibrations consultant Dr. Wilma Wigglesworth performs the analysis. WHEMCO may charge separately for the cost of Wigglesworth's analysis.

Stresslevel and Fabricant are long-term AWS members. With assistance from suppliers of welding equipment and consumables, they prepare most of Wonderful's welding procedures. They are proud of the large number of the company's qualified welding procedures and welders qualified to use them. When a question having to do with welding is beyond their knowledge, they contract for help with welding engineering consultant Mario Fusion.

Purchasing. Purchasing agent Will Bicheap reports directly to Jay Beancount and indirectly to the factory superintendent. He cooperates with the factory in operating the company's computerized inventory control and document management system.

Manufacturing. Superintendent Abel Fabricant runs the factory. After graduating from high school, this unusual 50-year-old began his career working for a competitor as a factory helper. In progressing from helper to factory superintendent, he learned most of the skills used in manufacturing heat exchangers. During this time, he took night courses in shop math, laying out for boilermakers, welding, and fitting. After mastering these subjects, he attended a local engineering school at night, where he earned the bachelor of industrial engineering degree. Because of this background, Fabricant supports the local vocational-education school program for apprenticeship training.

Fabricant's current stable workforce is approximately 60 production workers. Reporting directly to him are (1) a materials manager responsible for receiving, stowing, issuing, documenting, and tracking materials; (2) a quality control and testing manager who operates a small destructive testing facility in a sectioned-off area of the fabrication shop and maintains testing records; (3) the fabrication shop foreman; and (4) the machine shop forewoman. A welding foreman, responsible for qualifying welders and welding procedures, maintaining qualifications current, supervising the storage, documentation, and issuing of welding consumables, reports to the fabrication shop foreman. The welding foreman and quality control-testing manager maintain very tight control over storage and issuance of welding consumables, monitor recycling of flux for subarc welding, and make sure that welders keep at-hand electrodes in warmers, and return opened packages of electrodes for storage in constant temperature ovens.

Two testers report to the testing manager. One is qualified to the American Society for Nondestructive Testing (ASNT) level III radiographer requirements, and the other is qualified to perform level III fluid penetrant examinations, and halogen leak and helium leak tests. Wonderful subcontracts NDT for which it is not equipped or not competent to perform to Readem and Weep (R&W). R&W specializes in radiography, magnetic particle, eddy-current, magnetic flux, and ultrasonic examination. The testing manager has a staff-line responsibility to the chief inspector.

On receiving a factory order, and establishing a tentative production schedule, Fabricant reviews its requirements with the fabrication shop foreman, machine shop forewoman, and chief inspector.

The machine shop forewoman schedules all drilling, turning, milling, and other machining operations on the basis of availability of materials, machine time, and skilled workers. If possible, she assigns each worker to one or two machines. When a machine requires maintenance, she requires its operator to work with the maintenance foreman or mechanics he assigns to the repair. She also decides when it is more economical to subcontract a machining operation than do it in-house. She supervises job time entries on machine shop job time cards.

The fabrication shop foreman schedules all shearing, burning, plate rolling, and forming; coordinates use of welding and assembling equipment; and deploys movable equipment, such as turning rolls and assembly beams, and schedules testing. Through Fabricant, he advises purchasing when an operation intended to be done in-plant must be subcontracted. He supervises job time entries on fabrication shop job-time cards.

The chief inspector, Hank Eagleye, scrutinizes the purchase specification to acquaint himself with paragraphs that require the customer's inspector's presence during various operations. He enters these and operations that require the Authorized Inspector's acceptance on the "traveler," a document that accompanies parts and assemblies as production goes forward.

Fabricant assigns construction of an exchanger or group of identical ones to a senior or first-class fitter and senior or first-class qualified welder who regularly work together and whom he deems up to the job. He allocates floor space on the basis of exchanger size and weight and the capability of minimizing handling parts, subassemblies, and the main assembly. The fitter draws from the factory's labor pool second-class and apprentice fitters, welders and helpers, tube joiners, and other skilled personnel required to run the job.

Fabricant compares the job time cards with the estimated times for performing various operations. He hands them on to the bookkeeper, whose payroll clerk conforms the sum of daily times each worker reports with the times punched in and out for payroll. After resolving discrepancies with Fabricant, the clerk enters the times in the cost control record. Fabricant has experimented with having a full-time factory time keeper but found it unsatisfactory.

WHEMCO is currently investigating an advanced payroll and job timekeeping and labor cost-allocation system. It consists of terminals conveniently located throughout the factory, which are hard-wired to a central processing unit (CPU) with a battery-backed clock. The terminals have a keypad and magnetic reader slot. Each employee is issued a plastic card that has a magnetic identifying number. Workers insert the card on entering and leaving the premises and each time they start and finish applying time to a job. A panel on the terminal instructs the employee to type in *0 to indicate clocking in for work and #0 to indicate clocking out for the day. It instructs workers to punch in * and the job number to indicate start of applying time to a job and # and the job number to indicate clocking out of the job. Separate job codes are used to track time applied to nonproduction tasks. The CPU is programmed to keep track of the applied times and to reject time input for a new job until the previous one has been clocked out of. The program generates payroll records, paychecks, job time records, and records of times applied to nonproduction tasks.

Quality control. Chief Inspector Eagleye reports directly to President Steelbender and has a staff relationship with Fabricant. He deals with Authorized Inspectors on code inspections and other code matters and with customers' inspectors. Hank has two junior inspectors to help him make sure that the factory complies with required quality control standards. Eagleye works closely with the materials and testing managers.

Staff relationships. Steelbender believes administrative, engineering, marketing, and sales and factory personnel are equally important for the company's well-being. Therefore, he has established a policy of encouraging people to discuss their work with each other. He makes himself available to factory and office personnel, whom he encourages to use intelligence and initiative. He regularly tours the manufacturing facilities, can greet nearly all employees by name, and knows a little bit about their families and personal interests. However, he defers to Fabricant for final decisions about factory-personnel-offered suggestions. Eagleye reports directly to Steelbender any problems encountered with quality control that he is unable to resolve on the factory floor.

Bicheap coordinates with Plansales and Fabricant WHEMCO's effort to minimize inventory of plate, tubing, forgings, and other parts. When making a proposal, Plansales would like to be able to tell customers that all critical material is on hand. Bicheap is constantly aware of the cost of money tied up in raw-materials inventory. They balance current procurement time for materials against Plansales's needs. Fabricant has to offset the benefit of having materials available just when needed against cost of the space and handling required to maintain a stock of materials. All are aware that

Steelbender tracks the value of production per square meter (per square foot) of factory floor space and per factory worker-hour. It is a standing company joke that heat exchanger fabricators and other successful job shops were using "just-in-time production" long before it became an industrial byword.

Using Wonderful's scheduling software, they set up a preliminary critical construction path and try to coordinate it with Wonderful's current backlog. With this information they establish the delivery schedule to quote in responding to inquiries. A standard proposal clause states that all quoted shipping schedules are based on current availability of labor and materials.

When a proposal turns into an order, Selzam, Stresslevel, Bicheap, and Fabricant revisit the estimate and critical-path schedule. They establish a current engineering, drawing, procurement, and production schedule. Selzam advises the customer of the current timetable, which, he states, is based on approval of drawings in a specified time.

With Mario Fusion's help, Fabricant designed and constructed a portable bolted-together cleanroom for welding titanium, zirconium, and other metals that require a clean atmosphere. It can be set in place and assembled in approximately 48 worker-hours. Disassembly takes somewhat less time. This structure is fitted with a panel that has power, oxygen, propane, and compressed-air connections that connect with those in the main bay. It is supplied with clean, filtered air under slight positive pressure through flexible ductwork that draws air from outside the factory.

Office equipment and infrastructure

The offices are equipped with modern desktop personal computers (PCs) with color monitors, high-speed modems, and optical-disk readers and supplied with graphics, CAD/CAM, and communications software. The individual PCs are connected in a local area network (LAN). The desktops share black-and-white and color laser printing facilities, machines for scanning, and plotters. WHEMCO can electronically transmit drawings to customers equipped to receive such transmissions.

The company has high-speed reproduction equipment and facilities for binding manuals. Wonderful has a substantial investment in software, computer equipment, and wiring. It also has fax/modem-equipped laptops for its personnel's use when traveling. Steelbender has invested in a video camcorder and editing system that WHEMCO uses for training, documenting certain operations, and advertising. Videographic work is contracted to small local firm, Show N'Tell, who uses Wonderful's equipment.

Internal communication is interconnected with the telephone system and annunciators in the factory. Each individual has a direct-dial number, voice mail, and a recent version of an electronic-mail (e-mail) system. High-speed fax machines, modems, etc. complete a modern office environment.

Product line

Wonderful Heat Exchanger Company produces custom-designed pipe- and rolled-plate-shell tubular process and power-plant auxiliary heat exchangers, waste-heat boilers, and electrically and dowtherm-heated steam generators. They have built some low-pressure boiler feedwater heaters and replacement feedwater heater bundles but do not consider themselves to be in that business. Their size-range limits are approximately 270 mm (~10 in IPS) to 4.5 m (~15 ft) O.D. They are most competitive in larger pipe sizes and rolled plate shell sizes smaller than 3 m (~10 ft). They ship exchangers larger than 3.6 m (~12 ft) in diameter on special low-bed trucks over routings acceptable to the various jurisdictions through which they pass. Their length limit is approximately 20 m (~66 ft).

Wonderful takes pride in being competitive with North American, Asian, and European manufacturers in all the common heat exchanger materials such as carbon and low-alloy steels, high alloys, and nonferrous metals. Although its two bridge cranes can together lift only 30 metric tons (t) (~33 tons), WHEMCO has built exchangers weighing as much as 45 t (~50 tons), using a spreader beam between the two cranes to lift one end at a time onto flatbed railcars. WHEMCO's plate thickness range is 4.75 to 65 mm (~3⁄16 to 2½ in) depending on material of construction and shell diameter. They construct for vacuum and pressure services, with the uppermost pressure limited to 3000 psi (~20,700 kPa) and the limiting product of nominal diameter and pressure 60,000 lb/in (~10,508,000 kPa•mm).

In addition to manufacturing new tubulars and replacement bundles and parts, WHEMCO has a lucrative business retubing, repairing, and altering heat exchangers. For the most part they do this work in the factory or adjacent outdoor area. However, WHEMCO has made on-site repairs and alterations to accommodate good customers. Their capability for field operations is limited by lack of field service equipment and their nonunion status. Steelbender, Numbercruncher, and Fabricant have sketched out a plan for possible future entry into field services.

WHEMCO will not work on an exchanger or bundle to be retubed, rebuilt, or altered unless it has first been thoroughly cleaned. They require the customer first to remove all plugs from plugged tubes or to cut through plugged tubes behind the tubesheet. Customers must agree to provide a certificate attesting that the exchanger or bundle is free of noxious, flammable, explosive, or carcinogenous materials, and materials data sheets for the substances that flowed through the unit before WHEMCO accepts an exchanger for rework. These practices have somewhat dampened this business because some of WHEMCO's competitors are less cautious.

Fabricant, the fabrication shop and welding foremen, and the machine shop forewoman discuss any safety hazards with the crew selected to do the job and provide them with copies of the materials data sheets. Workers who are concerned about their health or safety are free to refuse to work on an exchanger that has been in service without prejudicial effect on their standing with the company.

Wonderful tries to apprise the scrap yard with which they deal of the potential hazards from entrapped substances when selling them parts of retubed or altered exchangers. They provide copies of the materials data sheets and recommend that air be blown through the shell and each tube before cutting or burning through the bundle.

Land and buildings

Figure 3.1 shows the overall layout of Wonderful's buildings and property. The plant is aligned approximately north and south with columns 1E and 1W at the north end and columns 18E and 18W at the south. It is built on 1.5 ha (~3¾ acres) of land. The main structure's west wall is located 3 m (9 ft 9 in) from the west property line, and the north wall is set back from the street 10 m (~32 ft 6 in). The building is as close to these property lines as the Board of Zoning permitted when the factory was erected.

The factory is a structural-steel frame surrounded by lightweight block and glass-window walls that carry the roof. The walls bear only wind, earthquake, and snow loads. It has one main bay 20 m (~65 ft) wide and 102 m (~335 ft) long, center-to-center of end columns. The floor is heavy-duty steel-reinforced concrete. Two floor-operated bridge cranes, which ride on common rails, span the main bay. Each can lift 15 t (~16.5 tons). The craneways are set on the tops of stub columns tied to the roof-support columns by steel connector plates. The maximum height, crane hooks to floor, is 9.5 m (~31 ft). The columns, footings, and craneways were designed and built to accommodate an additional 30-t (~33-ton) cab-operated crane, but the owners have not been able to justify the expenditure or amass the required capital.

On the west side a railroad/truck door is located midway between columns 16 and 17. A railroad track set flush with the main-bay floor connects with a siding and extends beyond a second railroad door set

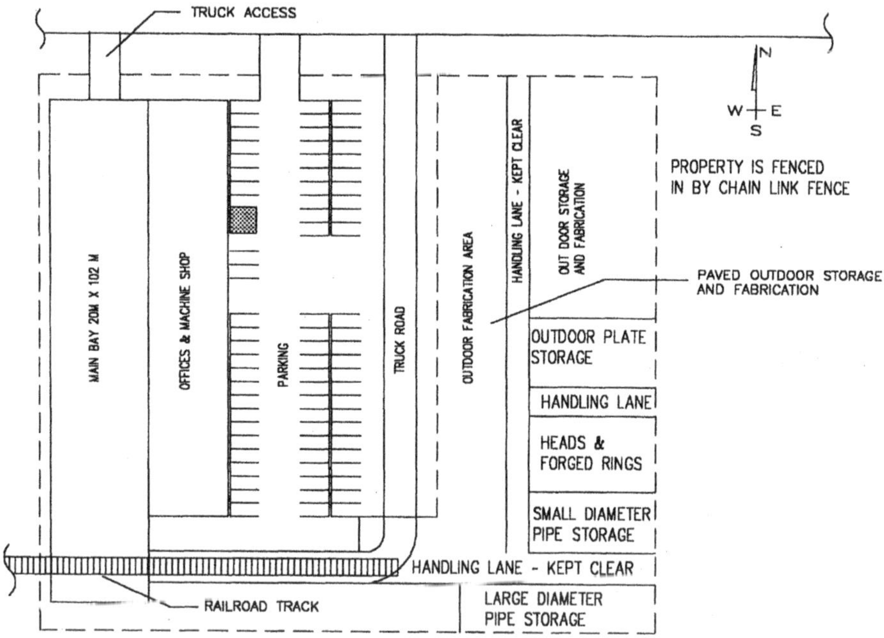

Figure 3.1 Layout of Wonderful Heat Exchanger Manufacturing Co., Inc. plant site, 1000 Tubular Exchanger Street, Oilfield City, Okla.

between columns 16 and 17 on the east. The east-side track extension can accommodate three railroad cars end to end between the factory wall and the deadhead bumpers. The north wall of the main bay has a truck loading door. Rails, set flush with the floor midway between the east and west walls, run from 10 m (~33 ft) north of the north wall exterior to one column before the railroad track that crosses east to west. Wheeled trucks can be set on these rails for rolling heavy sections along the shop. The trucks may also be used for loading bundles into shells and extracting bundles.

The building's columns are on 6-m (~19-ft 6-in) centers. Beginning with columns 2E and 2W, every second column is provided with fittings for a jib crane at an elevation of 5 m (~16 ft 4 in) above the floor. The jib arms are 6 m (~20 ft) long I-beams that can swing 180°. A manually positioned 3-t (~3.3-ton) electric hoist rides on the lower flange of the jib arm. The distance between the factory floor and the jib crane hoists is approximately 4.5 m (~14½ ft). Fabricant sets jib cranes and hoists where they may be used to best advantage.

Connected to the main bay alongside its east side is a 16-m (~52-ft)-wide structure running from column 1W to column 15W, a distance of 84 m (~273 ft). This building is divided into a single-story factory section open to the main bay and a two-story office section. The lower floors are walled off from each other at column 10W, 54 m (~175 ft 6 in) from the front of the building. The office section extends from column 10W to column 15W. Fireproof doors provide access from the executive and administrative offices to the factory floor.

The factory section houses a 38-m (~123-ft 6-in)-long machine shop. Tool room, factory testing and inspection office, and personnel facilities occupy an additional 16 m (~52 ft).

The machine shop is traversed by three 5-t (~5.5-ton) manually positioned, floor-operated bridge cranes on common craneways. Heavy plastic curtains separate it from the main bay. Blowers maintain a slight positive pressure in the machine shop to exclude fabrication dust.

The factory areas are illuminated by natural light and high-intensity mercury-vapor lighting. During cold weather, the factory temperature is maintained at approximately 15°C (~59°F) by natural-gas-burning radiant heaters except in the hydrostatic test area, where it is maintained between 21 and 23°C (~70 to 73°F).

The plant is supplied with 1000 A of 480-V, three-phase, 60-cycle power, transformed to 220 V three-phase and 110 V single-phase. Air-driven tools can be connected anywhere in the factory to a lubricated, 690-kPa (~100-psi) compressed-air loop. Propane for torch burning is piped to various burning areas from an outdoor spherical pressure storage tank.

The hydrostatic test area has a regenerative water demineralizer for supplying hydrostatic test water, a filtered air blower in line with an air dryer and warmer, and storage space for nitrogen purging and testing bottles. WHEMCO pumps hydrostatic test water from the demineralizer into an air-cushioned storage tank that maintains a pressure of 690 kPa (~100 psi). The test workers raise exchangers to the designated hydrostatic test pressure by means of Baker Oil-Field pressure testing pumps that can develop 69,000 kPa (~10,000 psi) of pressure. The factory conducts gas pressure tests with considerable reluctance, and then only at a maximum of 80 percent of the pressure permitted by the ASME Code but no higher than 6900 kPa (~1000 psi). WHEMCO follows to the letter the code's restrictions and requirements for incremental pressure increases. They use nitrogen for gas pressure tests and record all pressure tests on a strip chart.

The remaining, completely paved, usable area of approximately 1 ha (~1¾ acres) is devoted to personnel parking and a leveled area paved with load-bearing reinforced concrete for materials storage and fabrication. Built into the concrete are sockets for pipe columns. The socket pattern is arranged so the columns can support panels of lightweight fiberglass-reinforced corrugated coverings that can be set in place as required to provide protection from the weather. Serving the outdoor work area are two forklift trucks and a motor-driven "cherry picker" crane, each with 5 t (~5.5 ton) lift capacity.

Factory layout and operation

Figure 3.2 shows the general layout and flow path of the WHEMCO factory. WHEMCO receives most raw materials through the truck

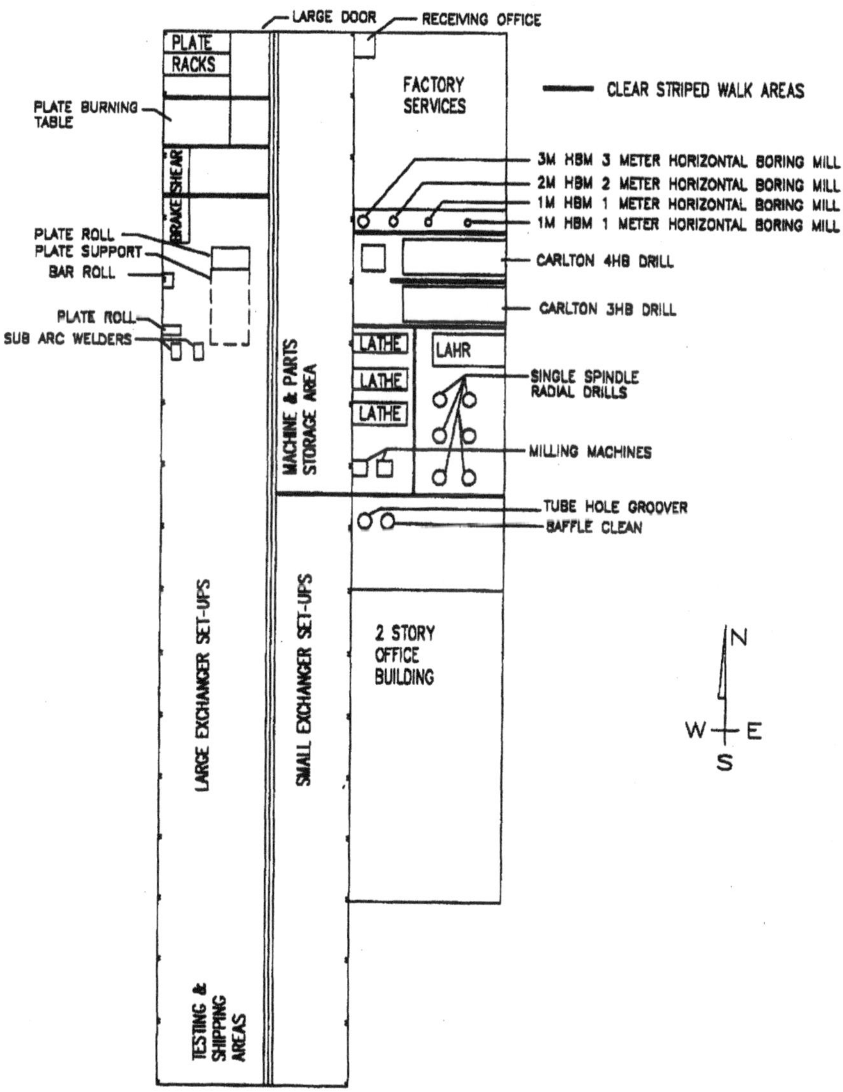

Figure 3.2 General layout and flow path of WHEMCO factory.

door at the north end. Eagleye or his people inspect the material and tag it *accepted* with a pressure-sensitive green tag or *rejected* with a red one. The tags are in two parts. One is placed on the material and the other, on the receiving document passed to purchasing. The red/green acceptance/rejection tagging system is used to designate acceptance and rejection throughout manufacture, with one part placed on what has been accepted and the other on the paperwork or traveler that accompanies fabrication. The red tag cannot be removed from the rejected piece unless a factory inspector signs acceptance on the traveler.

Mill test numbers are recorded, and the original certified mill test reports are filed in a central file with copies placed in the job files, engineering files, and inspection office. Job (serial number) identifications are marked on materials when received, after which they are moved directly to the site where they are to be used, parted and cut to size, or stored until required. No matter how the material is disposed of, it is entered into the computerized inventory. Inventory is relieved only when the person who draws material signs for its receipt. In this way, Wonderful maintains very tight inventory and materials tracking.

Bicheap buys plate and pipe that Fabricant must have on hand for smooth production. The materials handler receives and stores these stock plates outdoors. To avoid excess handling, plate is stored on its long edge in plate racks. Stocked plate is segregated by specification. Within each specification type, the plate is separated by thickness, width, and length. Bicheap and Fabricant select what they believe are the most advantageous thicknesses and sizes for plates to maintain in inventory.

Bicheap usually purchases double random lengths of pipe 457 mm O.D. and larger (corresponds with 18 in IPS and larger) for Fabricant's inventory. These are also segregated by specification, diameter, wall thickness, and measured length.

Wonderful does not carry an inventory of heads. Because they take up so much space, ones larger than 1675 mm (~66 in) received for jobs are stored outdoors. Other items stowed outside are large forgings, test fittings for floating-head bundles, and boxed or weatherproof wrapped bellows (thin-walled) expansion joints.

Wonderful is not equipped for U-tube bending. Steelbender, Bicheap, and Fabricant conducted a study showing that the dollar value of production per unit of floor space did not justify the required investment in equipment, training, and facilities. When possible, they purchase prebent, heat-treated, and tested austenitic and some ferritic high-alloy and nonferrous U-tubes from the tube mill. When they cannot procure mill-bent tubes, they subcontract the bending to nearby fabricators who have U-bending and U-tube heat-treatment facilities.

The factory layout is designed to require the least possible handling of parts and finished exchangers. In contrast to production-line factories, where parts flow toward a moving line on which exchangers are assembled, at WHEMCO exchangers are assembled at a stationary location to which parts and subassemblies are brought. They are moved only when required for testing, preparation for shipping, and loading on for transport.

As Fig. 3.2 shows, all burning, plasma cutting, tube preparation, bending, and forming operations are done on the northwest side of the shop. All machining operations are done on the northeast side of the shop. Finished machined parts for assembling bundles are loaded onto pallets and set adjacent to the bundle assembly beams.

Propane is piped into the factory and to the outside work areas from a storage sphere placed out of harm's way at the southeast corner where the machine shop's south wall intersects the main-bay east wall. Fabricant has set up a plate burning table near the plate storage area. He has equipped it with semiautomatic tracking torches for burning multiple patterns. He also keeps a pair of oxypropane portable burning outfits there, along with 200 and 400 amps DC and 600 amps AC welding machines there.

Subassemblies are welded on welding positioners, on tank turning rolls, or at the submerged arc welding stations. Depending on fabrication requirements, welded subassemblies may be returned to the machine shop for further machining after welding. Subassemblies that require heat treatment after forming and welding are trucked to Hotstuff Heat Treating, Blasting, and Coating Company on the other side of town. Cut and formed and edge-beveled parts are welded on the subarc girth and long seam welders, whenever economical, and subsequently moved to the assembly locations shown in Fig. 3.2 as large and small exchanger setups. These areas are provided with bundle assembly fixtures (Fig. 3.3) and turning rolls.

When a fixed-tubesheet unit is large enough, WHEMCO assembles the bundle in the shell. For fixed-tubesheet units too small to permit such assembly, and for all removable-bundle units, they complete the bundle on the bundle assembly fixture, complete (or nearly complete) the shell on the turning rolls, and stab the bundle into the shell. An additional operation, hanging the second tubesheet, is required for fixed-tubesheet units. After final assembly the exchanger is moved to the testing-shipping area at the south end of the factory, where it is hydrostatically tested, dried, and leak tested when required and its nameplate is applied and stamped in the Authorized Inspector's presence.

Wonderful may apply a coating of paint if prior sand or grit blasting is not required. However, when the coating system requires such surface preparation, they subcontract the work to Hotstuff. For that pur-

Figure 3.3 Bundle assembly fixture. (*Atlas Industrial Mfg., Clifton, N.J.*)

pose, the finished exchanger is trucked out of the east railroad door to Hotstuff and back in through the same door for final skidding, crating, or boxing.

Equipment

WHEMCO saved considerable capital by judiciously purchasing much of its heavy equipment, used, at auctions of bankrupt organizations, from companies selling surplus equipment, and dealers, and reconditioning it before installation. However, all its welding equipment and its large hydraulic plate roll was purchased new as were the cranes, hoists and forklifts, computer-controlled lathes, and drills. Following is a partial list of WHEMCO's major equipment.

Materials parting and trimming

(1) Plasma arc, computer-guided outfit: 130 mm (~$5\frac{1}{8}$ in)-thick high-alloy plate

(1) Eight-torch computer-guided burning outfit: 75-mm (3-in)-thick steel plate

(1) Plate shear, capable of shearing 20 mm (~$\frac{3}{4}$-in) × 3 m (9 ft 9 in) wide

(2) Manual, handheld plate edge bevelers capable of beveling austenitic plate 20 mm (~$\frac{3}{4}$ in) thick

Forming

(1) 350-t (385-ton) gap-type hydraulic press with 600-mm (~$23\frac{5}{8}$-in) platen

(1) Hydraulic, drop-end plate bending roll, 20-mm (~$\frac{3}{4}$-in) × 3-m (9-ft 9-in) capacity for cold, 475,000-kPa (~70,000-psi) yield steel

(1) Gear-driven, drop-end, initial pinch plate roll 6.5-mm × 1.25-m (~$\frac{1}{4}$-in × 4-ft) mild steel;

(1) Gear-driven, drop-end plate bending roll, 100-mm-thick × 2450-mm-wide (~4-in-thick × 96-in-wide) hot SA 516 Gr. 70 and 45-mm-thick × 2130-mm-wide (~$1\frac{3}{4}$-in-thick × 84-in-wide) cold

(1) 150 × 150 × 12-mm (~6 × 6 × $\frac{7}{16}$-in) Buffalo combination angle/bar bending-shearing machine

Machining

(1) Vertical-boring mill, 3-m (117-in) table

(1) Vertical-boring mill, 2-m (78-in) table

(2) Vertical-boring mills, 1-m (39-in) table

(1) Horizontal-boring mill (floor type) with 180-mm (~7-in) spindle and facing head and 3-m (~9-ft 9-in) swing

(3) NC lathes from 460- to 900-mm (~18- to ~36-in) swing × 2500 × 3650-mm (~8-ft 3-in × 12-ft) bed

(1) Semiautomatic facing and drilling machine

(1) Semiautomatic buffing machine for tubesheets and baffles

(1) Carlton 4HB computer-controlled drilling machine with milling capacity

(4) Single-spindle radial drill presses, with numerical or manual control
(2) Bokum grooving tool assemblies
(1) High-speed tube-hole grooving machine
(1) Portable milling machine

Welding and assembling

(9) Manual turning rolls 600-mm (~24-in) to 3000-mm (~120-in)-diameter capacity × 5-t (~5.5-ton) capacity
(8) Power turning rolls 600-mm (~24-in) to 3000-mm (~120-in)-diameter capacity × 5-t (5.5-ton) capacity
(6) Welding positioners
(2) Linde Unionmelt subarc welding machines
(1) Arcos overlay strip welder
(3) Gas-tungsten arc welders, 600 amp
(21) AC, DC 50- to 1000-amp welding machines
(2) 400-amp diesel-engine-driven DC portable welding machines
(2) Hobart Orbitig tube-to-tubesheet welding machines
(2) Gas-metal arc welding machines, 600 amp
(3) Vernon Oremco hydraulically driven and controlled tube expanding machines
(1) HydroPull hydraulic tube expanding machines
(1) HydroPull 500 hydraulic tube and stub puller

Nondestructive testing

(1) GE halogen leak detector
(1) Portable mass spectrometer (helium leak detector)
(1) Fluorescent-light fluid penetrant kit; miscellaneous dye penetrant equipment

Laboratory

(1) Research metallograph
(1) Binocular microscope
(1) 850-kN (~190,000-lb) tensile testing machine
(1) Stationary hardness tester
(1) Portable hardness testers

Miscellaneous camera equipment, impact testers, etc. are also included; not included are bundle assembly beams; tube-hole, tube-diameter, and thickness measuring equipment; and the large number of miscellaneous tools and equipment with which fabrication shops equip themselves.

Large Exchanger Factories

Large custom factories are much like intermediate-sized ones in layout and production flow and control. But they employ more personnel, occupy more space, and have equipment with greater capacity. Frequently they build pressure vessels, columns, and other large structures in the same facility. However, they customarily reserve a discrete area of the plant for exchanger and feedwater heater production. Manufacturers who also produce steam surface condensers either construct them in a separate facility remote from the main plant or set apart an area of the plant for condenser manufacture.

The buildings of one large North American exchanger manufacturer occupy approximately 20,600 m^2 (~218,000 ft^2) on approximately 4 ha (~10 acres). The company has a separate smaller facility where it builds steam surface condensers. Another plant is on approximately 22 ha (~55 acres), with its buildings occupying about 66,200 m^2 (~700,000 ft^2). This firm also builds large pressure vessels, towers and columns, spheres, and other types of equipment. A third manufacturer, who constructs process and power tubular exchangers and is a major feedwater heater manufacturer, has land and buildings midway between these two. All three have access to truck, rail, and barge shipping facilities.

Large manufacturers may build exchangers with proprietary channel-access designs for higher pressures and a greater pressure-diameter range than do intermediate-sized and small ones. Their products may include high-temperature waste-heat steam generators, large tubular catalytic reactors, feedwater heaters, and replacement feedwater heater bundles. Some large shops may exclude pipe-shell sizes altogether and others ones, smaller than about 600 mm (~24 in). Some typical attributes of the loosely defined large factories are described in the following paragraphs.

Large manufacturers employ more salaried people than do intermediate-sized ones, but may have more favorable ratios of production to nonproduction people. This competitive advantage is offset by the larger capital investment in land, buildings, and machinery than that of their smaller competitors. It is not possible to state a general rule about whether an intermediate- or a large-sized manufacturer will be more competitive for exchangers within the competitive range of both facilities.

Larger organizations usually employ full-time welding and quality assurance engineers. Project engineers, whose primary responsibility is to conduct jobs along their preestablished critical paths and eliminating bottlenecks, shepherd orders through the plant. Factory personnel are more likely to be unionized than at smaller works. The factory may have a formal apprenticeship training program.

Some physical ways in which large shops differ from small- and intermediate-sized ones are

1. They occupy substantially more land, usually located farther toward the outskirts of town and often adjacent to a river or canal on which barge shipments can be made, such as a facility one of the authors surveyed on 11 ha (~28 acres) adjacent to an estuary leading to a major port.
2. They usually have two or more main fabrication bays as wide and long as or wider and longer than the WHEMCO one.
3. Large factories can handle larger sizes and thicknesses and heavier weights than can intermediate-sized ones.
4. They have a greater range of drilling and machining capacities and some machines not found in smaller companies—for example, multispindle drilling machines with as many as 150 spindles, two-spindle gun drills, a plate edge planer, head spinning machine, bulldozer forming machinery for forming special shapes, and higher-capacity shearing and braking equipment.
5. They have more welding equipment, which may include plasma cladding equipment, additional tube-to-tubesheet orbiting-head welders, internal-bore welding apparatus, electroslag, electron- and laser-beam welding gear, and a separate cleanroom for welding exotic alloys.
6. They may perform explosive tube-to-tubesheet expanding and welding.
7. They may have in-house heat-treating furnaces, tube bending facilities, and a greater number of more sophisticated testing equipment housed in dedicated structures.
8. Company personnel use company-owned equipment to perform almost all nondestructive examinations.
9. They may have a concrete shielded room for radiography and developing and interpreting equipment.

Some larger factories are equipped to do hot forming, have car-bottom furnaces capable of heat treating large exchangers and parts, and have separate structures in which exchangers can be sand-, grit- and glass-bead-blasted and spray- and electrostatically painted or coated.

Repair and alteration facilities located near large oil-refining, petrochemical, and chemical complexes have distinctive equipment and procedures for rapid turnarounds and for quickly producing replacement bundles and exchangers. This may include hydraulic and mechanical bundle pulling equipment,* specially equipped trucks for bundle and exchanger transport, chemical cleaning tanks, and hydroblasting facilities. Their repair-alteration business often overlaps with designing and building new exchangers.

Feedwater Heater Manufacturers

The fifth (1992) edition of the *HEI Closed Feedwater Heater Standards,* widely used in North America, lists five members considered to be the major North American closed feedwater heater manufacturers: Foster Wheeler Energy Corporation (Clinton, N.J.), Marley Heat Transfer Company (Houston, Tex.), Struthers Industries, Inc. (Warren, Pa.), Senior Engineering Company (Los Angeles, Calif.), and Yuba Heat Transfer Corporation (Tulsa, Okla.)

Foster Wheeler Energy currently produces heaters at its factory in Danville, New York and cooperates with Foster Wheeler Limited (St. Catharines, Ontario, Canada) on some heater projects. Engineers and Fabricators (EFCO) of Houston, Texas builds Marley heaters and other tubulars in its main plant. They have a second factory that builds other larger tubulars. Struthers builds feedwater heaters in a factory at Gulfport, Mississippi. Senior Engineering (now part of Thermal Engineering) has a factory in Joplin, Missouri. The Yuba factory is in Tulsa, Oklahoma.

Reflecting market conditions over the years, other large North American heater manufacturers, such as ALCO Products, C. F. Braun, Griscom-Russell, and M. W. Kellogg, have come and gone in the feedwater heater and heat exchanger businesses. Undoubtedly other capable manufacturers will be included in the future and some now listed will drop out.

Manufacturers such as Nooter Corporation of St. Louis, Missouri, have built feedwater heaters from time to time. Some shops, such as Hydrodyne of Massilon, Ohio, change the mix of their work from mostly repairing and altering feedwater heaters to one more engaged in building new ones. Other heat exchanger manufacturers may be capable of building satisfactory feedwater heaters competitively.

*Examples are hydroextractors and bundle wagons.

However, purchasers are cautioned to examine their design and manufacturing experience before soliciting bids.

Not every organization physically equipped to manufacture feedwater heaters knows how to design and build them. Readers are advised to investigate thoroughly companies not widely known as a feedwater heater manufacturers who ask to bid on supplying feedwater heaters. If thorough research indicates competence and competitiveness, it would be prudent to allow the prospective manufacturer to bid only on small, low-pressure heaters until a record of experience has been established.

No factory devotes itself exclusively to manufacturing feedwater heaters. All produce other kinds of tubular exchangers. Although one of the best known has more than 5000 heaters installed worldwide, they also build shell-and-tube heat exchangers, steam surface condensers, moisture separator reheaters, deaerators, combustion air preheaters, flue-gas reheaters, dump condensers, submerged-tube evaporators, and recuperators in sizes from 610 to 4875 mm (~24 to 192 in) diameter. Their factory occupies approximately 19,000 m^2 (~200,000 ft^2).

They construct rectangular and round steam surface condensers and, depending on shipping and installation clearances, shop-tubed or field-assembled and modular condenser sections for retrofits.

This organization's engineering department uses CAD and proprietary programs to optimize their products. When appropriate, they perform finite element analysis for static, transient, and upset conditions to which the tubulars may be subjected. They have designed and built feedwater heaters for feedwater pressures as high as 34,500 kPa (~5000 psi) and other tubulars for pressures as high as 51,700 kPa (~7500 psi).

This manufacturer advertises that it has built equipment weighing 181,500 kg (~400,000 lb). In addition to equipment similar to that listed for WHEMCO, the factory has three multispindle drills that can drill as many as 41 holes simultaneously, numerically controlled four-, three-, and one-spindle drills that can drill through a 762-mm (30-in) forged-steel tubesheet with minimal drill drift and plasma cutting facilities with a 3 × 98-m (~11 × 32-ft) table. They have modern welding equipment, including electroslag equipment and orbital-head tube-to-tubesheet welders. Their workforce is skilled in mechanically, hydraulically, and explosively expanding tubes.

Some of the major firms that manufacture feedwater heaters also offer these field services for their own and other manufacturers tubulars:

Testing

Repairs

Tube plugging

Inspection

Troubleshooting

Maintenance

Removal and replacement

Turnkey installations

Field Service Organizations and Erectors of Exchangers Too Large to Ship

Diverse organizations service exchangers in the field and erect ones too large to ship from the factory. Some may have engineering departments well versed in analyzing the structures and their erection requirements. Others may simply be capable of doing the mechanical erection or other service work. These organizations buy almost all materials and parts needed for repairs or alterations against a job bill of material. Therefore, unless the organization is also a manufacturer, it probably does not carry an inventory of raw materials. However, it must have a system for receiving, issuing, and tracking materials.

Welding performed in the field on pressure parts or parts welded to pressure parts of code-stamped exchangers must be performed by welders qualified in accordance with the prevailing pressure-vessel code requirements for materials and thicknesses and in welding positions. They must maintain accountability by using a system of stamping welders' symbols adjacent to the welds. The NBIC and API 510, discussed in Chap. 6, require radiography and heat treatment in accordance with the rules of the pressure-vessel code that applied in the year the heat exchanger was built. The service organization must document its fulfillment of these requirements.

As noted above, some exchanger manufacturers work on their own and others' tubulars at job sites, and do turnkey installations of exchangers, feedwater heaters, condensers, and associated equipment. Frequently field service organizations' key personnel work with skilled workers and laborers hired for each job from the local labor pool. For the most part they rent heavy lifting equipment, welding machines, and air compressors from firms near the job site. However, they provide any special machines critical to the work and smaller tools. These may be portable flange facing machines, pass-partition groove millers, hydraulic tube pullers, tube expanders, orbital-head tube-to-tubesheet welders, etc.

Erectors of tubulars too large to ship prefabricate as many pieces as possible and match-mark the parts for assembly. A job superintendent may follow the course of production in the factory and carry the responsibility of the field assembly. The superintendent arranges for subcontracting operations such as field heat treating, blasting, and coating.

Work in the field takes much longer than in the factory. Field service organizations must mesh their safety and health practices with those of owners. Temperature extremes, wind, rain, sleet, and snow may impede outdoor work. The contractor may have to erect temporary shelters or "dog houses" to protect workers, equipment, and exchanger parts. Installations, repairs, or alterations during scheduled outages are subject to delays caused by the conflicting needs of the various contractors. Owners' work practices sometimes subject field service work to interruptions.

R-symbol holders, repairing or altering National Board–stamped and –registered heat exchangers in accordance with the NBIC, must coordinate the work with their own and the Owners' Authorized Inspection Agencies and Authorized Inspectors.

Production-Line Heat Exchanger Factories

Production-line factories do not manufacture tubulars to meet specific service conditions. Instead, they make graduated sizes of the same configuration, which they sell as commodities from catalogs. The selection of materials and available options is limited. Catalogs of exchangers for air-conditioning and refrigeration service may tabulate the capacities of the various sizes to enable purchasers to choose models and sizes that fit the service conditions. The design, purchasing method, and production scheme reflect these differences.

Design

Heat exchangers to be produced as a commodity must be designed for a specific market or group of markets. Each one in the product line must be able to provide thermal performance for a group of similar services at acceptable pressure losses over a range of flows, pressures, and temperatures. The range of capacities of incrementally sized units in the progression must not substantially overlap. Ordinarily, the whole series has the same maximum pressure and temperature ratings and similar tables of permissible coincident shell-and-tube metal temperatures.

The exchangers must lend themselves to mass-production techniques. Mechanically, the designs are based on readily available, standard-sized raw-materials components. Bills of materials consist of indented lists of assemblies, subassemblies, sub-subassemblies, and parts, each of which has its own assembly or part number and drawing. Manufacturers use parts identified by one number and dimensioned on one drawing in constructing as many models and sizes as possible. They engineer assembly drawings, which are retained on the factory floor, to cover several sizes. Shells and channels of rolled plate or sheet may be designed to be laid out and punched, perforated, or drilled in the flat before rolling and welding. Designers pay considerable attention to eliminating production costs by using jigs and fixtures to eliminate manual labor.

Marketing

Production-line shops may make sales through stocking distributors, wholesalers, and retailers, with prices based on a preestablished discount schedule. Factories frequently sell these exchangers to original equipment manufacturers for incorporation into their products. Commercial building contractors who erect refrigerated structures may also buy and install air-conditioning and refrigeration exchangers. The marketing department constantly reviews projected sales with its distributors and wholesalers and estimates the required numbers of each size of unit in the line to meet the projections. The marketing group writes factory orders on the basis of collations of orders from various sources, projections, and estimates.

Production and purchasing

Production-line factories are set up to produce a limited number of model sizes and types of exchangers. They equip their assembly lines with some of the same equipment as job shops for shearing, plate rolling, welding, and tube expanding. The arrangements are much like automotive ones, where raw materials enter one end and pass through the hands of workers who perform repetitive operations at fixed locations.

These plants manufacture inventories of identical parts that feed the streams of shells, bundles, and channels that move toward assembly of finished exchangers. Some parts may be castings cast in economical quantities. Production controllers initiate, control, and track production and procurement. Subcontractors, who may punch instead of drill holes, may provide baffles and tube supports when possible. Roll operators set their equipment to roll groups of same-sized shells and channel barrels rather than making adjustments to roll one or a few at a time.

The objective of the factory's layout is to ensure that materials, parts, subassemblies, and assemblies move smoothly to the various machines and assembly stations. Some factories build exchangers on carts that roll from station to station. Others make assemblies from parts that move along the production line on conveyors. Welding is likely to be computer-controlled gas-tungsten arc or gas-metal process and incorporated into robotic systems for positioning, handling, and joining. When brazing is done, it is by automatic methods.

Most production-line exchangers are built with roller expanded tube-to-tubesheet joints with the rolling done by multiple roll machines. Recently some production-line factories have begun using hydroexpanding.

The production controller establishes the quantities of identical-sized units to be built in economical production runs. The controller requisitions materials and subcontracted parts as required to maintain inventory and sets the earliest and latest dates acceptable for their arrival. Skilled materials handlers operate computerized materials control and tracking programs. Using a computerized inventory control system, they enter receipts into inventory and relieve the inventory as parts are used.

Quality controllers and inspectors are constantly present on the factory floor at all steps from raw-materials preparation to final testing and preparation for shipment. Testing methods are adapted to quick filling and emptying. Subcontractors are rarely used for testing. The exchangers may be automatically leak-tested. The Authorized Inspector may be a resident who inspects groups of identical units as they are built. The manufacturer's Data Reports may be preprinted on forms, with only the serial number and Authorized Inspector's signature not filled in.

On acceptance of all tests, surface preparation and coating may be done in closed booths or move through blasting and coating tunnels. The factory's packaging workers fabricate skids, crates, or boxes designed for the exchanger sizes moving through the line, apply protection to the connections, and bolt and strap the exchangers to the skids or load them into the skids or crates.

Surveying Manufacture-to-Order Factories (Job Shops)

Initial surveys and evaluations of custom heat exchanger factories usually are performed when considering procuring exchangers from organizations with which users have had no previous experience. Subsequent ones are done to update changes to ownership, financial status, management, engineering, production and quality assurance, and control personnel and facilities.

To a large extent, evaluating a manufacturing facility is subjective. Table 3.1, designed to assist readers in making dispassionate judgments, assigns a rating of 0 to 100 for each of the following: manufacturer-provided information; administrative procedures and practices;

TABLE 3.1 Factory Evaluation

Manufacturer-provided information	0	____________	100
Administrative procedures and practices	0	____________	100
Safety	0	____________	100
Engineering department	0	____________	100
Record-keeping practices	0	____________	100
Testing laboratory	0	____________	100
Workspace	0	____________	100
Lifting and shipping capacity	0	____________	100
Housekeeping	0	____________	100
Materials control, storage, and handling	0	____________	100
Control of welding consumables	0	____________	100
Age and condition of machinery	0	____________	100
Factory personnel	0	____________	100
Quality of fabrication	0	____________	100
Quality control and internal inspection	0	____________	100
Total	0	____________	100
Overall evaluation = total/15	0	____________	100

safety; engineering department; record-keeping practices; testing laboratory; workspace; lifting and shipping capacity; housekeeping; materials control, storage, and handling; control of welding consumables; age and condition of machinery; factory personnel; quality of fabrication; and quality control and internal inspection. We discuss these elements below.

Before embarking on a physical survey, request information about the manufacturer's ownership, management and financial structures, facilities, areas of expertise, engineering and manufacturing capabilities, and quality assurance and quality control. Supplement 3.1 is a form based on one the authors use for evaluating custom manufacturers. It excludes provision for references from manufacturers' customers. We believe manufacturers will not list dissatisfied ones and that none will fully disclose any negative experiences.*

The information that a manufacturing firm can fairly be asked to supply depends on its size. A small facility will cheerfully fill out a one- or two-page information request form, but will usually balk at one as detailed as that of the supplement.

Purchasers should review the responses to the survey form to decide whether to consider doing business with manufacturers with whom they have not had previous experience. Following are recommendations for conducting physical surveys.

*This survey form is available from the authors on a floppy disk. Using the diskette is cost-effective and efficient for manufacturers and owners and permits updating of subsequent surveys.

Administrative procedures and practices

Establish in advance the manufacturer's contact person and discuss the administrative procedures and practices the plant applies to visitors and its safety rules. Most require logging in and logging out of the premises. Depending on the kind of work in which the plant engages, you may have to provide proof of citizenship. Some shops demand visitors sign secrecy and nondisclosure agreements.

The management may insist, and we recommend, that you don protective clothing before entering the factory. These should include hard hats, shatterproof goggles or glasses with side-shield protection, hearing protection, steel-tipped safety shoes, work gloves, and full-length clothing with close-fitting long sleeves. Individuals with long hair should make sure that it cannot be caught in machinery. We urge all visitors to workplaces to follow these courtesies and safety precautions:

- Always keep the factory apprised of your whereabouts in the plant.
- Wear the protective clothing discussed above.
- Talk with workers only with the factory supervisor's permission. Wait until after an operation has been completed before distracting the person doing the work.
- Do not discuss labor times, costs, and prices with the workers.
- Unless previously arranged with the management, do not operate any machine or tool or perform any fabrication task.
- Do not enter an exchanger channel or shell without permission, and then only if it is adequately ventilated and you are under supervision of factory personnel.
- Do not ask for whom jobs for other companies are being done. The person accompanying you will usually say so when permissible.
- Do not voice opinions comparing the factory you are touring or the work it does with what you have seen elsewhere. Many workers take great pride in their work and are very loyal to their factory. Save your opinions for your confidential assessment.
- At all times, look where you are going. Watch out for uneven floors, drains, welding cables, air hoses, tools, and all the miscellaneous things that can impede safely walking on the factory floor.
- Keep your hands free and out of your pockets. Having your hands in your pockets is a sure way to increase your chance of being injured in a fall.
- Be alert for moving loads. Always step out of the way to a position in which the load will not pass over you.

- Never look directly at arc welding without appropriate eye protection. If you are passing a place where arc welding is taking place, shield your eyes. The ultraviolet light may permanently damage your vision. As a minimum, you can count on a day or so of red eyes and discomfort if you do not observe these precautions.
- Do not touch any structure on which welding has recently taken place. You may check to see if a part is cool enough to touch by moistening a fingertip and rapidly touching and withdrawing.
- Stand back from workers performing burning operations. Slag and molten metal may fly some distance from the part being burned.
- Do not approach a site where a worker is chipping or grinding without donning a full-face shield. Goggles and glasses with side shields are not adequate to prevent particles from entering eyes.
- Do not insert fingers into tubesheet tube holes or tube ends or run them over baffle edges and backfaces or other parts unless you are completely sure that all upset metal and sharp edges have previously been removed. Many fingers have been torn this way.
- Do not touch operating machinery. Be sure to keep loose clothing and long hair away from drills, lathes, boring mills, plate- and bar-bending rolls, shears and brakes, parts being rotated on turning rolls, etc.
- Do not hover about the pressure-testing area. Unless there is a specific agreement for you to witness pressure testing, stay clear of the area when tests are conducted.
- If you become aware of faulty work or misconduct by a member or members of the workforce, do not intercede to correct the situation. The factory management is responsible for dealing with such problems.

First visit and walk-through

Begin your first visit with a brief walk-through of the offices and works. Follow it with a more detailed inspection. Some places maintain a hospitality room for visiting inspectors, equipped with desks, telephones, facsimile transmission (fax) machines, and various other amenities. Others provide barely enough space for inspectors to perform the written part of their work and to wait before being called to perform inspections. Observe these facilities closely because they may indicate the factory's attitude toward inspectors and inspection.

Find out the shop's procedures for supplying documents to inspectors and other visitors. If possible, observe the degree of assistance—tools and personnel—and manufacturer provided inspectors.

During the first walk-through, get a general impression of the place and amount of work on the floor. We believe that an orderly facility, with clear aisles and a systematic work flow, is most likely to make the fewest mistakes and produce the best quality tubulars. Some apparent overcrowding is desirable because it generates workers' confidence in continuing employment and builds morale. However, if so much is going on that it is difficult to walk through the factory, the organization may be stretched beyond its capacity to produce quality work. Be aware that, when there is insufficient work and excessive unused space, workforce morale deteriorates and quality may suffer.

Note the workers' age mix. Most desirable is a spectrum spanning those of entry-level age to ones nearing retirement. The best work is usually done by a mixture of seasoned, skilled workers, leading and training those with less experience and skill. Try to get a feel for the relationships between people in the factory. Notice how supervisors and the people they supervise respond to each other. Despite the tensions involved, most workers in well-run, successful places soberly concentrate on their tasks and cooperate with each other. Skylarking, malingering, and argumentation indicate low morale. Look for knots of workers clustered around drawings, which is an indication of poor supervision and workers "dogging" their work. The condition of the factory washrooms is also a key to workforce morale.

Assessing the engineering department. Often informal discussions are more effective in assessing an engineering department than filling in a survey form. Look for easy or tense associations between mechanical and thermal engineers and their relationships with the drafting department and manufacturing engineers. Look at the computer equipment and what is on the screens to get a feel for computer capabilities. Find out which thermal and mechanical design programs the department uses. Determine whether the staff manually verifies program output often enough to make sure that computer-generated results are valid.

Examine how the company produces drawings. Some make drawings manually; others use CAD/CAM programs. Either is acceptable if the drafters recognize that drawings are abstractions intended to convey information and instructions concisely. Investigate the procedures for controlling drawings and bills of materials. These should include how drawings and bills of materials are issued and treatment of revisions and corrections, certifications, and as-built drawing revisions.

Manufacturers prepare fairly elaborate operating and maintenance manuals for feedwater heaters and simpler ones for other types of tubulars. The quality of these documents indicates how well the engineering department understands and discharges its responsibilities. If possible, briefly examine a typical manual.

An engineering department's physical arrangement may indicate how its members communicate with each other, the management, and the factory. Large, well-lit, unpartitioned, open rooms with people seated near windows and files and machines situated against inner walls promote cooperation. The loss of privacy and effect of others nearby is modest, compared with isolation and confinement in small interior cubicles. Senior personnel's offices, glass-walled for privacy, but accessible to junior staff, convey a sense of teamwork to junior staff.

Publications that should be on hand include updated and applicable versions of the codes and standards to which the company works, software manuals for their thermal and mechanical design programs, works such as the AWS *Welding Handbook,* recent issues of technical journals that relate to thermal and mechanical design and welding, and miscellaneous technical literature to support its work.

Despite well-thought-out procedures and best efforts, manufacturing mistakes happen. Correcting an error requires learning what caused it, how it occurred, and considering possible fixes and their effect on strength, utility, and merchantability of the exchanger. This makes coordination between sales, engineering, manufacturing, and factory inspection and quality control personnel essential.

Examining record-keeping practices. Look at the shop's record-keeping practices. Examine how the manufacturer records receipts and disbursements of materials and ties mill test reports of materials to its inventory control. See how it files and updates procedures, procedure qualifications, and qualifications of welders, welding operators, brazers, and nondestructive examiners. Most pressure-vessel codes require welders to stamp their symbol adjacent to their work. This requires parallel record keeping at which you should look.

Notice how the manufacturer meets code requirements for recording, filing, and retaining radiographs and reports of ultrasonic, magnaflux, fluid penetrant, eddy-current, and other nondestructive examinations. And examine its practices for noncode exchangers.

Look into the company's system for filing and retrieving drawings and practices for destroying outdated prints. Note also its practices for filing manufacturer's data reports, other code-required documents, repair and alteration reports, and nameplate rubbings.

Appraising the testing laboratory. Not all exchanger shops have testing laboratories. Those that do may be equipped with hardness testers, tensile testers, equipment for making bend tests on weld specimens, flaring, flattening and expanding tests on tubing, chemical and spectrographic analysis of materials, microscopes and equipment for polishing and microphotographing specimens, etc. Evaluate the laboratory's equipment and its maintenance. Ask about the qualifications and training of its personnel to use the equipment and interpret results. Look into how it takes, controls, and retains specimens.

Detailed factory inspection

Use the detailed factory inspection to evaluate

- Available workspace and lifting and shipping capacity
- Factory housekeeping
- Materials control, storage, and handling
- Control of welding consumables
- Age and condition of machinery
- Factory personnel
- Fabrication operations

Available workspace and lifting and shipping capacity. Compare what you see with the information provided on the survey form. Note the condition and utility of outdoor facilities. Check how the factory protects materials stored outdoors. Bearing in mind that in cold and snowy climates, materials stored outdoors may not be accessible when needed, examine the factory's provisions for moving materials from outdoor storage to the fabrication area in inclement weather.

Verify that the reported lift capacity actually exists. Note whether cab-operated bridge cranes are attended or operated most of the time. If your company is considering buying exchangers that must be shipped by barge or rail, make sure that the railroad and barge companies that serve the factory are still extant.

Housekeeping. Generally orderly factories efficiently and safely produce high-quality exchangers. Therefore, observe housekeeping in the machine shop, plate shop, and assembly and testing areas. Here are some specific questions:

1. Are the factory floors regularly swept clean?
2. Does the machine shop regularly collect, segregate by alloy, and remove to scrap storage metal turnings, shavings, and chips?
3. Are there procedures in place in the plate shop for clearing scrap away from work and materials storage areas and for moving parts to be scrapped out of the production area?
4. Is space allotted to each exchanger under fabrication clearly defined?

5. Is the flow of materials and parts to the assembly beams or fixtures and the path that units take toward testing and shipping apparent?
6. Does the factory have a system for detoxifying and removing filth from bundles and exchangers it receives for retubing, repairs, and alterations before setting them in a work area, or does it require customers to preclean and detoxify them before shipment?
7. Are procedures in place to ensure workers' safety when working on equipment that has been in service and to alert them to any potential hazards?

Materials control, storage, and handling. Observe how the factory controls inventory of raw materials and parts. Look into its system for tracking materials received into and withdrawn from storage and how it ties materials to the mill test reports. Find out whether carbon steel, austenitic, and high-alloy steels and nonferrous materials are stored and kept segregated from each other.

The way a fabricator stores and handles tubes affects the life of the exchangers in which they are installed. Iron contamination initiates corrosion in some tubes, and gouging and denting reduces the life expectancy of others. Therefore, well-run shops take protective measures. The authors have seen places that pile austenitic tubes on the floor where they are walked on by passersby.

Check to see if tubing is stored in its original packaging until just before tubing exchangers. Ideally, the materials receiver should open boxes or cartons of tubes for inspection when received, then reseal them. The tube loading crew should open the containers just before loading the tubes into the skeleton. Materials receivers should wrap and seal unpackaged tubes in heavy plastic sheeting, which should not be removed until loading. The assemblers should similarly wrap bundles until just before stabbing into shells.

Also observe how the factory protects austenitic tubing from iron contamination. It should not be in direct contact with steel racks. If it is not stored in closed packaging, it should be wrapped in plastic wrappings, sealed against the entry of iron dust.

See how the factory stores and handles expansion joints. Thick-walled expansion joints are generally sturdy structures. However, weld drags, dents, and corrosion all shorten their life. Thin-walled ones are extremely sensitive to the effects of weld drags and dents. Any twisting, angular bending, precompressing, or preextending for which they are not designed significantly reduces their cycle life. In coastal areas, salt air can predispose austenitic bellows to stress-corrosion cracking. Therefore, notice whether the factory closes the welding stub ends to exclude dirt and foreign matter. If shipped with bridges, to maintain length, they must remain in place until the exchanger is installed.

All hooks and plate-lifting clamps should be approved safety types. Watch how the factory lifts bundles and other parts. They must be lifted in ways that will not cause hidden damage and shorten exchanger life. Bundles should preferably be lifted on devices that support them for substantially their whole length. Wooden dunnage beams are acceptable if carefully placed to allow several tube supports or baffles to carry the load. Nylon slings should be spread apart sufficiently to avoid bundle sagging or flexing and bear on tube supports or baffles, not on the tubes.

Note how machined surfaces are protected. Parts waiting for or in transit from one part of the factory to another should be set and stored on wooden pallets or wrapped in plastic to make sure that the machined surfaces remain free of blemishes. Machined parts stored for long periods should be coated with a protective coating. After testing, nozzles, flanges, and threaded connections should be sealed and protected.

Control of welding consumables. Look at how the factory stores and controls welding consumables. Tracking weld rod analysis information is essential to controlling weld quality. Examine the factory's practice for recording analysis and traceability of weld rod from storage of unopened packages to final shipment. Check the temperature maintained in ovens for keeping coated rod dry; it should be 190 to 200°C (~375 to 400°F). Ask the person responsible for welding how the factory verifies that shielding gases are bone-dry. Notice whether welders are equipped with and required to use portable heaters in which to store coated rod ready for use.

Excess flux covers the joint during submerged arc welding. It is proper for the shop to recover the excess, provided it takes appropriate steps to maintain its integrity. Examine how the workforce does this.

Age and condition of plant and machinery. It takes special training to assess the condition of heat-exchanger manufacturing machinery. Well-maintained World War II vintage horizontal-boring mills are often acceptable for machining channels and tubesheets. Properly cared for milling machines of similar age are adequate for milling tubesheet grooves. Some old gap-bed lathes are useful for machining removable and fixed-tubesheet bundles. However, some equipment must be of relatively recent age. An example is tube-hole grooving tools which can wear in ways that cause the grooves to be eccentric relative to the centerline of the holes in the tubesheet. This causes the groove to be deeper than specified for about 180° of the hole circumference and barely below the surface in the remaining 180°.

To produce sound welds, it is necessary to control voltage, amperage, shielding gas flow, wire and ribbon feed rates, etc. within the ranges the welding procedures specify. Welding equipment should be free of accumulations of particulates, oily deposits, and other foreign matter that affect control. If it is covered with dirt and gummy deposits, you can assume that the factory has problems with weld quality.

Factory personnel. Assessing the quality of factory personnel requires some fabrication experience and an understanding of the work they do and skills required to do it. One factory the authors toured invites visitors to discuss with workers the specific tasks they are performing at the time of the visit, and ask about acceptance standards, training, experience, and any other questions that come to mind. Not all shops display such openness. Those that do inspire confidence in their capabilities.

If the shop has a procedure for screening entry-level workers, has an apprenticeship program, and certifies its workers to various levels in accordance with industry standards, its personnel will generally be competent.

Production system. Essential items to review in a factory's production system are (1) how drawings are issued to the factory and controlled, (2) how the factory feeds back information to the engineering department, (3) how supervision makes sure that approved procedures are made available to the workforce, (4) how management controls production to make sure that the work is performed in accordance with approved procedures, (5) how the workforce achieves the required level of quality, and (6) how the workforce goes about correcting of errors.

Pay particular attention to error corrections. There are two types: (1) errors correctable by repairs that do not change dimensions and configurations accepted on the approved drawings and (2) errors that require a design change. The former do not ordinarily delay delivery, and although the user's inspectors should note them in the inspection report, they are usually acceptable if they do not affect the code status and quality of the finished heat exchanger. The latter, however, may delay delivery or may be acceptable only if the manufacturer extends the warranty for the affected part.

Take, for example, the situation in which a factory accidentally taps all the stud holes in a channel flange oversize. The options for correcting the problem are limited to (1) filling the holes with weld metal and retapping the flange, which may not be acceptable for metallurgical reasons; (2) cutting off the mistapped flange and welding on a correctly tapped one; (3) tapping the flange to fit the next-larger-size stud and replacing the studs with ones larger on the end threaded into the flange and the original size at the other end; and (4) machining out the oversized threads, drilling deeper into the flange, and tapping the deeper-drilled section, leaving an annular space in the previously tapped section. Only options (1) and (2) do not constitute a design change, and of these two, only option (2) would be acceptable in most situations. However, if option (2) would unacceptably delay delivery, the purchaser would be forced to decide on one of the other options. Consider the factory's attitude to negotiating an extended warranty for the affected part in your evaluation.

Quality control system. The quality control system must ensure that the factory uses the correct materials, uses qualified workers to fabricate them to the agreed-on dimensions, performs the required destructive and nondestructive tests and examinations, inspects to verify quality during fabrication, and documents the work. The documentation must provide for traceability of materials and show which workers are accountable for the quality of the work.

The workforce must construct the pressure envelope of code-stamped heat exchangers to the pressure-vessel code's minimum requirements. If the Authorized Inspector rejects materials or work, the factory has no alternative than to replace the materials or redo the work until it is acceptable. However, an exchanger acceptable to the AI may function poorly or not at all because its nonpressure parts do not meet the required quality.

Therefore, factory inspectors must have the qualifications and authority to verify that the materials and work meet both code and noncode quality standards. Accordingly, it is important to explore the factory inspector's qualifications and authority to reject unsatisfactory materials and work.

It is also important to look into the factory's examination and testing practices and the qualifications of NDE subcontractors. Indications of how seriously the factory treats testing are their acceptance and rejection practices and the frequency with which they calibrate instruments.

Fabrication operations. Observing fabrication operations is the most fascinating part of a factory survey. The following lists some of the most important:

- Tubesheet drilling, grooving, machining, and preparation for tube-to-tubesheet welding.
- Baffle and support plate deburring.
- Setting up bundle cages (skeletons).

- Connecting zone enclosures to tubesheets.
- Welding zone enclosure flat-plates to semicylindrical parts.
- Tubesheet and tube end cleaning before joining.
- Tube loading—using guides and rubber-headed mallets.
- Tube expanding—look especially at equipment, methods, and control practices.
- Tube-to-tubesheet welding—check out the tube-to-tubesheet welding equipment and methods used.
- Sequence of tube joining—weld first, then expand?
- Stabbing bundles into shells.
- Bolting up and assembling.
- Final testing.

Assessing Field Construction, Repair, and Alteration Organizations

Following are suggestions for assessing field construction, repair, and alteration organizations.

Organization-provided general information

As with custom manufacturers, start the survey by requesting pertinent information about the organization. Establish whether the company engages purely in field operations or is connected with a heat exchanger factory. If connected, determine whether the field service organization operates within the factory's administrative structure, is semi-independent, or is a completely separate division or subsidiary.

Certificates of authorization, inspection agency, and coordination with owner's AI

Of prime importance for field assembly, repair, and alteration of code-stamped vessels are the Certificates of Authorization and Authorized Inspection Agency. If the organization is being considered for field erection of oversized, code-stamped exchangers built to Section VIII, Division 1 of the ASME Code, it must hold and be authorized to apply the U symbol. If the exchanger to be erected is to be designed and constructed to meet the requirements of Section VIII, Division 2, it must have a Certificate of Authorization to apply and hold the U-2 symbol. If the exchangers to be assembled on the site are steam generators that fall within the purview of Section 1, the field organization must hold and be authorized to apply the S symbol.

If you are considering a field service organization for on-site repairs or alterations to code stamped exchangers in accordance with the requirements of the National Board Inspection Code, make sure that it holds and is authorized to apply the R symbol issued by the National Board.

Find out how the prospective field service company coordinates its work on vessels to be code-stamped, or repaired or altered in conformity with the National Board Code or API 510 with the owner's Authorized Inspection Agency and Authorized Inspector responsible for in-service inspections.

Engineering

Establish whether the organization has the engineering background to produce the calculations required to satisfy pressure-vessel code requirements or whether it just erects what others have designed. Depending on the extent of the field work, the erector may need to be staffed with civil and mechanical engineers. If the purpose of the evaluation is to determine whether the organization is qualified to make repairs or alterations to code-stamped exchangers, they must have sufficient engineering capability to make sure that the work conforms with NBIC or API 510 requirements.

Purchasing and materials control

Investigate how the service company buys materials and parts needed for repairs or alterations, especially when the work is to be done on a cost-plus basis. Examine their arrangements for expediting delivery to the job site of parts purchased from a job bill of material. Make sure that there is a satisfactory system for receiving, issuing, and tracking materials delivered to the field.

Supervisory personnel

The individual who can either get your field work done on time or cause late-completion headaches is the job superintendent. Therefore, check the experience level of the supervisor who may be assigned to your job.

Field service workforce

Find out if the service company sends a crew of skilled workers and laborers to the site or if they hire from the local labor pool. Although travel and living costs of home-based crews is costly, their skill, teamwork, and work quality may outweigh these costs. If the owner is unionized, the union status of the workforce may affect how the field service crew and owner's personnel work together.

Welding performed in the field on pressure parts or parts welded to pressure parts of code-stamped exchangers must be performed by welders qualified in accordance with prevailing pressure-vessel code requirements. The service organization's welders must be qualified to weld the exchanger's materials in the thickness range of its parts, using the specified procedures and in the positions suitable for the installation. Note that, because the ASME Code does not provide for portability of welders', welding operators', and brazers' qualifications, these individuals must requalify whenever they change employers. However, the ASME Code allows a contractor to subcontract welding to another organization that has qualified its welders. The service company must maintain accountability by using a system of stamping welders' symbols adjacent to the welds.

The NBIC and API 510 require radiography and heat treatment in accordance with the rules of the pressure-vessel code that applied in the year that the heat exchanger was built. The service organization must document its fulfillment of these requirements.

Equipment and facilities

Economics dictates what a service organization brings to the field. Determine whether the company has traveling workshops, owns equipment for doing its work, buys it for the job and subsequently disposes of it, or rents it. For the most part, companies that work in the field rent heavy lifting equipment, welding machines, and air compressors from firms near the job site. However, they ordinarily provide smaller tools and special machines critical to the work. These may be portable flange facing machines, pass-partition groove milling machines, hydraulic tube pullers, tube expanders, orbital-head tube-to-tubesheet welders, hydraulic bundle extractors, and proprietary machinery.

Quality control

It is harder to monitor the quality of work done in the field than in an exchanger factory. When evaluating the field service company's capability, scrutinize their procedures for inspecting prefabricated parts before shipment to the site. Check how they make sure that only qualified workers do the work. Look into how they track welding consumables. Examine their procedures for performing and documenting nondestructive examinations and tests.

Contractors' procedures for dealing with impediments to performance

Some of the difficulties field service contractors encounter are similar to those described previously for shop construction. In performing an evaluation, find out how the prospective contractor deals with such problems.

Evaluating Production-Line Exchanger Factories

In some ways rating a production-line exchanger factory is simpler than appraising a custom fabrication shop. With few exceptions, the product line is limited, This means that drawing control is less complicated since quantities of identical parts are produced from standard drawings and assembled into a modest number of configurations. It also means that, except for application engineers, most of the manufacturer's engineering is directed toward improving manufacturing methods and machinery.

After gaining an understanding of the machinery layout and production scheme, investigate the company's purchasing and inventory control practices since these can affect quality and delivery times. Look for potential impediments to the smooth flow of parts into the production line, especially if their manufacture is subcontracted. Find out what the manufacturer does to make sure that the breakdown of a machine in the production line does not substantially delay shipment.

In production-line factories, plant inspectors check dimensions and quality of each lot of components and subassemblies against established criteria. They decide whether to accept, reject, or require recycling parts and assemblies. Consequently, there is often tension between production controllers whose mission is to pump out production and inspectors whose duty is to make sure that the exchangers are made correctly. For this reason it is important to learn who has the final authority.

References

1. *Standards of the Tubular Exchanger Manufacturers Association,* 7th ed., The Tubular Exchanger Manufacturers Association, Tarreytown, N.Y., 1988.
2. *Standards for Closed Feedwater Heaters,* 5th ed., The Heat Exchange Institute, Cleveland, Ohio, 1992.
3. *Standards for Steam Surface Condensers,* 9th ed., The Heat Exchange Institute, Cleveland, Ohio, 1992.
4. *Standard for Power Plant Heat Exchangers,* 2d ed., The Heat Exchange Institute, Cleveland, Ohio, 1990.
5. S. Yokell, "Understanding the Pressure Vessel Codes," *Chem. Eng.* (May 12, 1986); S. Yokell, *A Working Guide to Shell-and-Tube Heat Exchangers,* McGraw-Hill Book Company, New York, 1990.
6. *Welding Handbook,* 8th ed., 3 vols., The American Welding Society, Miami, Fla., 1987, 1991, 1996.

Supplement 3.1: Survey Form for Evaluating Custom Heat Exchanger Manufacturing and Repair Shops*

Date of First Report________ Date of This Update____________________

Company Name____

Parent Company____

Mailing Address____

____

____

Shop Location(s)____

Telephone No.(s) ________Telex ________ Fax ________ TWX ________

Years in Business__________ Approximate Annual Volume, $__________

Bank and Financial References ______________________________

____

Approximate Total No. of Employees:________ Shop________ QC ________

Engineering________ Drafting________ Administrative ____________

Principal Personnel and Titles:

Owner, Senior Partner, President______________________________

Financial Officer(s) ______________________________________

Head of Marketing ______________________________________

Head of Sales____

Head of Mechanical Engineering______________________________

Chief Mechanical Designer__________________________________

Head of Thermal Engineering ______________________________

Chief Thermal Designer ____________________________________

Head of Purchasing____

Shop Superintendent______________________________________

*MGT Inc., Boulder, CO 80303-3607.

Memberships: ______________________________

Codes and Standards to Which Units Are Built: ______________________________

____

Land and Buildings:

Open Land Areas - Paved ______________ Unpaved ______________

Nos. and Sizes of Plate Shop Bays ______________________________

____

Nos. and Sizes of Machine Shop Areas ______________________________

____

Nos. and Sizes of Assembly Areas ______________________________

____

Segregated Buildings or Areas for Work on Exotic Metals and Method of Isolation from Contaminants:

____

Shipping Facilities:

Size of Largest Door ____________ Railroad ____________ Barge Access? ________

Industries Served: ______________________________

____

Product Lines: ______________________________

____

Alterations, Repairs and Retubing:

Shop: Alterations __________ Repairs __________ Retubing __________

Field: Alterations __________ Repairs __________ Retubing __________

Inspection Codes ______________________________

Normal Competitive Fabrication Ranges: Maximum Weight ______________

Shell Diameters:

Pipes / Rolled Plates

Min. ________ Max. ________ Min. ________ Max. ____________

Thicknesses:

Shells/Channels / Tubesheets / Tubes

Min. _____ Max. _____ Min. _____ Max. _____ Min. _____ Max. _____

Tube O. D. Minimum ________ Maximum ________

Field Service/Erection

Union ____

Scope of Field Service Work ____________________________________

____

____

____

____

Engineering Capabilities:

Thermal Design Capabilities

☐ From Specifications ☐ Verify Only ☐ None

Describe scope and limits of thermal designs shop is capable of (for example liquid/liquid, thermosyphon and forced circulation reboiling, evaporation, two-phase flow, unfired steam generation, condensing pure vapors, condensing mixed vapors, dehumidfying, refrigeration and air conditioning evaporation and chilling, ammonia condensing etc.

____

____

____

____

Computer Programs:

☐ Proprietary ☐ HTRI ☐ HFS ☐ B-Jac/Aspen ☐ Other

Mechanical Design Capabilities:

☐ From Heat Exchanger Specification Sheet ☐ None

Describe scope and limits of mechanical designs shop is capable of (for example conventional and integral double tubesheet designs, wind and earthquake load analysis, fatigue analysis etc.

____

____

____

____

Vibration Analysis: ☐ Perform ☐ Guarantee ☐ None

☐ Subcontract at Customer Request

Drafting Capabilities: ☐ Manual ☐ Computerized

☐ Setting Plans ☐ Assembly Drawings ☐ Details

☐ Work to Customers' Drawings Only

Quality Assurance, Quality Control, Inspection and Testing:

Quality Assurance Manager________________________

Quality Control Manager ________________________

ASME Code Authorized Inspection Agency ____________________

____

Authorized Inspection Agencies for Codes Other than ASME Code

____

Agencies Where Manufacturers Data Reports are Filed: ____________

____

Nondestructive Examination Facilities and Levels of Qualification:

☐ Borescopy (type)________________________

☐ Dye Penetrant Examination; Examiner Qualification Level__________

☐ Fluorescent Penetrant Examination; Examiner Qualification Level________

☐ Radiographic Areas and Equipment (describe)______________

____

Examiner Qualification Levels ____________________

☐ Magnetic Particle Examination; Examiner Qualification Level ________

☐ Eddy Current Flaw Detection; Examiner Qualification Level ________

☐ Ultrasonic Flaw Detection (describe types) ________________

Examiner Qualification Level ________

☐ Ultrasonic Thickness Measuring ☐ Other ________________

____

Pressure Testing Capabilities:

Fluid Media Available ____________________

Hydrostatic Test Water Controls:

Purity (Distilled, Demineralized, Filtered etc.) ________________

____

Maximum Pump Pressure ________ Recording Method ________

Temperature Control System ____________________

Post Hydro Drying System ____________________

Air and Inert Gas Test Controls:

Maximum Pressure ________ Recording Method ________

Filtering & Drying Systems ____________________

Pressure Test Shielding for Personnel ________________

____

Leak Detection Systems and Equipment:

Bubble Former Used for Bubble Testing ________________

Examiner Qualification Level ____________________

☐ Halogen Leak; Examiner Qualification Level ____________

☐ Helium Leak; Examiner Qualification Level ____________

Laboratory Equipment:

☐ Metallographs (describe) ____________

☐ Microscopes (describe) ____________

☐ Tensile Testers ____________

☐ Hardness Testers ____________

☐ Impact Testers ____________

☐ Cameras ____________

☐ Video Equipment ____________

Lifting and Handling Capabilities:

Bridge Cranes:

No.	Capacity	Span	Floor to Hook

Jib Cranes:

No.	Capacity	Reach	Floor to Hook

Portable Cranes:

No.	Capacity	Reach	Floor to Hook

Shell Turning Rolls:

Manual				Power			
No.	Max. Load	Max. Dia.	Min. Dia.	No.	Max. Load	Max. Dia.	Min. Dia.

Manipulators:

No.	Max. Load	Max. Dia.	Min. Dia.

Welding Positioners:

No.	Max. Load	Reach

Welding Capabilities:

Metals for Which Shop has Qualified Welding Procedures, Welders and Welding Operators:

Fully Automatic ______________________

____

Semi-Automatic ______________________

____

Manual ______________________

Welding Equipment: (No., Type & Manufacturer, AC or DC, Amps)

Submerged Arc ______________________________

____

Gas Tungsten Arc ______________________________

Gas Metal Arc ______________________________

Plasma Arc ______________________________

____

Orbital Head Tube-to-Tubesheet GTAW ______________________________

____

Wire Overlay ______________________________

Strip Overlay ______________________________

Plasma Arc Overlay ______________________________

Spray Welding ______________________________

Other ______________________________

Cutting Equipment: (Max. Thickness, No., Type & Manufacturer)

Flame Cutting:

Manual Single Torch ______________________________

Electric Eye Tracer __________ Torches ______________________________

__________ ______________________________

Numerically Controlled __________ Torches ______________________________

__________ ______________________________

Carbon Arc ______________________________

Plasma Cutting ______________________________

Band Sawing ______________________________

Shears: No. ______ Thickness __________ Length __________

Other ______________________________

In-House Forming Capabilities:

Hot Forming Available? ______________________

Plate Rolls:

Thickness, Diameter, Width Ranges

____

Manufacturer	Type	Top Roll Dia.	Length
____________	____________	____________	____________
____________	____________	____________	____________
____________	____________	____________	____________

Press Brakes:

Manufacturer	Type, Tonnage, Window, Clear Width
____________	______________________
____________	______________________
____________	______________________

Bulldozers:

Manufacturer	Type, Tonnage, Window, Clear Width
____________	______________________

Head Spinning Machines:

Manufacturers	Types and Capacities
____________	______________________
____________	______________________

In House Tube Bending? _____ ☐ Manual ☐ Hydraulic

Range of Bending Capabilities:

Tube Diameter		Tube Gage		Bend Radius	
Min.	Max.	Min	Max	Min	Max
______	______	______	______	______	______

Heat Treating Equipment:

No.	Type & Controls	Width	Height	Length

Machine Tools:

Manually Operated Radial Drills:

No.	Drive	Arm	Column	Spindles	Manufacturer

Numerically Controlled Radial Drills:

No.	Drive	Arm	Column	Spindles	Manufacturer

Multispindle Drilling Machines:

No.	Maximum No. of Spindles	Maximum Stack Thickness	Maximum Diameter	Manufacturer

Numerically Controlled Post and Beam Drilling Machines:

No.	No. of Spindles	Maximum Hole Diameter	Maximum Stack Thickness	Manufacturer

Deep Hole Drilling Machines: ☐ BAT Drilling System

No.	No. of Spindles	Maximum Hole Diameter	Maximum Stack Thickness	Manufacturer

Tube Hole Grooving Equipment:

____

Post Drilling Baffle/Support Plate & Tubesheet Deburring System:

____

Flange Drilling Machines:

____

Vertical Boring Mills:

No.	Swing Table to Tool	Height	Manufacturer

Horizontal Boring Mills:

No.	Type	Spindle	Swing Diameter	Manufacturer

Milling Machines:

No.	Type	Capacity	Manufacturer

Lathes:

No.	Type	Swing	Bed Length	Manufacturer

Assembly & Repair Equipment

Tube Expanding Equipment:

Tube Rollers:

Type of Drive	Size	Speed (RPM)	Torque Control System	Manufacturer

Torque Analyzer:______________________________________

Hydroexpanders:______________________________________

Explosive Expanding Available?______________________________

Bundle Pullers/Pushers:__________________________________

____

Ultrasonic Bolt Stress Measuring Device:_________________________

____

Bolt Tensioners:

No.	Capacity	No. of Bolts Tensioned Simultaneously	Manufacturer
___	______	______	________________

Tube Pullers:

No.	Max. Tube Dia.	Max. Pulling Force	Manufacturer
___	______	______	________________

Field Service Equipment:

Portable Machining Tools:

No. ____ Type & Capacity ________________________________

Miscellaneous ____

____

Chapter

4

Maintenance Inspection and Inspecting Out-of-Service, Mothballed, and Used Exchangers

Introduction

This chapter is mostly about inspecting exchangers for maintenance but includes some discussion of inspecting out-of-service, mothballed, and used ones. The objectives of this kind of inspecting differ from those for inspecting in a heat exchanger factory. Instead of trying to make sure that manufacture meets minimum standards of quality, or requirements for factory retubing, repairing, or altering exchangers, its aims are to

- Appraise the present condition of an exchanger
- Determine causes of observed failures
- Find out what needs to be repaired
- Assess feasible kinds of maintenance, repair, or alteration procedures
- Evaluate proposed alterations
- Judge whether an exchanger can be used in a service different from the one for which it was designed

These appraisals and surveys cover structural condition and degree of fouling. To plan for maintenance and future replacements, you must make periodic evaluations and establish deterioration trends. The discussion in this chapter includes checklists followed by comments that apply to the listed items.

Safety Precautions When Inspecting Shutdown Exchangers

It is of utmost importance that inspectors be aware of the hazards of examining installed, shutdown equipment. This applies to inspecting exchangers temporarily out of operation during outages and ones that have been removed from service and stored.

To comply with applicable governmental, insurer, and plant safety requirements, inspectors must be familiar with regulations that apply in the jurisdiction and the company's safety standards. Before inspecting, assess the safety requirements for the work to be done, select the most suitable safety equipment, and hold safety meetings required to transmit information and provide education about safe practices.

Outline of basic safety-related practices when inspecting out-of-service exchangers

The following outline, intended for organizations and inspectors who do not have access to a formal safety program, lists some basic safety-related practices. It is not a complete guide to safe inspecting. Comments on the elements follow the outline.

I. *Log* in and out of the site.

II. *Cooperate* with personnel responsible for safety and follow their established safety requirements.

III. *Use and wear customary and usual safety gear that conforms with governmental standards* such as

   A. Hard hat
   B. Protective goggles
   C. Safety shoes
   D. Dust masks and breathing apparatus, when indicated by local conditions
   E. Gloves and other protective clothing
   F. Hearing protection in noisy areas

IV. *Depressurize, drain, and purge both sides* of the unit before opening any part for access.

V. *Clean* the unit completely free of irritants; detoxify before permitting entry into or contact with its surfaces.

VI. *Flush* units through which irritating, noxious, or lethal materials flow, and those that carry fluids that would interfere with the inspection.

VII. *Drain, dry, and purge with clean, fresh air* after flushing.

VIII. *Test the atmosphere* for oxygen and hydrocarbon content before entering channels and shells. In nuclear power stations, where radioactive fluids may be present, do not enter until the unit has been given a *Radiation Work Release.* Be sure to adhere to the time exposure limits as instructed.

IX. *Isolate piping* to and from the shell and tube sides.

X. *Cool down* units with metal temperatures warmer than 55°C (~120°F).

XI. *Measure temperatures* to make sure that all parts to be entered or contacted are cool enough for safe inspection.

XII. *Wear a safety harness* when entering a closed space.

XIII. *Have a safety observer standing by* when entering a closed space.

XIV. *Use eye protection* when observing open arc welding and burning.

XV. *Use hearing protection* in noisy areas.

XVI. *Observe the limits of roped-off areas* during radiography and take precautions to limit and protect against exposure to radiation.

XVII. *Demand information* about the use of, exposure to, and consequences of inhaling, ingesting, or skin contact with any substance used for cleaning and testing, and take the indicated safety precautions. Brief inspectors on any substances in or on the unit that may be

A. Flammable
B. Explosive
C. Irritating to tissues
D. Noxious
E. Lethal
F. Carcinogenic
G. Radioactive
H. Threatening to life, health, or property in any other way

XVIII. *Use sparkproof tools and equipment* when there is a hazard of explosion or fire.

XIX. *Ground* all electrical tools before use.

XX. *Require a fire watch* when it is necessary to burn into a part to provide access for inspection or maintenance.

XXI. *Isolate* the unit from operating areas of the plant.

XXII. *Have emergency and first-aid procedures and equipment* in place.

XXIII. *Have designated plant personnel trained in first aid procedures*

XXIV. *Designate* a responsible person to whom accidents must be immediately reported.

Comments on outline for safe inspecting

Logging in meets a basic safety rule in any plant, which is to have people in authority know where visiting individuals are and why they are there. Even the smallest facility should maintain a log that indicates where, when, and what people are inspecting. Inspectors should not fail to log in and out.

When testing the atmosphere in a closed space, use a tester appropriate for detecting hydrocarbons when the unit has carried organic materials. Use appropriate detectors for exchangers through which radioactive or other toxic materials have flowed. Always test for oxygen content after purging; the percentage should never be less than 19% nor more than 23%.

Inspectors should make sure that piping is either disconnected and the ends capped or otherwise blocked off, or that adequate double-block-valve protection is provided. Either the whole string in which a high-pressure feedwater heater is installed should be removed from service, or if it is bypassed, double-block valves should be closed off in the feedwater lines. The space between the block valves should be fitted with an unvalved, open vent and drain.

Metals at a temperature of 55°C (~120°F) are about as hot as a person can touch without extreme discomfort. The confined space must be cool enough and have sufficient ventilation to permit the work to be done efficiently and safely. This applies to entering a full-opening channel and especially to entering a hemihead or elliptical head through a manway.

Plants and systems of which exchangers are a part often continue operating during preplanned partial outages. This is common practice in power generation. Here, individual feedwater heaters or a string of heaters may be bypassed and taken off the line while other units continue to run. The only safe way to inspect under these conditions is to isolate the heater being entered with line blinds or double-block valves. Similar isolation is required in other industries.

All supervisory personnel should be thoroughly grounded in the organization's procedures for reporting accidents. Note that reports of accidents in which property is damaged or people are injured are legal documents.

Planning for Maintenance Inspections

Ideally, the frequency of heat exchanger inspections should be determined from records of rates of deterioration and fouling from the time of installation. On-line inspections may not serve this goal, no matter how extensive the inspection.

Most heat exchangers are part of an equipment network which may not be capable of being shut down at the times most suitable for exchanger maintenance. Consequently, engineering, operating, and maintenance personnel must cooperate in planning for cleaning and maintenance.

Plan to inspect at these times: (1) before regularly scheduled outages; (2) whenever a partial shutdown makes an exchanger available; (3) during the tense period of an unforeseen shutdown; and (4) after maintenance, repairs, or alterations are complete.

Collect or have available all pertinent data before making an inspection. Assemble the following:

- Nameplate data
- Heat exchanger, feedwater heater, or condenser specification sheets
- A complete set of manufacturer-supplied drawings, procedures, and records including the items listed in Checklist 4.1
- Records of alterations and repairs
- Records of previously performed maintenance listed in Checklist 4.1
- Manufacturer's recommended installation, maintenance and repair package (if provided)
- Operating logs that show both normal and deteriorated performance

For steady-state conditions, the log of temperatures and pressures at the four fluid terminals reveals changes in the unit's duty capability and increases in pressure losses. However, when feedstock properties, stream flow rates, inlet temperatures, and inlet pressures vary, a

CHECKLIST 4.1 Manufacturer-Supplied Documents and Previous Exchanger Inspection and Maintenance Records to Assemble before Inspecting

- ☐ Setting plan
- ☐ Assembly drawings with as-built dimensions
- ☐ Detail drawings of the components as built
- ☐ Procedure specifications
- ☐ Records of nondestructive test results and interpretations
- ☐ Mill test reports of critical parts, e.g., tubing, tubesheets. and any corrosion-resistant metals
- ☐ Manufacturer's data reports and code documentation
- ☐ Records of previously performed inspections
- ☐ Tube plugging records, plug maps, and failure locations
- ☐ Tube replacement records and location maps

thermal design engineer or rater should analyze how much the actual performance deviates from the performance to be expected in a new, clean unit.

Planning for inspection during scheduled outages and partial shutdowns

Review the documents listed in the checklist with the Inspector. If there has been no significant performance deterioration since the last inspection, prepare a plan to evaluate the exchanger's structural condition. If performance has declined, prepare a plan that includes inspection to verify or disprove speculations about the cause of the deterioration.

The plan for inspecting structural condition should cover:

- A *budget* for the kinds of nondestructive examinations and tests to be performed
- The *sequence* in which nondestructive examinations and tests will be done
- The *personnel* who will do the examining and testing, and qualification levels of examiners and testers
- The *response time* of the contractors who may be asked to perform nondestructive examination
- The *equipment* required for the scheduled inspection
- *Materials* required for closing up the unit after inspection
- *Preparations* for safely inspecting the unit
- *Coordination* with other work being done in the area
- *Contingency plans* for dealing with what the inspection may disclose

Examples of equipment required for inspections are video camcorder, borescope probes, TV monitor, and VCR. Action plans should include how to coordinate and make sure that trained contract or plant inspection personnel, materials, and equipment will be on hand when needed.

During outages, all kinds of equipment, piping, electrical lines, etc. may be worked on. Therefore, it is very important to coordinate with the maintenance superintendent the availability of plant personnel and facilities for inspecting and maintaining the tubular equipment. This includes coordinating with the plant safety department the providing of safety harnesses, safety watches, and other required safety equipment and checking atmosphere and taking whatever other precautions are required. Other groups' plans must also be known and coordinated with. An example is erecting staging for and performing work that might interfere with inspecting and working on the tubular equipment. Therefore, submit the inspection-maintenance plan to the maintenance superintendent as far in advance as possible to facilitate coordinating all the work being done during the outage.

A major purpose of inspection is to uncover hidden incipient failures and problems and possible causes of performance deterioration. Examples are vibration damage and metal loss from corrosion, erosion, or erosion-corrosion. There are a large number of possible problems, for many of which there may have been no indications during normal operation. Therefore, have contingency plans for dealing with the most frequent types of failure in place before inspecting the exchanger. Include estimates of times required to perform any maintenance procedures, repairs, and/or alterations that may be required. Checklist 4.2 is a partial list of things that maintenance inspection may reveal.

Plan to procure beforehand any replacement parts, gaskets, bolts and nuts, and other equipment that will be needed. If it appears that pressure parts will require welded repairs or alterations, discuss the proposed work with the Authorized Inspector responsible for in-plant inspection. This should include agreeing on the acceptability of the proposed work and the testing required when it is complete. Be sure to notify the appropriate authority in jurisdictions where the laws and regulations governing pressure vessels require such notification.

CHECKLIST 4.2 What Maintenance Inspections May Reveal

- ☐ Leaking tubes that require plugging or replacement
- ☐ Possible impingement on adjacent tubes of high-pressure fluid from leaking tubes requiring insurance plugging
- ☐ Leaking tube-to-tubesheet joints that may have to be repaired by reexpanding, rewelding, or seal-welding previously expanded-only tube-to-tubesheet joints
- ☐ Cracked, eroded, or wormholed ligaments
- ☐ Excessively fouled bundles, which may require mechanical, chemical, or hydraulic cleaning or combinations of these cleaning methods
- ☐ Wasted surfaces requiring repairs by welding and shop or field machining
- ☐ Pinholes in welds, requiring repair
- ☐ Indications of flow-induced vibration, requiring staking the tubes
- ☐ Thinned or "feathered" tube ends, requiring sleeving or ferruling
- ☐ Baffle wear or damage indicating a need for retubing or rebundling
- ☐ Bent or otherwise damaged pass partitions, requiring repair or replacement

Planning for inspection during unscheduled outages

Unplanned shutdowns usually happen when there are indications of leaks or sudden performance deterioration. Fluids may leak to the atmosphere through flanged joints or corroded-through or failed parts. Decreased performance or product deterioration may indicate air leaks into vacuum units or higher-pressure stream leaks into lower-pressure ones. Another reason for an unplanned shutdown is an abnormal pressure or temperature excursion that requires a shutdown to examine the integrity of the unit or system.

Planning for unscheduled outages is similar to planning for scheduled ones. But troubleshooting the problem and restoring production as quickly as possible allows very little time for planning. Therefore, include an inspection section in contingency plans for such events.

Planning for inspection on completion of maintenance

Plan for inspecting maintenance work before the exchanger is buttoned up. Include requirements for inspection on the work order. Describe what to inspect; by what means; who is to perform the inspection; and the necessary materials, tools, gages, etc. State who is to witness air and hydraulic pressure tests, the recording format, and filing requirements. Jurisdictional agencies concerned with pressure vessels often require filing forms documenting the work. In North America, many jurisdictions and most insurers require owners to keep pressure-vessel status current with the National Board. Whenever a pressure part has been worked on, be sure to coordinate pressure-testing requirements of the prevailing inspection code with the Authorized Inspector responsible for in-service inspection, the Authorized Inspection Agency, and the local jurisdiction.

Instruct the maintenance force to have all necessary materials at the site when the unit is ready for inspection.

Inspection before Maintenance

Start with an on-line external inspection. When the exchanger is shut down and properly prepared, inspect off line to (1) ascertain present condition and compare with condition at the previous inspection and (2) find deteriorations and failures that require immediate and future repair and maintenance. Depending on what the on- and off-line inspections uncover, you may have to implement maintenance procedures and inspect the completed work.

On-line inspection before maintenance

Operating units cannot be fully inspected. Insulation hides most of what one would like to observe, and access is limited. However, a preliminary on-line inspection can reveal leaks, cracks, and improper or unauthorized installation of parts and appurtenances. With the right kind of inspection equipment, you may be able to find out whether harmful tube vibration is occurring. The Electric Power Research Institute (EPRI) has furthered the development of an ultrasonic device for on-line monitoring of tube and tube-to-tubesheet joint leaks in feedwater heaters. This technique, pioneered in Europe in the 1970s, can alert operators early of a tube leak. The advance warning can indicate the need for action well before performance deterioration becomes apparent. Following are suggestions for making an initial on-line survey followed by off-line inspection.

Examine operating logs. Examine operating logs to see if pressure-relief devices were tripped. Trips may indicate that a unit was exposed to excessive pressure. Look for a record of an excessive pressure or temperature episode that may have damaged the unit. Investigate the causes of deviations from normal conditions and their effects. Thoroughly inspect for possible damage to the pressure-bearing parts.

Inspect the external parts. Inspect ladders, platforms, foundations, connections to piping, and protective coatings for broken and loose parts, holidays, and discontinuities. Use inspection hammers and flat and pointed scrapers. These are some specific things to examine:

- Safety valves
- Insulation condition
- Condition of vents and drains
- Appurtenances and supports
- Piping to the exchanger

Observe the condition of safety relief devices. Regularly test-trip safety valves. Take notice of safety valves encrusted with corrosion or deliberately rendered inoperative and report the deficiency to the appropriate authority for corrective action.

Examine insulation for cracks that indicate possible exposure of the underlying part to excessive pressure and stains that might betoken hidden leaks. When either is visible, remove enough insulation to expose the part for further inspection.

Inspect all accessible vents and drains. Try shutoff valves in vent and drain lines to make sure that they are operative. Also, make sure that vents and drains intended to operate continuously are not closed off.

Look for appurtenances not shown on drawings or maintenance records fastened to the unit by operators. If they are acceptable and do not violate the prevailing inspection code, update the as-built exchanger drawing. Otherwise, recommend corrective action to be taken at the next scheduled shutdown. This is especially important if the undocumented addition can affect safety.

Inspect supports and foundations for cracking, bending, and corrosion. Make sure that no corrosion has taken place between an exchanger and the cradles on which it rests. Verify the integrity of sealant of exchangers to cradles. When saddles are continuously welded to exchanger shells, make sure that the holes slotted at the hold-down bolts to accommodate expansions are not plugged. Check to be certain that the hold-down bolts are not excessively corroded. Measure the thicknesses of the various parts of saddles that appear to have corroded.

Take scrapings for analysis from any places where water or process fluids have stained the surfaces. Examine these sites with a magnifying glass. Liquids entrapped under insulation are frequent causes of corrosion.

Look for external corrosion, signs of leakage, angularity between pairs of flanges, cracking, strains, and stains or drips indicating leaks. When there is evidence of such conditions, schedule the connections for sand, grit, or water blasting and closer inspection by magnetic particle or fluid penetrant examination when magnetic particle examination is not feasible.

Off-line maintenance inspection

Here are some things to look at during off-line maintenance inspection.

General external appearance. After the unit is off line and prepared for premaintenance inspection, look for the following:

- Structural distortions
- Wear or pounding of support boltholes and slots
- Distortions indicated by nonuniform gasket compression
- Flange and tubesheet distortions

Condition of parts. Inspect inside the heads or channels and, as much as possible, bundles and shell interiors for the following:

- Corrosion and erosion
- Fouling on both sides
- Tube damage and deterioration
- Tubesheet face damage
- A tube plug pattern in which the plugs are all in one group of tubes
- Damage to machined surfaces
- Distortions in the pressure envelope
- Corrosion

Checklist 4.3 is a partial listing of structural distortions to search for. Checklist 4.4 lists places to examine for corrosion and erosion that may require investigation and repair.

Compression marks on the gaskets and the compressed thicknesses are good indicators of how a flanged joint is functioning. Check to see if the gasket surfaces are capable of bearing on the gasket. Permanent warping may require remachining and investigating the cause.

CHECKLIST 4.3 Structural Distortions to Look for

- ☐ Bowed covers
- ☐ Bulged or collapsed parts
- ☐ Cupped flanges and tubesheets
- ☐ Squirmed or angularly deformed expansion joints
- ☐ Bent bolts
- ☐ Bent, twisted, or out-of-square nozzles

CHECKLIST 4.4 Locations to Examine for Corrosion and Erosion

- ☐ Opposite shell inlets and outlets
- ☐ Adjacent to baffles, supports, and tubesheets on the back side of a tubesheet and tubes where they emerge from the hottest tube pass*
- ☐ In the shell at baffle and support plate locations
- ☐ Anywhere sludges or other sediments may have accumulated
- ☐ On the underside of the top tubesheet of vertical units
- ☐ On metals less noble than the ones they are near or to which they are attached
- ☐ Tubes in which wall thickness may have corroded or eroded below that required for the operating condition

*Tubes in the hottest pass are most sensitive to corrosion. Water leaving hot-pass tubes may flash and erode the tube ends as it enters the channel space.

Using articulated mirrors or borescope probes, examine the condition of the exterior of the tubes in way of the inlet connection for impingement erosion and corrosion.

Vertical steam generators with insufficient flow velocity across the bottom tubesheet may build up flocculent materials corrosive to the tubes and tubesheet. The water-level control in the shell must be adequate to provide for a sufficiently high blowdown flow rate to prevent solids from building up on the shell side of the bottom tubesheet. If flow rates to the shell fluctuate widely, the top tubesheet may be shocked as cold feedwater strikes the hot surface. If the underside of the top tubesheet is inadequately vented, the accumulation of noncondensible gases wet with the shell-side fluid may cause severe corrosion of the tubes in the vicinity of the tubesheet and damage the tubesheet surface.

Material lower on the electromotive scale will be preferentially corroded. An example is flanged carbon-steel channels or bonnets, bolted and gasketed to brass tubesheets.

During shell-side inspections of removable-bundle units, look for signs of vibration damage such as fretting and pounding where the tubes penetrate supports and baffles, and interspan collision damage.

It is not possible to determine how badly the shell side of a fixed-tubesheet exchanger is fouled. Estimate it by (1) examining the condition under all shell nozzle connections with mirrors, lights, and bent rods; (2) scraping samples of deposits from the nozzle and shell for analysis of corrosion products; and (3) making water-flow tests through the shell.

Checklist 4.5 lists the kinds of damage and indications of deterioration to search for on the tube sides of fixed tubesheet exchangers.

It is a good practice, after water or solvent washing, to blow out the tubes with plant compressed air before inspecting their interiors. Collect the material blown out for chemical analysis. A substantial content of metal ions representative of the exchanger's material of construction may indicate previously unknown deterioration. It may also indicate potential problems downstream.

A difference between the expanded and unexpanded tube-wall thicknesses of more than 10 to 12 percent may indicate any of the following conditions: (1) excessive flow velocity (the tubes may also have "feathered" ends), (2) insufficient turnaround area, and (3) overrolling during expansion.

Make relatively quick tube-condition checks with a borescope. Scan the tubes for etching and pitting. Measure visible pits that can be reached with a pit gage to determine whether remaining wall thickness is adequate for the pressure and service. If time and the budget permits, measure the tube-wall thickness by eddy-current examination or ultrasonically.

CHECKLIST 4.5 Damage and Indications of Deterioration to Inspect for on Tube Sides of Fixed-Tubesheet Exchangers

- ☐ Bowed or collapsed tubes, indicating excessive differential thermal expansion
- ☐ Twisted tubes indicating improper rolling technique
- ☐ Split tube ends
- ☐ Thin tube ends
- ☐ Deposits of scale and foulants
- ☐ Signs of vibration damage
- ☐ Excessive difference between I.D. of tube behind the tubesheet and in the expanded area
- ☐ Interior corrosion, wear, and damage
- ☐ Etching, pitting, and/or gross corrosion-erosion revealed by ridges in tubesheet face surfaces running into tube holes, tubes, and pass-partition grooves (when present)
- ☐ Cross-ligament cracks
- ☐ Ligament width smaller than design

Always locate the position along the length of the tube of any tube leaks. If the leak position cannot be found with a borescope, use an axial tube leak position finder like that shown in Fig. 4.1.

Examine the ligaments with fluid penetrant, strong light, and magnifying lens. Stains and spots indicate through leakage. Widths smaller than design indicate overrolling or excessive numbers of retubings. Look for erosion of the tubesheet and tube ends in the

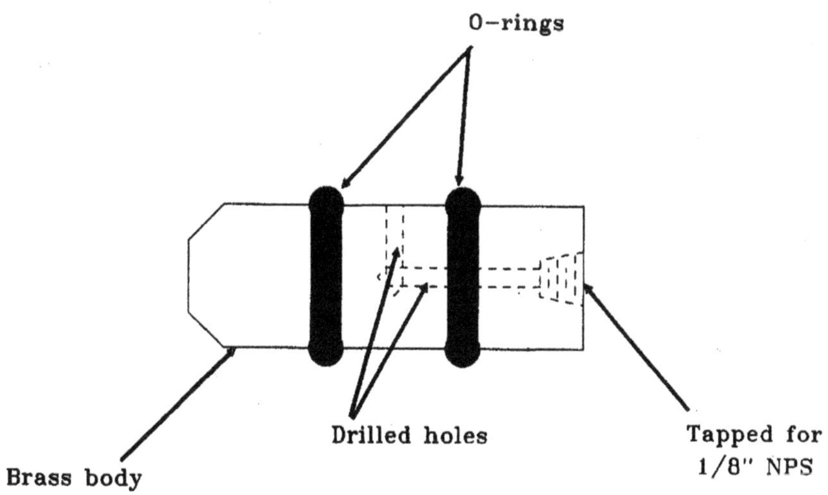

Figure 4.1 Axial tube leak position finder. (*MGT Inc., Boulder, Colorado.*)

region of the tube-side inlet nozzle. The cause(s) of tube failures may be determined by pulling out the plugs and locating the failures.

Measure flatness across machined tubesheet surfaces with a straight-edge and dial indicator. Waviness indicates probable exposure to excessive interpass temperature differences.

Check the tubes for bowing, which can indicate inadequate provision for differential expansion between the shell and tubes and excessive interpass temperature differentials.

Assessing structural integrity of pressure parts. The most common way to examine the structural integrity of the pressure envelops is to hydrostatically test the unit at 1.5 times the maximum allowable working pressures (MAWPs) shown on the nameplate, adjusted for allowable stress at ambient relative to design temperatures.

Hydrotesting is not the most effective way to determine the leak-tightness of an exchanger. Depending on the required degree of leak-tightness, check for leakage by bubble testing, helium leak mass-spectrometer sniffer testing, or halogen-diode halogen sniffer testing.

Check for corrosion that exceeds allowable wasting limits by drilling test holes or trepanning sections and making micrometer measurements of thicknesses or ultrasonically measuring the thicknesses of the parts.

Off-line inspection of removable-bundle exchanger shells. Outage time permitting, disassemble removable-bundle units. Sand-, grit-, or hydroblast the shell interior. Inspect for the internal and external corrosion and erosion of the shell. Examine the most susceptible areas in Checklist 4.6.

Off-line examination of removable-bundle exteriors. Before you can safely inspect a removable bundle, it must be clean of materials hazardous to the inspector inside and out. Achieve this by chemically cleaning or high-pressure lancing. After cleaning, examine the following:

CHECKLIST 4.6 Areas of Removable Bundle Shells Most Susceptible to Corrosion and Erosion

- ☐ Internal welds where nozzles and couplings penetrate the shell
- ☐ Places where hot fluid strikes (usually inlet regions)
- ☐ Regions in which there is lack of fluid circulation
- ☐ Locations where cross-flow baffles and tube supports rest on or near the shell
- ☐ Regions between the inlet-end tubesheet and the first baffle or support and between the last baffle or support and the bundle end

- Cross-flow baffle edges for damage resulting from bundle insertion and pulling, erosion, and corrosion
- Thickness of cross-flow baffles and tube supports
- Tubes and baffle holes for signs of vibration damage
- Baffle and tube support penetrations for crevice corrosion
- Tubes between tubesheets and adjacent baffles for dead-spot corrosion
- Tubes in way of the shell inlet and outlet connections for erosion

As a general rule, a bundle's useful life is over when cross-flow baffles and tube supports are half their design thickness. Checklist 4.7 identifies places to examine for indications of vibration damage.

CHECKLIST 4.7 Places to Look for Indications of Vibration Damage

- ☐ Between cross-flow baffles and tube supports for interspan collision damage
- ☐ Between the tubesheet and first baffle or support for fatigue cracking
- ☐ In the vicinity of the rearmost U-tube bundle support for abrasion of the tubes
- ☐ Where the tubes pierce the baffles or supports for fretting and wear
- ☐ Just behind the inner face of the tubesheet for tube cracking

Inspecting during Performance of Maintenance Work

There are two aspects to inspecting maintenance work: (1) inspecting to see that the work being done conforms with the procedures issued for its performance and (2) inspecting flaws uncovered as the work progresses. Coordinate inspecting for conformity with procedures with the AI responsible for on-site inspection. It is efficient to inspect jointly with the AI who will establish hold points.

Often, as repairs or alterations proceed, the people doing the work encounter previously hidden flaws and failures. For example, (1) a fluid penetrant examination of a tube-to-tubesheet weld repair may disclose ligament cracking adjacent to the weld or (2) disassembly of a fixed-tubesheet shell for retubing may reveal wasted baffles and cage parts. Before deciding what to do about these reported problems, the parts must be inspected.

Inspecting on Completion of Maintenance

Make a through inspection of completed maintenance work before restoring an exchanger to service. This should include visual inspection, fluid penetrant examinations, and ultrasonic and radiographic examinations. Things to inspect and types of inspections to make are

- Rewelds of eroded and corroded parts such as nozzle necks, flange faces, pass partitions, and ligaments by visual and fluid penetrant examination
- Taper and expanded-in thimble-plugged tubes by bubble testing, pneumatic testing, and hydrostatic testing
- Welded plugged tubes by fluid penetrant examination, bubble testing, and hydrostatic testing
- Integrity of the pressure parts by hydrostatic testing
- Rewelds of pressure seams and seams in which shells and channels have been cut by radiography and ultrasound
- Tube-to-tubesheet weld repairs by visual, fluid penetrant examination, and bubble and pneumatic tests followed by hydrostatic testing

The AI may require hydrostatic testing after any weld repairs or alterations. Coordinate the inspector's inspection with the AI's inspection.

Recording and Documenting Maintenance Work

Record and document repairs to the pressure envelope on for-the-purpose forms in accordance with the prevailing inspection code, for example the NBIC form R-1. Make these records part of the permanent maintenance file for each exchanger. In addition to these formal documents, be sure to record the work done in conformity with established company practices. The current trend is to computerize records of maintenance work performed.

Bar coding

The advent of readily available bar-coding systems has provided an excellent tool for storing and retrieving maintenance documents. Bar coding, widely used in commercial transactions, is now commonly used in libraries. It is highly suitable for filing, tracking, retrieving, and restoring to files, drawings, and documents in general, and those applicable to maintenance, repairs, and alterations in particular.

Videorecordings

Small videocameras are widely available and relatively inexpensive. This provides an opportunity to make dated visual and verbal records of the performance of maintenance procedures. Inexpensive editing equipment permits adding titles, eliminating unwanted parts of the original recording, and adding vocal and printed comments. Videorecording is not only a good way to document implementation of the procedures followed but also an excellent tool for training maintenance personnel. When videorecording the maintenance performed, the authors follow this procedure:

1. Videotape the nameplate of the exchanger. Record the time and date.
2. Videotape the region to be worked on before any work commences. During the prework videotaping, describe the condition of the parts on which the camera is focused and the repair or alteration to be done.
3. Videotape the work as it is done.
4. Videotape the final appearance of the completed work.
5. If a pressure test is performed, videotape the pressure gage(s) and or pressure recorder (if one is connected).
6. Edit the video and make copies for distribution in the form of a video report.

Assessing and Inspecting Mothballed and Used Exchangers

Assessing and inspecting units removed from service is similar in many ways to inspecting shutdown units during a scheduled outage. The differences depend on the purposes of the assessment and inspection. These may be to (1) determine whether a decommissioned exchanger can be restored to service (2) determine whether a decommissioned unit can be used in another service, and (3) evaluate the condition of a used-equipment dealer's unit before purchase.

Provide the inspector with access to and time to become familiar with the pressure-vessel code that applied when the exchanger was built. Insurance or legal considerations may necessitate the inspector's becoming familiar with the codes and requirements of the locality where the used or mothballed exchanger is to be reinstalled.

Wherever the NBIC and *ASME Boiler and Pressure Vessel Code* (ASME Code) prevail, do not consider reusing exchangers built to Section VIII, Division 2 and stamped U-2. Division 2 restricts their use, design, and portability. Furthermore, they are designed for out-of-the-ordinary pressure-diameter criteria or cyclical services. Therefore, reuse would require extensive nondestructive examination and engineering evaluation. Some suggestions follow.

Researching applicable information

The starting point for considering a mothballed or used exchanger is to gather as much information as possible. Following are items to research:

- Nameplate data
- Manufacturers' data reports
- Purchase specifications
- Manufacturers' drawings
- Records of repairs and alterations, usually attached to manufacturers' data report forms
- Maintenance records.

Pressure-vessel code requirements around the world require nameplate stamping. Exchangers built to the ASME Code must be stamped with the maximum allowable pressures and temperatures for each side. Do not consider putting into service a heat exchanger from which the manufacturer's nameplate has been removed. Generally, such units have been condemned and should have been destroyed. Most often, documentation of laid-up units or ones sold to used-equipment dealers are no longer available. However, nameplates provide some immediate information and are the key to obtaining more data.

Manufacturers' data reports provide basic information about the construction. This includes the original use, maximum allowable working pressures and design temperatures, materials of construction, tube counts, diameters and wall thicknesses, and schedule of nozzles. They give no insight into flow arrangements and do not reveal the number of passes on each side or the TEMA configuration.

It is unlikely that original purchase specifications will be available for mothballed or used exchangers. However, when they are, they can give some insight into their original uses.

Manufacturers' drawings are among the most valuable documents to acquire before inspecting used or mothballed exchangers. When they are available, use them as the basis for an orderly examination of all critical parts.

Maintenance records will almost never be available for used exchangers. For mothballed exchangers, stored in the company's storage facility, there is more likelihood of obtaining a history of the unit before inspection. If this shows a repetitive set of mechanical problems, caused by inherent defects in the exchanger, the potential uses of the unit are limited to moderate-pressure, moderate-temperature conditions and flow rates that will not lead to service failures. The maintenance record can provide a guide to specific parts to be carefully examined for current condition.

Assessing the condition of used and mothballed exchangers

Assessing the condition of used and mothballed exchangers is similar in many ways to evaluating an operating unit during a scheduled outage. However, thorough cleaning and more complete disassembly and nondestructive examinations are generally required. The following suggests how to go about this kind of inspection.

Examine the physical appearance before disassembly and cleaning. Inspect as you would for a unit shutdown for a scheduled outage. However, look for damage and deterioration that might have caused the unit to be mothballed and for handling and storage maltreatment. Set the unit on horses to permit a first examination of its physical appearance. Here are some additional things of which to be wary:

- A marked difference between the appearance of one part of a main-weld seam and that of another may indicate repair of a leak. This is especially probable if one part of the seam has been automatically or semiautomatically welded and another part welded manually.
- Patches to the pressure envelope may indicate that repairs or alterations were made. Look for repair or alteration stampings, and try to obtain records that indicate the nature of the work.
- Frozen-open pressure relief valves. Suspect excessive pressure application if you observe frozen-open relief valves, blown-out gaskets, or distorted flanges, bolts, and tubesheets.
- Poor appearance of weld repairs. Poor appearance indicates that the welders who did the work were probably not qualified, and the welding probably was not done in conformity with the NBIC, the API Inspection Code, or other required inspection code. Look for square or rectangular patches and lap welds, blow-throughs, undercuts, excessive weld-bead crown heights, wide welds, craters in welds, and other irregularities.

If there are any peculiar colors, odors, etc., chemically analyze scrapings of the surfaces that were in contact with the process fluids for hazardous residues.

Chapter

5

Tests and Nondestructive Examinations

Introduction

All pressure-vessel codes require tests and nondestructive examinations at various stages of construction. Requirements for nondestructive examinations of the ASME Code, the NBIC, and the *API Pressure Vessel Inspection Code* typify requirements in other codes for tests and nondestructive examinations. The American Society for Nondestructive Testing (ASNT), discussed below, is typical of these organizations and is devoted to nondestructive examining and testing in industry. Its constitution and bylaws state that its mission is to

- Provide the forum for exchange of NDT information
- Provide NDT educational materials and programs
- Provide training, standards, and services for qualification of and certification of NDT personnel
- Promote the discipline of NDT as a profession
- Facilitate and promote NDT research and application of NDT technology

In carrying out this calling, it establishes standards for NDT. Its SNT-TC-1A, *Recommended Practice,* lists the many standards it has set. ASNT establishes qualification levels for NDT personnel, offers training courses in various NDT areas, and administers tests in order to certify NDT examiners at skill levels I, II, and III.

Examiners who are certified by the ASNT or similar organizations are not *inspectors*. Tests and nondestructive examinations required by pressure-vessel codes should be performed and the results interpreted by individuals qualified by the ASNT or other reputable organizations.

The ASTM (formerly the American Society for Testing Materials) has prepared many widely accepted standards for performance and acceptance of the results of nondestructive examinations. The various standards of Section V of the ASME Code, titled *Nondestructive Examination,* list the ASTM standards as referenced documents.

Guidelines to ASME Code, Section V, "Nondestructive Examination"

Section V, "Nondestructive Examination," of the ASME Code has the information on examinations and tests required by ASME Code rules for constructing new pressure vessels. The NBIC Code requires, and the API Inspection Code implies, that the same kind of nondestructive testing (NDT) used in constructing a pressure vessel should be used when the vessel is repaired or altered. Section V has most of the general information needed for nondestructive examinations and tests, but it does not have procedures applicable to specific circumstances. These must be prepared by a professional in the NDT field.

Section V, Subsection A describes the methods for performing nondestructive examinations called for by other code sections. The ASME Code section that requires a nondestructive examination has the acceptance standard specific to the section. Subsection A also has evaluation standards, gives generally acceptable ranges of results, and tells how to deal with unacceptable ones. It even describes post-examining cleanup. Subsection B lists standards that cover nondestructive examination methods that the ASME Code has accepted by reference. Most are identical with ASTM Standards for the methods listed.

For people peripherally involved with NDE, but whose background and work are not specifically NDE-related, the tables, sketches, and photographs included in Subsections A and B are of inestimable help in understanding what NDE discloses. Another feature helpful to laypeople is Section V's glossary of terms used in nondestructive examinations and its sets of standard definitions of terms relating to various testing methods. The latter cover

AE	acoustic emission
ET	electromagnetic (eddy-current) testing
LT	leak testing
MT	magnetic particle testing
PT	liquid penetrant inspection
RT	γ and x radiography

UT ultrasonic testing
VT visual testing

Judicious use of nondestructive examinations and tests during inspections can reveal hidden flaws, indicate the degree (and sometimes the rate) of deterioration occurring with use, and provide a check on the quality of maintenance work.

Section V's methods, which are intended to detect surface and internal discontinuities in materials, welds, and fabricated parts and components, include visual inspection, radiographic examination, ultrasonic examination, liquid penetrant examination, magnetic particle examination, eddy-current examination, and leak testing.

Visual Inspection (VT)*

The first inspection to make is visual. The *unaided* human eye can tell whether

- Additional cleaning is needed before a more detailed inspection can be made.
- Rough corrosion, erosion, distortion, bulging, buckling, or misalignment is occurring.
- Severe cracking or other surface defects have developed.
- Tools are needed to aid in more detailed visual inspection.
- Additional types of inspection and more sophisticated tools are necessary for a more detailed inspection.

For tubular equipment, this is predicated on the observer's having considerable experience in looking at exchangers and making judgments. Unfortunately, the years in which experience is acquired take a toll on eyesight. The places and conditions for inspecting visually and the need for corrective lenses and eye protection also impede visual inspection.

An experienced observer can tell a fair amount about an exchanger by looking at its various parts without special cleaning. However, thorough visual inspection requires clean surfaces. For simple, localized work, use files, wire brushes, power brushes, emery cloth and sandpaper, and power sanders. Suspicious regions, coated or encrusted with process fluid residues and foulants, may require hydrocleaning. With proper precautions against excessive erosion, sand, grit, or hydroblasting may also adequately expose the surfaces.

*See Articles 9 and 28 of Section V of the ASME Code.[1]

Acid cleaning is another way to prepare an exchanger for visual inspection. It has the advantage of highlighting and sometimes enlarging surface defects. Its disadvantages are the hazards associated with using corrosives and the need to dispose of the waste acid in an environmentally sound way.

Good practice, before completely inspecting a disassembled removable-bundle exchanger, is to hydroclean the bundle inside the tubes and on its exterior and to use hydrocleaning or hydroblasting to clean the inside of the shell, channel head, and other parts. The tube side of straight-tube equipment can also be effectively cleaned by mechanical methods such as drilling, brushing, forcing cleaning plugs through the tubes with air or water, or combinations of these techniques. However, for deposits that these methods do not remove, hydrocleaning or hydroblasting is recommended.

It is not possible to clean the shell side of a fixed-tubesheet unit except chemically. Therefore, the possibilities for visually inspecting inside fixed-tubesheet exchanger shells are limited.

Aids to visual inspection

The human eye has limited capability to discern small details (especially as one gets older). Furthermore, what can be seen is only what is illuminated by sufficient light. For the best and safest lighting, use focused, commercially available, high intensity battery-powered or low-voltage (usually 12-V) light sources. A variety of devices are available to extend unaided vision. Simplest and least expensive are articulated mirrors and reflectors. Always use the largest one that will fit into the site. To enhance the ability to see small defects, a convenient 10-power pocket magnifier is excellent. When a flaw is found, it is a great help to examine it further with a portable microscope. These are available in magnifications of 10, 20, 30, and 40 diameters. Recently, small, 30-power battery-lit, pocket microscopes have become available as inexpensive commercial instruments. They are excellent for examining tube-to-tubesheet welds, ligament cracks, weld seams, and other places where cracks are suspected.

Borescopes, or flexible fiber-optic probes, extend the reach of eyes. They may be fitted to a VCR and monitor for recording and subsequent study. A keyboard permits labeling the tape, which can subsequently be edited and comments added by voicing over. Interpreting what borescopy reveals requires practice on standard defect specimens because the probes substantially magnify the surfaces on which they are focused. Interpreting the seriousness of defects is difficult without such advance practice.

Supplement visual inspection with photographs and videorecordings. We recommend a combination of instant-camera, 35-mm photographs and handheld camcorder videography to provide a permanent record of conditions found during a visual inspection. You can then review the appearance of defects and conditions with repair and maintenance personnel when planning and performing the work.

Radiographic Examination (RT)*

Radiographic examinations give pictures of the insides of parts. They are made either by using x-ray-generating equipment or a radioactive isotope, most frequently of iridium.

Spot radiography

Under the ASME Code rules, spot radiography consists of a manufacturer's radiographing one spot, chosen by the AI, for every 50 ft of butt welding *performed by a given welder.* Its purpose is to verify the welder's ability to produce sound welds. The ASME Code presumes that the welding of a spot-radiographed unit will have been performed by a welder whose work has been demonstrated to be acceptable. Therefore, it assigns an 85 percent joint efficiency to the weld.

Full radiography

The purpose of radiographically examining the full length of *all* the butt welds of the pressure envelope is to show that they conform with the required quality standard. Welds that meet the code's acceptance standards for full radiography are assigned a joint efficiency of 100 percent.

Partial radiography

The ASME Code's nameplate stamping requirements indicate when the full length of the welds of a *part* of a pressure vessel is fully radiographed. The 100 percent joint efficiency is applied only to the fully radiographed welds of the part.

For exchangers in critical services, it is prudent to require full radiography of the welds of those parts that can be radiographed. Radiography can be used to examine filet welds and is sometimes, but rarely, used to examine some tube-to-tubesheet joint welds. It is hardly ever used to examine the welds of welded tubing because the wall thickness is generally so close to the size of detectable flaws that defect resolution and interpretation are inconclusive.

*See Articles 2, 3, and 22 of Section V of the ASME Code.[1]

When radiography is used to examine pipes and other circular sections, it is necessary to understand the projection geometry to interpret the size of the flaws it discloses.

Ultrasonic Examination (UT)*

Ultrasonic examination is excellent for disclosing flaws and measuring thicknesses in seamless and longitudinally welded pipe and tubing. The techniques and practices have evolved rapidly.

Discontinuities in longitudinal welds and their heat-affected zones are discerned using an angle projection of a pulsed ultrasonic beam by either the surface contact or immersion method. This method, which may be used by mills during tube manufacture, may reveal indications that ET and water-submerged air testing do not. This is especially so for tubing that requires very high ET frequencies, such as copper-nickel, alloy 25-6MO, and titanium.

The principal maintenance uses are measuring the thicknesses of various parts of an exchanger to determine remaining thickness and examining welds that cannot be radiographed. For these purposes the surface contact method is preferable since it eliminates the need to immerse the part to be measured in water.

The *Internal Rotary Inspection System* (IRIS), developed to measure tube-wall thickness and reveal the depth of pits in tubes, is not suitable for these purposes. The IRIS has a probe that carries its own source of water into the tube. It can examine 100 percent of the tube surface without distortions even where the tube penetrates baffles or supports.

The *Quantitative Ultrasonic Analysis System and Recorder* (QUASAR)† is a computer-based system developed from the IRIS system (Fig. 5.1). An ultrasonic transducer lies axially inside the tube and the pulse emitted bounces off a 45° mirror onto the wall of the tube. An echo of the signal is returned from both the inside and outside tube-wall surfaces. By calculating the time between echoes, the wall thickness is determined.

The mirror is rotated 360° by a water-driven turbine, thus scanning the entire tube surface as the probe travels down the tube. Successive pulses from a single revolution are displayed on the computer screen, giving a pictorial view of the tube wall at that location (Fig. 5.2).

---

*See Articles 4, 5, and 23 of Section V of the ASME Code.[1]

†The QUASAR system was developed by Anser Corporation, 8 Cedar Street, Suite 67, Woburn, MA 01801 (USA).

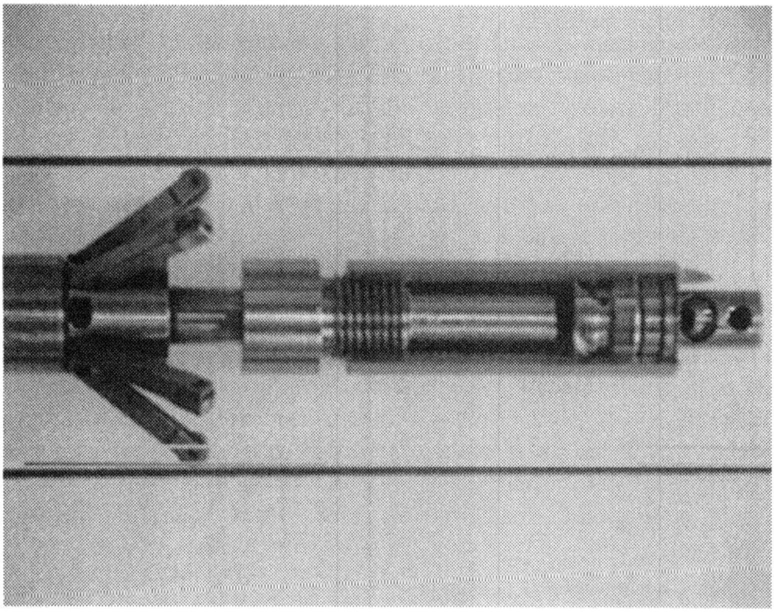

Figure 5.1 Turbine for IRIS examination. (*Anser Inc., Woburn, Mass.*)

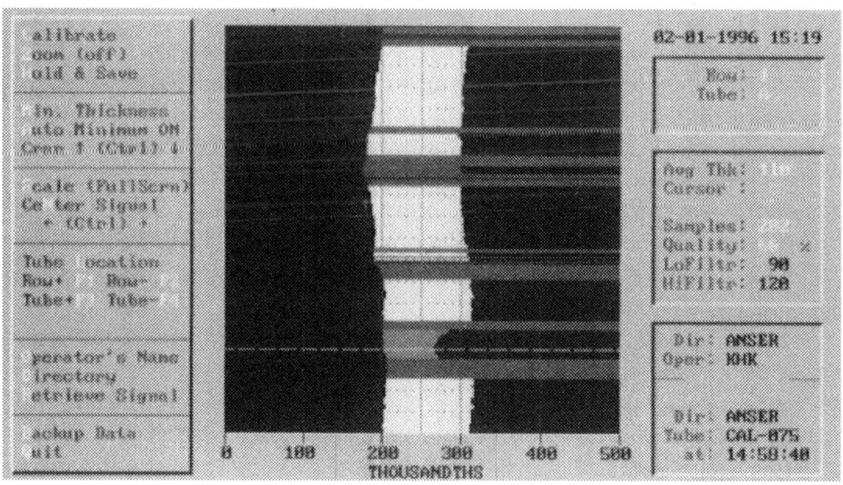

Figure 5.2 Output of QUASAR internal rotary inspection scan. (*Anser Inc., Woburn, Mass.*)

The principal disadvantage of IRIS is that it is a relatively slow process when compared with eddy-current examination—about 2.44 m (~8 ft)/min compared with 18.3 to 27.45 m (~60 to 90 ft)/min. This slow speed enhances the signal clarity, giving the system much greater accuracy than other methods. Tube-wall thickness can be measured to within 0.127 mm (~0.005 in) with the QUASAR system as compared with the average signal tolerance of 20 percent or more with eddy-current systems.

Sonic pulse examination

The slower IRIS systems are often used to verify critical wall loss indications found by other systems such as eddy-current systems or sonic pulse. The sonic pulse system transmits an audible sound wave down the inside of the tube. Echos from this wave are returned from any anomaly in the tube, whether a hole, through-crack, dent, obstruction, or internal corrosion of the tube.

Liquid Penetrant Examination (PT)*

Liquid penetrant examination reveals *surface* discontinuities such as cracks and porosity. Indications of these discontinuities are assumed to be evidence that the flaw extends below the surface. During construction, the ASME Code rules require liquid penetrant examining roots, interpass deposits, and surfaces of welds used to join some materials. It is used as an alternative to magnetic particle examining nozzle-to-pressure envelope welds when the materials are nonmagnetic. But, in contrast to magnetic particle examining, which can disclose discontinuities for a moderate depth beneath the surface, liquid penetrant examining will not expose any flaws beneath an unbroken, continuous surface. Flaws revealed by liquid penetrant examination must be ground or chipped out until sound metal is reached, rewelded, and reexamined.

A commercially available system, in widespread use, consists of three aerosol or pressurized cans used in sequence as follows: can 1 contains a cleaning agent, can 2 contains the liquid penetrant, and can 3 contains a developer. The cleaning agent is sprayed onto the surface and allowed to reside for at least 5 min, after which all excess is wiped off. The liquid penetrant is a low-viscosity red dye that searches out and stains any discontinuities in a surface. It is sprayed onto the part(s) to be examined and wiped clean of any excess on the surface. The developer is a dispersed white, powdered, chalk-talc mix-

*See Articles 6 and 24 of Section V of the ASME Code.[1]

ture. When it is sprayed onto the surface, it provides a smooth continuous layer. Capillary action and the properties of the developer are such that some of the dye is extracted from any discontinuity that it has penetrated. This provides a red trace of the flaw against the white background.

This system is so convenient and easy to use that workers not trained in fluid penetrant examination and having no formal qualifications often do the work. This practice is unsound because training in properly preparing the parts to be examined and interpreting the results are essential. It is strongly recommended that personnel who perform fluid penetrant examining be formally trained and certified to the level pertinent to the work.

A disadvantage of systems that use dry developers is that the chalk-talc mixture may obscure the surface immediately next to the flaws. When this is a concern, use liquid penetrants or oils that fluoresce under high-intensity ultraviolet light (black light) in a similar fashion. Note that some lubricating oils will fluoresce under black (ultraviolet) light. Use black light to examine surfaces in units handling such materials by wiping away the visible surface oil with a soft, clean rag and shining the high-intensity ultraviolet light onto the surface.

Practice is required to interpret what you see when using fluid penetrants. Training should include examining specimens with artificial flaws.

These are some of the places to examine by fluid penetrant:

- Shell and channel main-seam weld roots
- Replacement shell flange, shell-cover flange, and channel flange welds
- Repair-weld roots
- Flange-to-nozzle weld surfaces
- Nozzle-to-shell and -channel weld surfaces
- Tube-to-tubesheet joint welds

Magnetic Particle Examination (MT)*

*Magnetic particle examination** consists of dusting the area to be examined with colored iron dust and then creating a magnetic field in the parts. The iron dust arranges itself uniformly along the lines of magnetic flux. When there is a discontinuity, the defect distorts the

*See Articles 7 and 25 of Section V of the ASME Code.[1]

magnetic field. The particles are attracted to the point where the lines of magnetic force are deflected out from and back to the surface, arranging themselves so as to outline the defect. The magnetic field must be strong enough to create this effect. The dye used to color the particles may be fluorescent to enhance seeing indications of defects.

Because the iron particles will not line up along cracks parallel with the magnetic flux lines, vary the direction of magnetizing to pick up cracks that run in different directions. This process may produce undesirable residual magnetism in the parts examined. Therefore, usual and customary practice is to demagnetize after the examination.

Various devices may be used to produce the magnetic field in the workpiece. Among them are DC arc welders and electric storage batteries, both of which produce the required high current and low voltage. Their disadvantage is the potential for current arcing and damaging the surfaces. Magnetization may also be produced using strong permanent magnets. However, the most common means is to pass direct current through a set of coils wrapped around the work. The magnetic field generated in the coils magnetizes the part being examined.

Because magnetic particle examination is limited to materials that can be magnetized, it is useless for most austenitic stainless steels and nonferrous materials. It gives erratic results on porous materials. It can detect subsurface defects; however, it does not indicate the depth to which they extend. The depth to which it can penetrate depends on the strength of the magnetic field. The usual limits are approximately 6.5 to 12.5 mm (~¼ to ½ in).

Eddy-Current Testing (ET)*

Eddy-current examination measures the variations in electrical characteristics induced in a part by alternating-current coils or probes. Variations and discontinuities cause changes in eddy-current strength in the part being examined. The eddy current and its variations are detected by a variety of electronic devices and displayed on a cathode-ray terminal. The output may also be recorded on a continuous strip chart.

Considerable training in using eddy current to examine structures is required. Comparison specimens with standard flaws are used to establish the signatures of various indications.

A relatively recent development is to use regular eddy-current examination to estimate the remaining thickness in exchanger tubes and to trend tube thickness loss over time. Purchasers who intend to

---

*See Articles 8 and 26 of Section V of the ASME Code.[1]

monitor deterioration using eddy-current testing are advised to have a 100 percent examination performed at the exchanger factory to establish the baseline thicknesses. Be aware that the baseline data generated with internal probes in the exchanger factory is different from that of the encircling coil eddy current developed at the tube mill for acceptance purposes.

It is extremely difficult to interpret eddy-current signatures at locations where tubes penetrate baffles and supports. Therefore, the process is not reliable for detecting erosion, corrosion, and erosion-corrosion at such locations. For this purpose, IRIS is superior.

In order to develop reliable trending data, the tube examination method must be repeatable and relatively accurate. Accurate evaluation of test data requires the signal analyst to be highly experienced in the type of tube failure or degradation mechanisms that may be encountered in specific tube material and service being examined. The evaluation of the test data can be further enhanced by pulling a few sample tubes with typical signal indications and comparing the eddy-current response with the measured wall loss. With this reference base, important components of the signal such as amplitude, phase, and duration can be incrementally related to specific wall loss percentages.

The most widely used eddy-current testing techniques* are

- Eddy-current examination (ECE)
- Partial-saturation eddy-current examination (PSECE)
- Saturation eddy-current examination (SECE)
- Remote-field eddy-current examination (RFECE)
- Magnetic flux leakage examination (MFLE)

Eddy-current examination (ECE)

In ECE, the tubes are scanned internally by probes with two coils. One is wound around a bobbin such that the coil almost touches the inside wall of the tube. The two coils are parts of a Wheatstone bridge.

The probe is held in an undamaged section of tube to establish a reference voltage across the bridge. As the probe is pulled along the tube, plus and minus variations in the X and Y coordinates produce signals on the oscilloscope screen characteristic of the type of failure/degradation and severity of the defect.

---

*For a detailed discussion of these techniques, see the article on heat exchanger reliability by Bell and Strauss.[2]

An *absolute* probe contains a sensing coil close to the tube wall and a second, nonsensing reference coil away from the tube wall. In a *differential* probe, in which both coils are connected to the bridge, an abrupt discontinuity will result in one coil signaling a wall loss, which is compared with the signal from the other coil which does not sense the discontinuity. The probes are physically similar, but connected differently (Figs. 5.3, 5.4*a*, and *b*).

The absolute probe is sensitive to gradual tube degradation, such as inside or outside erosion-corrosion, stress-corrosion cracking, fretting, or pitting. The differential probe is more sensitive in evaluating pitting, and consequently is used more often when pit attack is suspected.

Saturation eddy-current examination (SECE)

SECE uses a probe that has the coil wound around a permanent magnet specific for the tube material, diameter, and thickness to be inspected.

Figure 5.3 Probe for eddy-current testing may be an absolute or differential probe depending on how it is connected. (*Scientific Technologies, Inc., Madison, Conn.*)

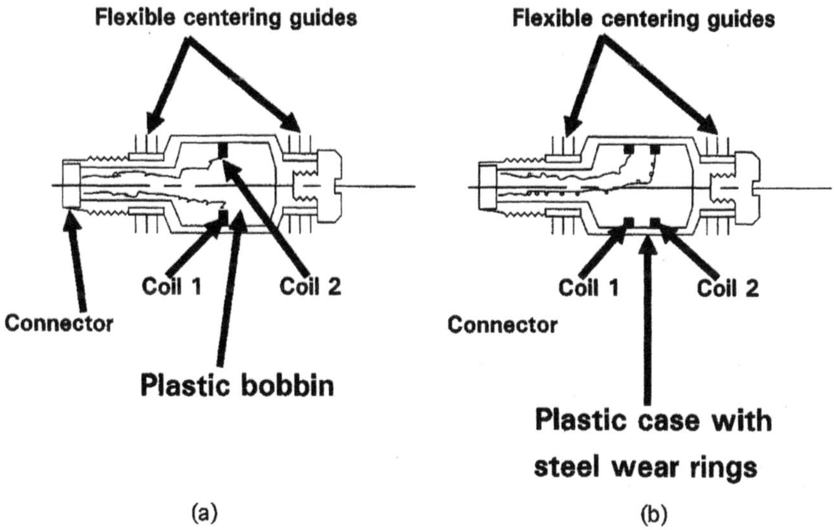

Figure 5.4 (*a*) Absolute probe and (*b*) differential probe.

Magnetic flux leakage examination (MLFE)

MLFE uses a probe that has two coils that sense the magnetic flux leakage through the tube wall. The probes for SECE and MLFE are physically similar in appearance (see Fig. 5.5).

Remote-field eddy-current examination (RFECE)

RFECE uses a probe that has one coil that induces the eddy-current field and another that senses the field (Fig. 5.6).

Choice of technique

The eddy-current technique best suited to producing reliable, repeatable results is determined by the tube material to be examined. Table 5.1 lists tube materials and recommended eddy-current techniques.

Leak Testing (LT)*

All pressure-containing structures leak to a greater or lesser degree. The leakage rate depends on the integrity of the pressure envelope

*See Articles 10 and 27 of Section V of the ASME Code.[1]

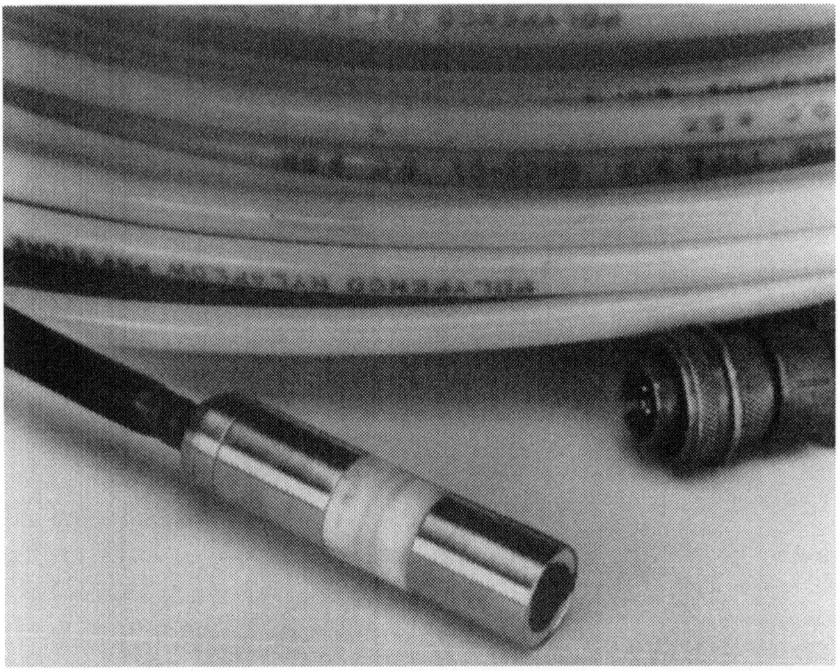

Figure 5.5 Usage for probe for partial-saturation eddy-current and magnetic flux leakage testing depends on how it is connected. (*Scientific Technologies Inc., Madison, Conn.*)

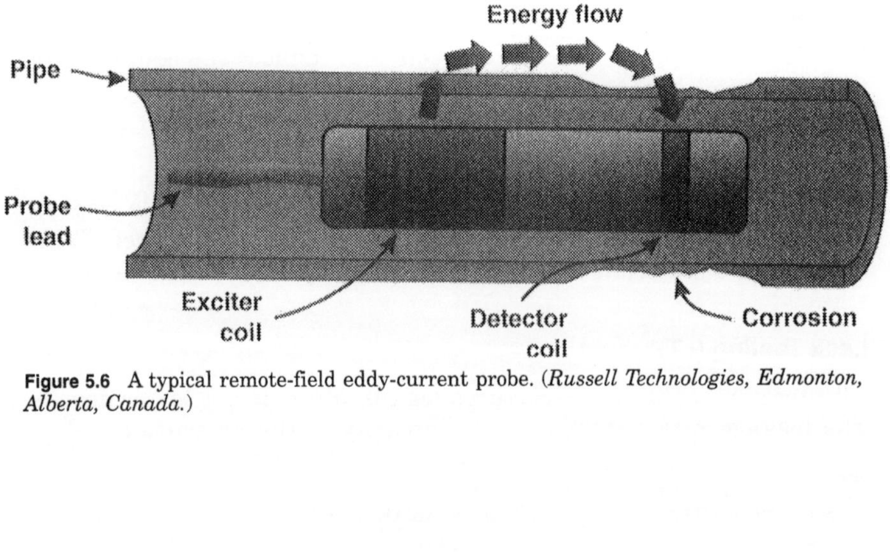

Figure 5.6 A typical remote-field eddy-current probe. (*Russell Technologies, Edmonton, Alberta, Canada.*)

TABLE 5.1 Recommended Eddy-Current Techniques for Various Tube Materials

Tube material	Eddy-current technique
Copper alloys and austenitic stainless steel	Eddy-current examination
Ferritic stainless steel and carbon steel	Partial-saturation eddy current; remote-field eddy current; magnetic flux leakage (should be verified by ultrasonic testing)

and the fluid it contains. Pressurized hydrogen, for example, will leak by diffusing at measurable rates through solid steel enclosures. Therefore, in testing for leaktightness, one must first set some limit for leak-rate tolerance and then find a way to measure the leak rate.

Obvious leaks to consider are ones from inside the exchanger to the atmosphere. For vacuum operation, consider leaks of atmosphere into the vacuum side, and leaks between the sides. Tolerable amounts of overall leakage out of or into the exchanger are determined by safety, process, and environmental considerations. Those between the sides are determined primarily by process conditions. However, if the material flowing through one side is hazardous when mixed with or dispersed in the material flowing through the other, safe operation also determines the tolerable leak rate.

Some of what follows is about tube-to-tubesheet joint leak testing. However, the methods and principles apply generally. When considering a leak-test method, look into the ASME Code, Section V, Article 27 SE-432, "Standard Recommended Guide for the Selection of a Leak Testing Method." Table 5.2 lists ASME Code acceptable leak rates for helium and halogen testing systems. Always bear in mind that leak rates measured at atmospheric conditions may be much different from the actual ones during operation.

The following discusses bubble formation testing, gas-lake testing, and detector probe methods. The detector probe methods include halogen diode and helium mass-spectrograph sniffer methods. Use these

TABLE 5.2 Acceptable Leak Rates for Helium & Halogen Leak Test Systems

System	Maximum acceptable leak rate, standard cm^3/s*
Helium sniffer	1×10^{-4}
Halogen sniffer	1×10^{-4}
Helium tracer probe	1×10^{-5}
Helium hood	1×10^{-6}

*Standard cubic centimeters per second.

methods when a unit is to handle lethal or noxious fluids and to ensure that air-conditioning and refrigeration equipment refrigerant gas leaks are within tolerances. None of these methods is quantitative.

Bubble testing

Bubble testing is a very effective way to find tube-to-tubesheet joint and other kinds of leaks. To test, pressurize the structure with dry air or nitrogen and squirt or paint a bubble former on the unpressurized face. Test solution must not break away from the surfaces or break down rapidly because of air drying or low surface tension.

Although a suitable bubble-forming solution may be a soap or detergent, household ones are not acceptable, nor does Section V permit their use. Use a commercial bubble former developed specifically for testing such as "Snoop."* Always investigate the chemical composition of the bubble-former solution for its effect on the metals it contacts.

For most purposes, a gas-side pressure of 345 kPa (~50 psi) is adequate. (See below for leakage rates and safety precautions.) A typical sequence for testing tube-to-tubesheet joints is

1. Fill the shell with inert gas or air at the lower of 345 kPa [~50 psig (lb/in^2 gage)] or the MAWP.
2. Allow a pressure dwell time of at least 15 min
3. Flow, paint, or squirt bubble former over the areas to be tested.
4. Observe for bubble formation.

Tube-to-tubesheet joint gas-lake testing

The gas-lake testing procedure uses a "lake" of liquid (usually water) to surround the region to be inspected. It is most suitable for testing the tube-to-tubesheet joints in the top tubesheet of a vertical, head-up unit. The procedure is as follows:

1. Pressurize the shell side with air or nitrogen to the lower of 345 kPa (~50 psi) or the MAWP at room temperature.
2. Set the unit vertically and seal a 2.5 or 5.0-cm (~1- or 2-in) -high ring dam to the tubesheet.
3. Fill the tube side with water to the level of the dam.
4. Observe for bubbles of gas in the lake.

---

*Snoop is a product of Nupro Company, 4800 East 345 St., Willoughby, OH 44094 (USA). Nupro states that it meets the performance requirements of MIL-L-25567C Type 1 and has no chlorine.

Detector-probe methods

Detectors or "sniffers" suck air through a tubular probe into an instrument sensitive to small amounts of tracer gas. To use a sniffer on tube-to-tubesheet joints, traverse the tube ends with the probe at a distance of approximately 3 mm (~⅛ in) at a scanning rate predetermined on a standard leak. Also insert it into the tube ends for about 6 mm (~¼ in). If the tubes are welded to the tubesheet, the welds may be checked by encapsulating each tube end with a funnel connected to the probe.

The mass-spectrograph *instrument* sensitivity is considerably higher than that of the halogen diode sniffer: 1×10^{-9} cm³/s compared with about 1×10^{-6} cm³/s for a 1% concentration of halogen tracer gas. However, the halogen diode sniffer *system* sensitivity is higher than that of the helium sniffer system.

Helium tracer probes and helium hood testing (the only one deemed quantitative by the ASME Code rules) are used too infrequently to discuss here, but consider them when a service is so critical that very small leaks cannot be tolerated. The ASME Code, Section V, "Nondestructive Testing," is an excellent place to get an overview of these methods.

Mass-spectrometer helium probe leak testing

Helium probe testing makes use of a (generally portable) mass spectrometer to detect leakage of helium used as a tracer gas. Typically the side opposite to the one being probed is cleaned, dried, and filled with bone dry air or inert gas to a pressure of ~345 kPa (~50 psi). Tracer helium is added until it occupies 10% of the space by volume [a total gage pressure of ~380 kPa (~55 psi)]. The places examined for leakage are probed. The presence of helium in the spectrograph indicates a leak.

The ASME Code requires the sniffer to be calibrated against a capillary tracer-gas standard that has a leakage rate of 0 to 1×10^{-5} std cm³/s before use and again at use intervals of not more than 2 h.

Include the manner and speed of probing in procedure specifications. Generally follow the guidelines prescribed in the ASME Code, Section V.

Halogen leak detector probe testing

Halogen leak detector probe testing uses an instrument that has a heated platinum anode and a collector plate (cathode). Air from the probe passes over the anode. This element ionizes any halogen gas present in the air. The ion formation rate and flow are indicated by a meter. The ASME Code requires the halogen sniffer to be calibrated against a standard leak, similar to calibrating helium sniffers.

Halogen leak detector probe bag testing. In the manufacture and maintenance of air-conditioning exchangers, common practice is to enhance the sensitivity of halogen probe leak detection by encapsulating the whole tubesheet in a plastic bag. After a preestablished soak time, the bag is perforated with a hole just large enough to permit inserting the probe. The detector indicates the combined leakage of all of the tubes and tube-to-tubesheet joints. When a leak is found, the individual tubes are probed to find out which one(s) leaked.

Typical procedure. Here is a typical procedure for halogen sniffer testing tube-to-tubesheet joints:

1. Provide tracer gas, preferably refrigerant 12; however, refrigerants 11, 21, 22, or 114 or methylene chloride may be substituted.
2. Clean and dry the shell side.
3. Purge and fill the shell with clean, bone-dry nitrogen or inert gas and add 10% by volume of tracer gas. An alternative is to prepare a mixture of the gas externally and fill the purged exchanger shell with the mixture. The volume of tracer gas in the prepared mixture must be adjusted to provide a final percentage of 10% of tracer gas. Typical fill pressures are 207 to 345 kPa (~30 to 50 psi).
4. Allow a soak time of at least 30 min for tracer-gas dispersion.
5. Determine the appropriate scanning rate by passing the probe of the calibrated instrument over the leak standard orifice that detects leakage of 1×10^{-5} std cm^3/s.
6. Traverse each tube joint, keeping the probe tip within 3 mm (~⅛ in) of the test surface at the interface of the tubes and tubesheet.
7. Check for cracked tube ends by inserting the probe into each tube end to a minimum depth of 20 mm (~¾ in) and holding it there for about 3 s.

Pressure Testing

The purposes of pressure testing* are to (1) stress the exchanger parts under *internal* pressure to some value below the yield stress so as to verify that the structure can withstand the applied *internal* pressure and (2) disclose leaks that penetrate the wall of the pressure

*See the ASME Code, Section VIII, Division 1, Paragraph UG-99 for standard hydrostatic test requirements.

envelope, whether at tube-to-tubesheet joints, main-seam welds, nozzle-to-shell and -channel welds, or flanged, gasketed joints.

Ordinarily, except for units designed for differential-pressure operation, testing is done with test pressure on one side and atmospheric pressure on the other and then again with test pressures applied simultaneously to both sides. Units designed for vacuum on one side may be evacuated on that side.

The most frequent pressure-testing inspection is simply to scan very closely for any leaks from the pressurized side to the atmosphere. This may be facilitated by adding fluorescent dye to the test water and scanning under black light. A variation is to determine if there is a pressure loss over a preset dwell time. Here are some pressure-testing considerations:

- Test media should be noncorrosive, free of slime-producing biological substances and suspended solids.
- Algicides used with test water should not harm the unit.
- Each side of an exchanger should be tested with the other side at atmospheric pressure except for exchangers designed for differential pressure operation.
- In addition to pressurizing each side with the other at atmospheric pressure, both sides should be tested simultaneously.
- Porosity and inclusions in filet welded joints may create hidden partial leak paths not revealed by hydro testing.
- Joints that do not leak under liquid pressure testing may not be sufficiently tight for the required service, such as in exchangers connected to the refrigerant side of a refrigeration system.
- Before performing leak tests on exchangers with expanded tube-to-tubesheet joints, a pressure test should be done to prove that the exchanger is structurally sound.
- If the tube-to-tubesheet joints are produced by welding followed by expanding, the tube-to-tubesheet welds should be leak-tested before expanding and hydrostatically testing.

Under shell-side pressure, leaks through tube-to-tubesheet joints are readily visible. When the tube side is pressure-tested, leaks can be seen only in removable bundles. The locations of leaks deep within a tube field may not be discernible. The only way to find out if there is a leak through the tube-to-tubesheet joints in a fixed-tubesheet unit when testing the tube-side is by pressure loss.

Testing both sides at the same time tests the ability of the tubesheet to meet this condition. Determining the effects of such a test requires observing differences in pressure on both sides.

Fluid penetrant and hydrostatic testing will probably not disclose the existence of hidden partial leak paths. For critical services, perform a series of tests in which test pressure is applied and reduced to atmospheric. The cycles may cause the partial leak paths to be completed and reveal the leaks.

Test and safety gages

Use at least two gages or pressure recorders on each side to be tested with ranges and subdivisions appropriate for the test pressure. Calibrate them against a standard test gage or dead-weight tester. When pressure testing exchangers fitted with safety valves and pressure-burst disks, be sure either to remove the safety devices and blank off the openings or to provide other means of ensuring that they are not damaged by the test pressure.

Media

These are the media commonly used for testing:

- Water
- Kerosene, diesel oil, and similar materials
- Air and inert gases
- Helium
- Refrigerant gases

The standard hydrostatic test medium is water. The local water supply is seldom acceptable because (1) it must be free of chlorides and other dissolved substances that can predispose the exchanger to corrosion, (2) it must contain no organic substances that will cause corrosion or fouling later on, and (3) suspended solids must be filtered out. Test water that causes the least problems is either distilled or demineralized and filtered free of any suspended solids.

When pressure testing with water would cause problems later on, it is generally permissible to use kerosene and similar materials. The medium must be compatible with the use to which the exchanger will be put. There may be advantages to using these media in place of water when they are more searching, are less viscous, and permit leaks to be more readily detected. Also consider materials that have distinctive odors. With such materials, take care not to spill any test medium. Pipe all vents far enough away from the exchanger to guard against false indications and remove any containers, pumps, etc. of the odoriferous material from the area.

Air and inert gases are used mostly for leak testing. However, when you cannot use a liquid medium to test hydrostatically—possibly because no residue can be tolerated, or because the weight of the medium would permanently distort the exchanger—you may use air or inert gas in a pneumatic test in place of the standard hydrostatic test.

Helium leaks are detected with a mass spectrograph. It is generally harmless to the metals with which it makes contact.

Refrigerant gases are the most searching, least viscous common test fluids. Most of them are halogenated aliphatic hydrocarbons. Any dissociation and production of free chloride ions can lead to stress-corrosion cracking of austenitic and some other materials. This is especially likely in the region of tubes expanded into tubesheets. Consequently, either avoid leak testing such units with halogenated gases, or make sure the surfaces being tested are bone-dry and subsequently fully purged with bone-dry air or inert gas.

Equation (5.1) shows that leakage varies directly with the square of the pressure differential across the leak and inversely with the test medium's viscosity:

$$R = \frac{K(P_2 - P_1)^2}{\mu} \tag{5.1}$$

where* K = proportionality constant (similar to an orifice constant for flow across an orifice)

P_1 = pressure on lower pressure side of the leak path

P_2 = pressure on higher pressure side of the leak path

R = leakage rate

μ = viscosity of the leaking fluid

Table 5.3 lists the viscosities of some common fluids used in leak testing. Accordingly, for the same size of leak path and pressure dif-

TABLE 5.3 Viscosities of Some Common Test Media

Fluid	Viscosity, poise
Air, nitrogen, helium	1.9×10^{-4}
Refrigerant gases	1.2×10^{-4}
Water	1.0×10^{-2}
Lubricating oils	1.0×10^{0}

*In consistent units.

ferential from end to end of the leak path, the leak rate of air and inert gases will be about 52 to 53 times as great as for water at 20°C (~70°F). Therefore, testing with air or inert gas is an excellent way to detect leaks.

Test pressures

In the United States, newly constructed pressure vessels are tested in accordance with the requirements of the ASME Code. Worldwide, ones tested after alterations or repairs are generally tested in accordance with the requirements of the NBIC or the API Inspection Code. For most new pressure vessels, this means that the unit is tested at 1.5 times the maximum allowable working pressure (MAWP)* to be stamped on the nameplate, adjusted upward for the corrosion allowance and for the difference in allowable stress at design temperature relative to test temperature. However, the test pressure for heat exchangers may not be corrected for corrosion allowance because of the effects of shell-side test pressure higher than MAWP on the tubes.

Do not include corrosion-allowance thickness in test pressure calculations for fixed-tubesheet units in which no corrosion allowance is provided for corrosion-resistant tubesheets and an allowance for corrosion is provided in a carbon-steel shell or channel. This is because of the common design method used to select the tubesheet thickness. In this calculation, you assume a thickness. Using the assumed thickness, you calculate the actual thickness and compare it with the assumed one. You reiterate until the calculated and assumed thicknesses match each other to within a specified minimum difference. When the tubesheet has corrosion allowance, you do the calculations with and without corrosion allowance.

Repaired units that must be hydrostatically tested are usually tested to only 80 percent of the MAWP stamped on the nameplate. However, if you plan to test exchangers which have been in service at pressures 1.5 times MAWP and to correct for corrosion allowance, use measured remaining thicknesses to calculate how much allowance for corrosion remains. For close inspection, be sure to reduce the test pressure to the MAWP.

Because the stored energy makes it hazardous, pressure testing with air or inert gas should never exceed the MAWP stamped on the nameplate.† If a unit is to be leak-tested with air or compressed gas,

---

*See the ASME Code,[1] Section VIII, Division 1, Paragraph UG-98 for a definition of MAWP.

†See the ASME Code,[1] Section VIII, Division 1, Paragraphs UG-100, "Pneumatic Test" and UW-50, "Nondestructive Examination of Welds on Pneumatically Tested Vessels."

the test pressure is usually kept at the lower of 345 kPa (~50 psi) or the MAWP stamped on the nameplate.

Test temperatures

The ASME Code specifies a minimum test temperature of 70°F (~21°C). Because the ASME Code applies to new work only, and because it is unlikely the metal temperature of work in the factory will be permitted to be much above 40°C (~100°F), it does not specify an upper limit. Manufacturers must make sure not only that they meet the ASME Code's minimum test temperature but also that metal ductility is adequate at the temperature of the test.

The ASME Code indicates the maximum thicknesses allowed for various materials at diverse minimum design metal temperatures for which it does not require Charpy impact testing to verify that the metal is ductile. Materials, not exempted from impact testing listed in ASME Code, Paragraph UCS-66 or shown in Fig. UCS-66, must be impact-tested before being used in fabrication. The impact test must demonstrate that the material retains its ductile characteristics at a temperature 40°F (~5°C) below the minimum design temperature.

In recognition of conditions encountered with installed exchangers, the NBIC specifies a minimum test-water temperature of 60°F (~16°C) and a maximum of 120°F (~49°C). The test-water temperature may be lower if the owner provides information on the metal ductility that shows that the lower temperature is acceptable. You can show this using ASME Code, Fig. UCS-66.1. When an owner requires a test temperature higher than 120°F (~49°C), the pressure must be reduced to the MAWP and the temperature to 120°F (~49°C) for close examination.

Temperature maintenance

It is often desirable to have test pressure reside longer than the minimum period. However, unless the exchanger is kept at constant temperature, the test pressure can vary substantially with time. If it drops somewhat, there is no hazard to the structure. But if the temperature increase is substantial, the test pressure may increase to the point of damaging the exchanger.

Dwell time

The minimum dwell time for hydrostatic testing to the ASME Code rules is ½ h. This may be insufficient for thick welded structures, especially exchangers with thick tubesheets in which the tube-to-tubesheet joints are made by welding and full-depth expanding.

Cycle testing

Most exchangers designed for high-pressure and/or high-temperature service are constructed with the tubes strength-welded to the channel side of the tubesheet.* In nearly all such exchangers, the tubes are subsequently expanded into the tube hole after welding, beginning about 10 to 13 mm (~⅜ to ½ in) inward from the weld root, and extending to about 3 mm (~⅛ in) in front of the shell-side face.

Neither liquid penetrant examination nor bubble testing will reveal subsurface porosity, cracks, and inclusions in the welds that join the tubes to the tubesheet. Pressure application deflects the tubesheet, which stresses the welds, in which discontinuities are stress raisers. Therefore, cyclically applying and relieving pressure in the manufacturer's plant may induce failures. This permits making appropriate repairs before the exchanger is shipped.

Most of the time, 10 pressure cycles are applied. However, in some circumstances a higher number is used. Since the higher design pressure is usually in an exchanger's tube side, that is the side on which most cycle tests are performed.

Because the channel is closed and full of high-pressure test medium, you cannot see leaks through the welds. Consequently, joint failures are indicated indirectly by holding pressure long enough to make sure that there is no pressure falloff (~30 min). After each pressure holding period, pressure is relieved to atmospheric for 5 to 10 min before repressurizing and holding. The leakage rate through a small crack in a tube-to-tubesheet weld may be very small. Therefore, it is desirable to use a pressure-recording instrument that indicates pressure in small increments and provides a graphic record of pressure.

Cycle testing the shell side of exchangers that have substantially higher shell-side than tube-side design pressure permits directly examining the tube-to-tubesheet joints. However, it requires considering how it will affect the whole shell-side structure. Most suitable for shell-side cyclic pressure testing are fixed-tubesheet exchangers without shell expansion joints. If a fixed-tubesheet unit has a shell expansion joint, examine the effect of the pressure cycling on the expansion joint. Cycle testing the shell side of removable-bundle units requires looking into how it will affect bundle-to-shell attachments.

Test sequence for welded and expanded tube-to-tubesheet joints

Most of the time, the preferred procedure for joining tubes to tubesheets by welding and expanding is to weld the tubes to the

*See the ASME Code,[1] Section VIII, Division 1, Paragraph UW-20 for requirements for strength-welded tube-to-tubesheet joints.

tubesheet before expanding. Expanding usually begins about 10 to 13 mm (~⅜ to ½ in) beyond the weld. This presents a testing problem. If the weld is the primary joint, you must test its strength hydrostatically to satisfy the ASME Code requirements. When you do this, the crevice between the tubes and holes may contain test water. Subsequent expansion may entrap test water in the unexpanded 10- to 13-mm (⅜- to ½-in) gap.

This is undesirable even when the test water is demineralized or distilled. Therefore, the test water should be drained and the unit heated and dried before the tubes are expanded. Because it is difficult to drain and heat to dry the unit, manufacturers prefer to leak test the welds and hydrostatically test only after expansion is complete.* If this practice is followed, tube-to-tubesheet strength welds, as defined in ASME Code, Paragraph UW-20, may not be stressed by direct application of pressure if the shell-side design pressure is higher than that of the tube side.

Effect of external pressure on tubes

Although the purpose of pressure testing is to determine whether or not the parts can withstand the applied internal pressure, it is necessary to consider the shell-side pressure's effect on the tubes. As pointed out above, the ASME Code hydrostatic test applies only to internal pressure testing. For an exchanger shell, this means that the pressure-vessel code's purpose is to test the capability of the shell and tubesheets under internal pressure, but not the tubes under external pressure. The ability of a tube to resist collapse under external pressure is a function of not only its physical properties but also the ratios of diameter to thickness and length to diameter. Large-diameter, thin-walled, long tubes of some metals may collapse under external test pressures of 1.5 times the shell MAWP. Therefore, before testing the shell, make sure that the tubes meet pressure-vessel code requirements for shells under external pressure.

Testing differential-pressure-design exchangers

When testing units designed under differential-pressure design rules, it is essential never to exceed the design differential pressure. This means following procedures for pressure testing that permit building up to the test pressures on each side in tandem with the other. Whenever possible, use a differential-pressure controller for this purpose.

---

*Under the ASME Code rules, the hydrostatic test is the final test.

What to look at and how

The places to inspect an exchanger under test pressure are all main-seam welds, all penetrations of the pressure envelope by connections, and all flanged joints. Inspect the tube-to-tubesheet joints very carefully because they are the most frequent source of trouble in an exchanger.

Be sure that all stray moisture has been removed before inspecting. This may mean wiping all surfaces with moisture-absorbing cloth or paper and blowing out the tubes with clean, dry air. The light that shines on the places being observed should be bright enough to permit reading fine print without special correction. Small, high-intensity pencil lights are excellent for examining tube-to-tubesheet joints.

If a fluorescent dye has been added to the test water, scan all the surfaces examined with the black-light source.

Acoustic emission (AE) testing during pressure testing*

When pressure is applied inside an exchanger shell or channel, structural discontinuities grow in size. The growth generates acoustic emissions that appropriate sensing equipment can detect. The acoustic emission system consists of sensors, signal processing, display, and recording equipment.

Sensors are located at critical points such as the junctions of feedwater nozzles in feedwater channels, or wherever there is a critical section, such as welds, manways, repaired sections, and regions of structural discontinuities.

The pressure is applied at a rate specific to the examination procedure. The rate must be rapid enough to expedite the examination with the least extraneous noise. Hold points are established beforehand. These are usually at 50, 65, 85, and 100 percent or maximum test pressure. Dwell time at the hold points is 10 min, with a 30-min dwell at maximum pressure. External noise (e.g., rain, mechanically transmitted sound through connected piping and supports and pressure-pump noise) must be kept below the examination threshold of the system.

Time and pressure are recorded. Pressure recording must be within ±2 percent of the maximum pressure. Background noise such as liquid splashing into the exchanger, noise from too-rapid a pressurizing rate, pumps, motors, and mechanical devices; electromagnetic interference; and environmental noise sources such as rain and wind must be identified, minimized, and recorded. Before examination, background noise must be monitored to provide a basis for the examiner to evaluate its effects during examination.

---

*See Articles 11 to 13 and 29 of Section V of the ASME Code.[1]

Because of the nature of measuring the signal, the acoustic emission testing during pressurization should be backed up by another means of nondestructive testing.

Continuous acoustic emission (AE) testing

In addition to testing during pressurization to determine if flaws exist at critical points in the structure, you can use acoustic emission testing to monitor the development of leaks during operation. The objectives of this kind of AE testing are to detect leaks, locate the sources, interpret the response to the noise signals, and evaluate their significance. Such a system, used on closed feedwater heaters at the Eddystone Generating Station (Eddystone, Pa.), has detected leaks in tubes and tube-to-tubesheet joints before they became evident in operation.

Because of the many tubes in a heat exchanger and the nature of the tube-to-tubesheet attachments, locating very small leaks that generate high-frequency noise are hard to identify. Nevertheless, when the AE testing equipment indicates the presence of such leaks, the advance warning should trigger a search during the next planned outage.

The sensors are either mounted directly on the exchanger locations to be monitored or, if this is not feasible, are attached to waveguides that bear on the exchanger. The system, equipment, and techniques must be capable of canceling out noise and interference inherent in the operation of the tubular.

When continuous acoustic emission testing reveals a leak in a primary pressure boundary, such as a channel-barrel weld, evaluate the indication by another means of nondestructive examination.

Destructive Sampling and Trepanning

How to take a sample for examination depends on why the sample is wanted and where it comes from.

Tube specimens

When nondestructive testing methods indicate tube deterioration, it may be worthwhile to extract tubes or parts of tubes from an exchanger for laboratory analysis and metallurgical examination. This is relatively simple when the tube-to-tubesheet joints are expanded only into bare tube holes, somewhat more difficult if the annular holes are grooved, and most arduous if the tubes are welded and expanded into the tubesheets. The welds of welded joints must be carefully trepanned away without damaging adjacent welds before a tube can be extracted.

If the shell side of straight-tube equipment is clean, you may extract whole tubes with a hydraulic tube puller. It is impractical to extract whole tubes from severely fouled straight tubes and not possible to extract through the tubesheet tubes from U-tube equipment. In these circumstances, extract specimens behind the tubesheet. Cut the desired tube length with an internal tube cutter. Then extract with the hydraulic extractor. Make sure to stabilize the tube beyond its cut position before plugging the hole in the tubesheet.

Boat specimens

"Boat" specimens, so named because of their shape, are taken from thick tubesheets during postmortem examining exchangers for the purpose of finding out how and why tubes failed in the portion embedded in a tubesheet. The chunk of tubesheet may be removed by carbon arcing, by torch burning, or with portable milling cutters. Carbon arcing and torch burning require removing a specimen large enough to encompass the area to be studied without damage from the removal method.

Trepanning

Trepanning is a destructive method for determining the thickness of a part and for examining welds. It consists of cutting a hole with a circular saw through the part to be examined. For determining thickness, the hole must be large enough to insert a device for measuring the thickness. For examining welds, the core must be large enough for the required specimen analyses.

Interpreting Test and Examination Results

Always precede radiography, eddy-current, ultrasonic, and other nondestructive examinations by visual examination. The weak spot of visual inspection is that one individual rarely gets to see all the various types of erosion, corrosion, cracks, pits, fractures, etc. that befall exchangers. The API's *Guide for Inspection of Pressure Vessels*[3] has some excellent photographs of such sights. There is currently no generalized photograph album of heat exchanger failures in the literature. Until such a work is published, it will be necessary to rely on photographic records maintained in-house and from cooperating organizations and specimens from previous failures and some specimens prepared to show what failures look like. Consultants are also available to help interpret the meaning of the observations.

Interpreting the output of highly technical examination procedures such as radiographic, eddy-current, and ultrasonic examinations requires special training. Organizations devoted to nondestructive testing, such as the ASNT, offer training and courses and qualify examiners in the various procedures discussed. Generally, there are three levels of qualification. These are based on training, including passing tests of knowledge, and experience. Roughly speaking, the levels are as follows: level I qualification authorizes the examiner to assist a level III examiner in setting up and performing an examination, but not to interpret the results; level II examiners may set up the examination under the supervision of level III examiners and interpret results under the guidance of examiners qualified to level III; and level III examiners have worked at lower levels under supervision of level III people, have undertaken training in performing the examinations in which they are qualified, and have passed tests to demonstrate their proficiency. To maintain one's level of qualification, it is necessary to continue to use the procedures and keep current with new developments.

Always discuss the results of nondestructive examinations with the person who signs the report of examination. Make sure to get an explanation and location sketch of indications reported and the cited paragraphs of the governing code.

If examinations are performed pursuant to repairs or alterations, make sure that the workers whose work has been examined are informed about what the examination disclosed.

Records of Inspections and Tests

For visual examinations, the best records are videorecordings and photographs, suitably labeled with the date, time, and location of the part photographed; serial number; National Board number (when so stamped); tag number of the exchanger; references to the manufacturer's drawing number; and the owner's purchase order number. Cross-reference the inspection report and tubesheet sketch or computer printout of the inspection record with the photographs, videos, and sketches or comment sheets.

When contractors or trained and qualified staff perform other kinds of inspections, be sure that the same kind of information is recorded on the examination report. Radiographs should show as a minimum the serial number, tag number, and location(s) for each film in addition to the usual penetrameter information. Radiographer's written reports should readily be identified with the films.

Precautions for Hydrostatic and Pneumatic Testing

Hydrostatic testing is a generally safe procedure for stressing an exchanger to disclose hidden weaknesses. A principal concern is to make sure that all deflections resulting from applying the pressure are elastic ones so that no part will be permanently distorted.

Nil ductility precautions

Considerable energy is stored in a vessel subjected to internal hydrostatic pressure because of the elastic deformation of the parts. Therefore, a brittle failure is hazardous. Recognizing that at low temperature many ductile materials become brittle, the ASME Code and NBIC set minimum test temperatures. Although these are generally safe minima, whenever the thickness of any pressure part of the exchanger is substantial, investigate the ductility-temperature relationship of the material for the thickness of the part. Make sure that the test is conducted at a temperature at which the metal remains ductile.

Safety precautions for testing with air and gases

Because of the energy stored in compressed gas contained in an elastic shell deflected by pressure, under no circumstances during pneumatic testing should air, inert gas, or nitrogen pressure in the shell exceed the MAWP stamped on the nameplate. Before pressurizing the shell with a compressible fluid, read the material in Section VIII, Division 1 of the ASME Code on pneumatic testing. Pay special attention to pressure-vessel code warnings such as those in ASME Code, Paragraphs UG-100 (c) and (d), which read

> (c) The metal temperature during pneumatic test shall be maintained at least 30°F above the minimum design metal temperature to minimize the risk of brittle failure [see UG-20 and General Note (6) for Fig. UCS 66.2]; (d) The pressure in the vessel shall be gradually increased to not more than one-half of the test pressure. Thereafter, the test pressure shall be increased in steps of approximately one-tenth of the test pressure until the required test pressure has been reached....

References

1. *The ASME Boiler and Pressure Vessel Code,* American Society of Mechanical Engineers, New York, 1995.
2. Robert J. Bell and Sheldon D. Strauss, "Advancing Heat Exchanger Reliability," *Power,* July 1991.
3. *Guide for Inspection of Pressure Vessels*, 3d ed., The American Petroleum Institute, Washington, D.C., Dec. 1978.

Chapter

6

Inspection Codes

Introduction

This chapter discusses the two principal inspection codes used in the United States and Canada: the *National Board Inspection Code**[1] and the *API Pressure Vessel Inspection Code,* API 510.[2] Throughout, we will call these codes the *NBIC* or *National Board Code* and the *API 510* or *API Code.* Neither code is prepared or administered by the national government of the United States or Canada or any state, provincial, or territorial government.

The National Board of Boiler and Pressure Vessel Inspectors (National Board; Columbus, Ohio) produces and issues the NBIC. The American Petroleum Institute (API; Washington, D.C.) produces and issues API 510.

Provinces, states, cities, and counties which have laws and regulations governing inspection of boilers and unfired pressure vessels are likely to adopt the National Board or the API in whole or part and incorporate it into their laws and regulations. Depending on the dimensions and design pressures, heat exchangers are pressure vessels as defined in pressure-vessel safety and construction codes and these inspection codes. In this chapter, we are concerned only with the in-service inspection of condition, repairs, and alterations required to maintain the integrity of heat exchangers. Therefore, we

*There are repair/alteration organizations holding certificates of authorization to apply the National Board's R symbol in most of Canada (Provinces of Alberta, Manitoba, New Brunswick, Nova Scotia, Ontario, and Quebec) and in Belgium, Brazil, England, France, India, Israel, Italy, Korea, Malaysia, Mexico, The Netherlands, Saudi Arabia, Scotland, Singapore, Venezuela, and Germany.

may substitute the terms *heat exchanger, exchanger,* and *tubular* for *pressure vessel* when citing the codes.

ASME Code, Section XI, "Rules for In-Service Inspection of Nuclear Power Plant Components," addresses the requirements of in-service inspection programs that define the boundaries of the nuclear and nonnuclear systems in power plants. If the jurisdictional requirements at the plant site mandate using the NBIC, heat exchanger repairs, alterations, and rerating must be in accordance with the NBIC rules for R and NR certification and stamping. The rules for R stamping apply for heat exchangers outside the Section XI confines; the rules for NR stamping apply to ones within its boundaries.

The purpose of this chapter is to provide guidance to the codes. In this effort we may paraphrase and abstract the codes' texts or cite portions out of context of the original documents. When you employ the inspection codes, rely only on the original texts in the applicable editions.

In the following, we call the individual authorized to inspect nonnuclear pressure-retaining parts the *Inspector*; we call the Inspector authorized to inspect nuclear pressure-retaining items the *Authorized Nuclear Inspector* (ANI). Be aware of the distinction between the *Inspector,* whose duties the construction and inspection codes define, and a *quality control inspector,* who works for a manufacturer, purchaser, architectural-engineering firm, owner, or user not authorized to inspect by the applicable construction or inspection code.

We caution that some standards, to which the ASME Code refers, use the terms *inspector* and *examiner* and *inspection* and *examination* interchangeably. However, under the ASME Code definitions, *inspection* is conducted only by an *Authorized Inspector* and *examination* is done by an *examiner* qualified to the appropriate level.

Other in-service inspection codes

At present, there are no other in-service inspection codes in general use. If others are produced elsewhere, they will probably be written, administered, and enforced by government bodies and may be similar to and more or less restrictive than NBIC and API 510.

Relationship of the ASME Code to the NBIC and API 510

Many people do not know that the ASME Code applies only to new construction by the manufacturers ASME accredits. It often surprises purchasers of ASME Code–stamped vessels that the ASME Code does not govern poststamping use or cover in-service inspection and nameplate stamping of repairs and alterations of in-service pressure vessels. Users not familiar with its scope may be astonished that the ASME Code does not address repairs and alterations to existing units and that the ASME Boiler and Pressure Vessel Code Committee does not monitor repairs and alterations. The ASME accepts no responsibility after a pressure vessel has been accepted by the manufacturer's AI, the nameplate stamped, and the manufacturer's data report signed and filed, except when a violation comes to light.

The rules and regulations of the political jurisdiction where a unit is installed govern responsibility for continuing inspection of condition at regular intervals and for inspection of repairs and alterations after a pressure vessel is placed in operation. Almost all North American jurisdictions and many jurisdictions elsewhere either require or accept the NBIC or the API Inspection Code.

Confusion about the relationship of the inspection codes to the ASME Code may arise because NBIC applies the original construction code's rules to repairs and alterations and API 510 applies the *principles* of the ASME Code's rules. Another source of misunderstanding is that the National Board offers training, testing, and qualifying of Inspectors and Authorized Nuclear Inspectors, defined under the ASME Code rules as *Authorized Inspectors* (AIs) and *Authorized Nuclear Inspectors* (ANIs). The National Board conducts these activities in cooperation with jurisdictions responsible for pressure-vessel and boiler safety.

The NBIC and API 510 are similar to but different from each other. For example, API 510 states, "A Repair Organization means: a. The holder of a valid ASME Certificate of Authorization for the use of an appropriate ASME Code symbol stamp...," whereas the NBIC definition is "A Repair Organization is a holder of a valid National Board certificate authorizing the use of the R, VR, or NR stamps."

Relationship between NBIC and API-510

There is no conflict between the NBIC and API 510 or between these codes and the ASME Code. The introduction to the 1995 edition of the NBIC states

> The American Petroleum Institute Code, API-510, covers the maintenance, inspection, repair, alteration and re-rating procedures for pressure vessels used by the petroleum and chemical process industries. Similarly, API-570 covers piping used by the petroleum and chemical process industries. It is the intent that this Code cover installations other than those covered by API-510 and API-570 unless the jurisdiction rules otherwise.

Jurisdictional Requirements*

The following is excerpted from the NBIC:

> Reference is made throughout this Code to the requirements of the "jurisdictional authority." *Where any provision herein presents a direct or implied conflict with any jurisdictional regulation, the jurisdictional regulation shall govern.*

Unlike other countries' national pressure-vessel codes, the ASME Code does not have the force of law unless a political jurisdiction adopts all or part of it by reference in its laws and regulations. The U.S. and Canadian laws, rules, and regulations for construction, installation, and inspection of boilers and unfired and fired pressure vessels vary widely in the various states, provinces, territories, counties, and cities. Most heat exchangers are pressure vessels within the definitions of these laws, rules, and regulations and those of the ASME Code, the NBIC, and API 510. Therefore, they are within the jurisdictional requirements for in-service inspection of unfired pressure vessels.

The 1995 edition of the NBIC defines *jurisdiction* or *jurisdictional authority* as "A governmental entity which administers and enforces one or more sections of the ASME Code, one of which shall be Section I (Power Boilers)."

In most North American jurisdictions, the laws, rules, and regulations require such inspections to be made by Inspectors employed by the jurisdiction, by the insurance carrier that covers the facility, or by an *Inspector* defined as an Owner-User Inspector. These individuals generally must hold commissions issued by the jurisdiction, the National Board, or both. Users cannot inspect exchangers' integrity except where the jurisdiction has a provision that permits owner-user inspection. Owner-User Inspectors are commissioned by the National Board and under most circumstances must have qualifications and certification equivalent to those required for jurisdictional inspectors.

This work cannot provide a complete, jurisdiction-by-jurisdiction summary of governing laws, rules, and regulations.† However, the

*This work deals only with requirements in the United States and Canada. In the United States, information on existing International Standards and their availability can be obtained from the American National Standards Institute, 1430 Broadway, New York, NY 10018 [tel. (212) 354-3379]. Americans who need information on foreign standards should register with the National Institute of Standards and Technology-NCS-CI, Administration Building, Room A629, Gaithersburg, MD 20899 [Tel. (301) 975-4040].

†Be cautioned that many cities and counties have rules of their own or adopt the NBIC or API 510, even when the parent jurisdiction has no rules or has less stringent ones than the local authority.

Uniform Boiler and Pressure Vessel Laws Society Inc. [UBPVLS; 308 N. Evergreen Road, Suite 240, Louisville, KY 40243; tel. (502) 244-6029, fax (502) 244-6030] publishes, by subscription, a regularly updated synopsis of boiler and pressure-vessel laws, rules, and regulations.* It also summarizes this information for U.S. states, cities, and counties and Canadian provinces and territories.

The jurisdiction-by-jurisdiction information published by the UBPVLS provides the following information: (1) the responsible department, (2) the date of the passage of the law, (3) the rules that apply to construction and stamping, (4) objects subject to rules for construction and stamping, (5) objects subject to rules for field inspection, (6) inspections required, (7) insurance and owner-user inspection requirements, (8) requirements for certificate of operation, (9) jurisdictional inspection fees, and (10) miscellaneous other information. Many jurisdictions either accept the National Board "Rules for Repairs and Alterations to Pressure-retaining Items by Welding" in the NBIC or those in API 510 or require one or the other. Others have their own rules for repairs and alterations.

Guide to the *National Board Inspection Code (NBIC)*

The NBIC is an American National Standard, ANSI/NB-23. Its purpose is to maintain the integrity of pressure-retaining items after they have been placed in service by providing rules for inspection, repair, and alteration, thereby ensuring that these objects may be safely used.

The National Board publishes NBIC editions at 3-year intervals. Publishing dates parallel those of ASME Code editions. Between editions, it publishes addenda annually. The July 1, 1995 edition made extensive revisions to the format and some provisions of previous editions. The National Board publishes interpretations in response to questions about NBIC rules, but cautions that they are provided for information only and are not part of the NBIC. A paragraph in the 1995 edition foreword states, "It is intended that this Edition of this NBIC and any subsequent Addenda not be retroactive. Unless the jurisdictional authority imposes the use of an earlier edition, the latest effective Edition and Addenda is the governing document."

*The synopsis does not provide specific details about the rules that have been adopted. Note that in the NBIC and API 510, not all alterations are *welded* alterations. However, many jurisdictions adopt one or the other inspection code specifically for welded repairs; others, for welded repairs and alterations; and some accept or adopt one of the inspection codes in its entirety.

NBIC has rules for in-service inspection, repairs, and alterations to boilers and pressure vessels. For brevity, this discussion omits references to boilers and to pressure-retaining items not part of shell-and-tube heat exchangers.

NBIC disclaimers

The NBIC has the following disclaimers:

> The rules established by the NBIC Committee are not to be interpreted as approving, recommending, or endorsing any proprietary or specific design or as limiting in any way the repair organizations's freedom to choose any method of repair or alteration that conforms to the NBIC rules.
>
> The NBIC does not contain rules to cover all details of repair or alteration. Where complete details are not given, it is intended that the repair organization, subject to the acceptance of the Inspector, provide details for the repair or alteration which will be as safe as otherwise provided by the rules in the original code of construction.
>
> There are instances where the NBIC serves to *warn a repair organization or Inspector* against pitfalls; but the Code is *not a handbook* and cannot substitute for education, experience and sound engineering judgment.

NBIC definitions

There are subtle distinctions between the definitions in the NBIC glossary and API 510 definitions. These are some of the definitions in the NBIC glossary.

alteration Any change in the item described on the original Manufacturers' Data Report which affects the pressure containing capability of the pressure-retaining item. Nonphysical changes such as an increase in the maximum allowable working pressure (MAWP) (internal or external) or design temperature of a pressure-retaining item shall be considered an alteration. A reduction in minimum temperature such that additional mechanical tests are required shall also be considered an alteration. Here are some examples of alterations:

- Nonphysical changes such as an increase in the MAWP or the design temperature of the shell or tube side
- Adding new nozzles or openings except for those that the NBIC classifies as repairs
- Changing dimensions or contours
- Replacing a pressure-retaining part with a material of different nominal strength or composition from that used in the original design, except that replacement with a material of composition different from an allowable stress equal to or greater than that used in the original design may be considered a repair provided the replacement material

satisfies the material and design requirements of the original code of construction

This means that retubing with tubes of a nominal composition and lower allowable stress, and/or gage different from that used originally is an *alteration.*

Owner-User Inspection Agency An owner or user of pressure-retaining items that maintains an established inspection program, whose organization and inspection procedures meet the requirements of the National Board rules and are acceptable to the jurisdiction or jurisdictional authority wherein the owner or user is located.

Owner-User Inspector An Inspector who holds a valid National Board Owner-User Commission.

pressure-retaining item (PRI) Any pressure vessel, piping, or material used for the containment of pressure, either internal or external. The pressure may be obtained from an external source, or by application of heat from a direct source, or any combination thereof.

repair The work necessary to restore pressure-retaining items to a safe and satisfactory operating condition. Following are examples of repairs that meet the NBIC definition:

- Weld repairing or replacing pressure parts or attachments that have failed in a weld or base metal such as tube-to-tubesheet joint welds
- Replacing tubesheets in accordance with the original design
- Replacing tubes, where welding is involved
- Replacing pressure-retaining parts identical with those shown existing on the original exchanger and described on the manufacturer's data report (e.g., rewelding a circumferential or longitudinal seam in a shell or channel and replacing nozzles).

Repair Organization A holder of a valid National Board certificate authorizing the use of the R, VR, or NR stamps.*

rerating The increase of the maximum allowable working pressure (MAWP) or temperature of a pressure vessel regardless of whether physical work is carried out on the pressure vessel. Under NBIC rules, Re-rating is considered an alteration.

Organization of the NBIC

The 1995 edition of NBIC is organized in these parts and appendixes:

Foreword

Introduction

Part RA. "Administrative Requirements"

Part RB. "Inservice Inspection of Pressure Retaining Items"

*Jurisdictions may also authorize Repair Organizations. However, such Repair Organizations are outside the scope of the NBIC.

Part RC. "Repairs and Alterations"

Part RD. "Repair Methods"

Mandatory Appendixes

Appendix 1. "Preparation of Technical Inquiries to the National Board Inspection Code Committee"

Appendix 2. "Stamping and Nameplate Information"

Appendix 3. "Steam Locomotive Firetube Boiler Inspection and Repair"

Appendix 4. "Glossary of Terms"

Appendix 5. "National Board Forms"

Appendix 6. "Examples of Repairs and Alterations"

Appendix 7. "Procedures to Extend the `VR' Certificate of Authorization Stamp to ASME `NV' Stamped Pressure Relief Devices"

Nonmandatory Appendixes

Appendix A. "Standard Welding Procedures"

Appendix B. "Recommended Preheat Temperatures"

Appendix D. "Recommended Guide for the Design of a Test System for Pressure Relief Devices in Compressible Fluid Service"

Appendix E. "Recommended Procedures for Repairing Pressure Relief Valves"

Appendix F. "Pressure Differential between Safety or Safety Relief Valve Setting and Boiler or Pressure Vessel Operating Pressure"

Appendix G. "Safety Valves on the Low Pressure Side of Steam Pressure Reducing Valves"

Interpretations

Index

In this work, we discuss only those parts and appendixes pertinent to inspecting, maintaining, and repairing shell-and-tube heat exchangers that qualify as pressure vessels under the ASME Code and other construction code rules. The sequence of our review follows the organization of the NBIC.

NBIC Part RA, administrative requirements

The principal paragraphs of Part RA are RA-1000, "General"; RA-2000, "Accreditation of Repair Organizations"; and RA-3000, "Accreditation of Owner/User Inspection Organizations." Each has subparagraphs appropriate to the major subject, dealing with matters such as prerequisites and procedures for accreditation and administrative requirements.

Under the NBIC rules, only holders of the National Board–authorized and –issued R stamp may make repairs and alterations to pressure-retaining items. Holders of authorization to apply ASME Code symbols cannot register repair and alteration forms with the National Board unless they also hold authorization to apply an R stamp. This recognizes that the ASME undertakes responsibility only for new construction. Otherwise, a U symbol holder who does not also hold R-stamp authorization and who did unacceptable repair or alteration work could not be held accountable by either the ASME Boiler and Pressure Vessel Code Committee or the National Board.

The National Board administers four specific accreditation programs: R stamp, covering repairs and alterations to pressure-retaining items; VR stamp, covering repairs to pressure-relief valves; NR stamp, covering repairs, modifications, and replacements of nuclear items; and Owner-User Inspection Organization, covering in-service inspection of pressure-retaining items owned or used by the organization which is accredited as an Owner/User Inspection Organization.

The R stamp accreditation program provides requirements for an organization that performs repairs and alterations to pressure-retaining items. It offers six types of authorizations: (1) "Repairs Only—Shop Only," (2) "Repairs Only—Field Only," (3) Repairs Only—Shop & Field," (4) "Alterations & Repairs—Shop Only," (5) "Alterations & Repairs—Field Only," and (6) "Alterations & Repairs—Shop & Field." The distinctions among these types are self-apparent. When you contract with a Repair Organization be sure that its type of authorization fits your situation.

The NR stamp accreditation program provides requirements for organizations that perform repairs, modifications, and replacements of nuclear power plant components.

NBIC Part RB, in-service inspection

Part RB has guidelines for in-service inspection of pressure-retaining items. It begins by stating requirements for Inspector safety. Users requiring in-service inspection are responsible for determining that pressure-retaining items may be entered safely. The Inspector must decline to make an inspection if the vessel has not been properly prepared for an internal inspection. Paragraph RB-2030 details some necessary preparations.

Paragraph RB-3200 deals with inspection of pressure vessels. It requires a preinspection review of operating conditions, normal vessel contents, date of last inspection, current jurisdictional inspection certificate, ASME Code symbol stamping or mark of whatever other code of construction prevailed, National Board and/or jurisdictional registration number, and records of wall thickness checks, especially where corrosion is a consideration. Many of these are similar to the recommendations given in Chap. 4.

Subparagraphs cover assessing the installation, making an external inspection, and making an internal inspection, and lists the various nondestructive examinations (NDEs) that may be performed. If there is doubt about the extent of a defect or detrimental condition, NBIC asserts the Inspector's right to require additional NDE. It discusses when in-service inspecting requires pressure testing and the conditions of conducting the test.

Part RB deals extensively with remaining life and inspection intervals. It provides the rational way to estimate the remaining life, as follows:

$$\text{Remaining life (years)} = \frac{t_{\text{actual}} - t_{\text{required}}}{\text{corrosion rate}} \tag{6.1}$$

where t_{actual} is the measured actual thickness and t_{required} is the code-required minimum thickness.

You can determine t_{actual} by any suitable NDE method accurate to within a tolerance of $0.10t$ for thicknesses $^5/_{16}$ in (~8 mm) or less and the greater of $^1/_{32}$ in (~0.8 mm) or $0.05t$ for thicknesses greater than $^5/_{16}$ in (~8 mm), or you may make thickness measurements through available existing openings. If you cannot measure the thickness nondestructively, you may drill through the metal wall and measure the thickness. Give consideration to an appropriate way to plug the hole.

The owner or user may establish the corrosion rate from data collected on exchangers in the same or similar service, or the Inspector may estimate it from knowledge and experience with pressure vessels in similar service. If the probable corrosion rate cannot be determined in either of these ways, NBIC requires making on-stream thickness determinations after about 1000 h of service and making subsequent sets of thickness measurements after each similar interval until the corrosion rate is established.

The maximum interval between internal inspections or a complete on-stream evaluation must not exceed the lesser of 10 years or one-half the estimated remaining life. Where remaining life is less than 4 years, the inspection interval may be the full remaining safe operating life up to a maximum of 2 years.

Part RB discusses these aspects of remaining life evaluation:

- Deterioration
- Life evaluation when corrosion rate is not a consideration

- Construction with corrosion-resistant linings
- Different rates of corrosion in two or more zones (e.g., shell side and tube sides in heat exchangers and feedwater heaters, desuperheating, condensing and subcooling zones in feedwater heater shells, and tube-side flow passes)
- Interrupted service
- Circumferential stresses
- Longitudinal stresses
- Pitting
- Weld joint efficiency factor
- Ellipsoidal and torispherical heads
- Adjustments of corrosion rate

The NBIC list of deteriorations that affect remaining life includes bulging, sagging, stress-corrosion cracking (SCC), creep rupture, thermal and/or mechanical fatigue, hydrogen blistering, high-temperature hydrogen attack, carburization, graphitization, erosion, thermal shock, cyclic temperature changes, vibration, pressure surges, excessive temperature, external loading, and faulty material and fabrication.

The following definitions and comments paraphrase and may inadvertently modify the NBIC. In applying the NBIC refer to the original.

creep The distortion over time of metals under stress. Creep rates increase with increasing temperature. Temperature–allowable stress tables in construction codes are based on tolerable long-term creep rates. Creep increases from practically none at code-permitted temperature-stress levels to the degree that parts subjected to temperatures higher than design may distort and eventually fail. High-temperature exposure may also alter the chemical and physical properties of a metal, leaving it in a permanently weakened state. When you evaluate expected remaining life, examine the creep rate versus time at the temperature of exposure and the stress level applied.

fatigue Failure of parts resulting from high stresses and frequent reversals of stress direction, especially where secondary stresses are elevated. Examples are cyclically loaded expansion joints, and regions where the shell is joined to a fixed tubesheet that endures rapidly changing start-up and shutdown temperatures. Cyclic temperature and pressure changes may provide the stress reversals. Look for thermal fatigue where metals with different thermal coefficients of expansion are joined, such as the weld of an austenitic tubesheet to a carbon-steel shell.

galvanic corrosion Corrosion resulting from two dissimilar metals in contact with each other and an electrolyte which creates an electrolytic cell. An example is a film of water containing dissolved oxygen, nitrogen, and carbon dioxide.

Electric current flowing through the circuit may cause the metal with the greater electrode potential to corrode rapidly. Galvanic corrosion is most active when there are large differences between the electrode potentials, but may also exist with relatively small changes of alloy composition such as might exist between the weld deposit and base metal. Natural protective coatings, such as oxides on reactive metals, or applied ones may inhibit galvanic corrosion. However, metals and alloys are usually selected for their intrinsic corrosion resistance. Look for galvanic corrosion at weld regions and on flange faces.

general corrosion Corrosion covering a considerable area. When it occurs, evaluate the safe working pressure, which is directly related to the remaining material thickness.

grooving Metal deterioration caused by localized corrosion that may be accelerated by stress concentration. Look for it adjacent to welds and on flanged surfaces.

hydrogen attack* Degradation of mechanical strength of carbon and low-alloy steels in a water-free hydrogen environment at temperatures above 450°F (~230°C), caused by hydrogen permeating the metal and reacting with the carbon to form methane (CH_4). The strength of carbon and low-alloy steels is related to carbon content. Therefore, loss of carbon reduces strength. In addition entrapped methane at high pressure can form bubbles, cracks, and/or blisters. Hydrogen attack damage follows an incubation period in which there is no noticeable change in properties. The length of the period depends on the temperature and pressure of exposure and the alloy content. The damage is reversible during the incubation period, but once the metal begins to degrade it is irreversible.

hydrogen embrittlement Loss of strength and/or ductility in steels caused by solution of atomic hydrogen, also called *sulfide stress cracking, wet H_2S cracking, hydrogen-induced stress cracking* (HIC), *stress-oriented hydrogen-induced cracking* (SOHIC), etc. Hydrogen embrittlement is a low-temperature phenomenon that results most often from hydrogen evolved from aqueous corrosion reactions, but also results in welds from hydrogen evolved in the welding process (weld underbead cracking). Its appearance is different in the different environments in which it occurs. SOHIC, weld underbead cracking, and sulfide stress cracking require applied or residual stress. In HIC and blistering, the internal pressure from the recombined hydrogen is sufficient to cause the damage. Susceptibility to sulfide stress cracking and similar forms increases with the strength of the steel and severity of the hydrogen environment. In weld underbead cracking, hydrogen dissolved in a hard, high-strength heat-affected zone may be the cause. You can minimize the phenomenon by using low-hydrogen weld wire and postweld heat treatment. You can reverse hydrogen embrittlement as long as no physical damage has taken place by removing the hydrogen from the steel by heating in the absence of hydrogen to about 150 to 200°C (~300 to 400°F). To detect wet hydrogen sulfide cracking, which is not always visible to visual inspection, use magnetic particle, wet fluorescent, of fluid penetrant.

---

*See API 941[3] for more detailed information on hydrogen attack.

line corrosion A condition in which pits are connected or nearly connected in a narrow band or line that frequently occurs in the area of intersection of the support skirt and bottom of a vertical exchanger or at a liquid-vapor interface.

low-temperature effects Water and some process fluids expand at subfreezing temperatures, which may cause failure. The ductility and notch toughness of carbon and low-alloy steels is a function of thickness and temperature. In some, brittle failure may occur at ambient temperatures. Some failures have been attributed to brittle fracture of steels exposed to temperatures below their transition temperature under pressures greater than 20 percent of the required hydrostatic test pressure. However, most have occurred at the code-required hydrostatic test pressure.

pitting Shallow, isolated, scattered pitting over a small area that does not substantially weaken the vessel, but which could eventually cause leakage.

stress-corrosion cracking (SCC) Cracking of specific metals subjected to tensile stress in an environment to which the metal is sensitive. The tensile stress may be applied or residual. Typical are the transition from rolled to unrolled expanded austenitic tubes exposed to substances with chloride content above 15 parts per million (ppm), season cracking of brass tubes in ammonia solutions, and cracking of carbon steel in hot caustic solutions.

temper embrittlement Loss of ductility resulting from postweld heat treatment or high-temperature service. Be cautioned that low-alloy steels are prone to temper embrittlement.

When an exchanger has two or more zones, and the required thicknesses, corrosion allowances, or corrosion rates in each are so different that Eq. (6.1) would calculate substantially different remaining lives for the zones, the period between inspections may be established zone by zone.

NBIC requires exchangers installed aboveground to be inspected externally after the lesser of 5 years or 25 percent of the calculated remaining life. You do not have to remove the insulation if (1) the operating temperature is sufficiently high or low to preclude the presence of water, (2) if the remaining life is 10 years or more, and (3) if the exchanger is protected from exposure to corrosion. However, you must periodically observe the condition of the insulation and insulation jacketing and make necessary repairs.

The required inspection intervals are based on continuous service interrupted only by normal shutdowns. If you take a unit out of service for an extended period, you have to consider the environmental conditions inside and out when you revise the date of the next inspection.

NBIC gives means for calculating thicknesses used for determining corrosion rates when MAWP is governed by circumferential stress, and when longitudinal stresses would be important. It states that you may disregard widely scattered pits if (1) their depth is not more than one-half the required thickness exclusive of corrosion allowance,

(2) the total area of the pits does not exceed 7 in^2 (~45.2 cm^2) within any 8-in (203-mm) -diameter circle, and (3) the sum of their dimensions along any straight line within the 8-in (~203-mm) circle does not exceed 2 in (50.8 mm). You must take into account weld joint efficiency in determining MAWP. When efficiency is less than 100 percent and there is corrosion both on the weld surface as well as metal remote from the weld, you have to take into account the joint efficiency to determine which governs. In this calculation include metal adjacent to the weld 1 in (25.4 mm) or 2 times the minimum required thickness on either side. Because each code has formulas for calculations that apply to knuckles and spherical sections, the NBIC requires separate calculations for governing thicknesses when you measure the corroded thickness of ellipsoidal and torispherical heads.

If the measured thickness indicates a rate of corrosion different from that which you assumed, adjust the inspection frequency to conform with the measured rate. In-service inspection also includes examining the condition of appurtenances.

The NBIC devotes a special paragraph to in-service inspection of liquid ammonia vessels because the medium is so corrosive. If you have a liquid ammonia system, we suggest that you look into NBIC Paragraph RB-3700.

Paragraph RB-3800 deals with inspecting exchangers with quick-actuating closures, such as might be found on channel-opening exchangers and feedwater heaters.

Restamping or replacing a nameplate

Paragraph RB-4000 tells how to proceed when the stamping on a pressure-retaining item becomes indistinct or the nameplate is illegible or lost, but traceability to the original exchanger is possible. The Inspector instructs the owner or user to have the item restamped; the owner or user requests permission to restamp or replace nameplates to the jurisdiction where the pressure-retaining item is located on the "Boiler or Pressure Vessel Identification" form shown in Appendix 5, providing proof of the original stamping and such other data as is available. If these satisfy the jurisdiction, it may grant the request.

Replacement nameplates for ASME-stamped exchangers must be clearly marked "replacement" and must be identical with the original stamping. The Code symbol and National Board number restamping may be done only by an employee of the original manufacturer in the presence of an Inspector in the employ of the Authorized Inspection Agency who employed the Inspector who signed the original Manufacturer's Data Report, or an Inspector employed by the jurisdiction. Similar rules apply to items not ASME Code–stamped.

When restamping is complete, the owner or user must file with the jurisdiction a notice of completion together with a facsimile of the stamping or nameplate, signed by the Inspector who witnessed it.

NBIC requirements for records review

Paragraph RB-3270 sets forth requirements for records that must be maintained and the Inspector's records review. This includes examining the heat exchanger log, record of maintenance, corrosion-rate record, or any other examination results. It requires maintaining a permanent record for each heat exchanger that includes an ASME Manufacturer's Data Report or equivalent specifications if the unit is not ASME Code–stamped. You may use NBIC's form NB-5, "Boiler or Pressure Vessel Data Report—First Internal Inspection" for this purpose.

In addition to complete pressure-relief device information, the progressive record must include, but is not limited to, the following:

- Location and thickness of monitor samples and other critical inspection sites.
- Limiting metal temperature and its location on the heat exchanger when it is a factor in establishing the minimum allowable thickness.
- Computed required metal thicknesses and MAWP for the design temperature, and pressure-relieving device opening pressure, static head, and other loadings.
- Test pressure if tested at the time of inspection.
- Scheduled date of next inspection (approximate).
- Date of installation and dates of any significant changes in service conditions such as pressure, temperature, character of contents, or corrosion rate.
- Drawings that show sufficient details to permit calculating the service rating of all components on heat exchangers used in process operations subject to corrosive conditions. If drawings are not available you may use a combination of detailed data with sketches.

Conclusions of in-service inspections

At the conclusion of an in-service inspection, the Inspector will discuss with the owner or user any defects or deficiencies found in the condition, operating and maintenance practices, and, if necessary, make recommendations for corrections.

NBIC Part RC repairs and alterations

Part RC is subdivided into Paragraph RC-1000, which sets out general requirements for repairs and alterations; and Paragraphs RC-2000 and RC-3000, which have additional requirements for repairs and alterations, respectively.

General requirements. Only accredited organizations may perform repairs and alterations. When the ASME Code governed the original construction, repairs and alterations must conform, insofar as possible, to the section and edition most applicable to the planned work. When a different code or standard governed the original construction, repairs or alterations must conform, insofar as possible, to the edition of the construction standard or specification most applicable to the work. When this is not possible or practical, you may use other codes, standards, or specifications, including the ASME Code, provided the R Certificate Holder has the concurrence of the Inspector and jurisdiction where the unit is installed.

Materials and replacement parts. Materials used in making repairs or alterations must conform with the requirements of the applicable construction code rules used in the original construction. Carbon steel with carbon content more than 0.35% may not be welded.

NBIC classifies replacement parts subject to internal or external pressure as "Material or Preassembled Welded Parts." *Material* is parts that have been formed to the required shape, but on which no welding has been performed. These parts must be marked with the material and identifications. The markings are considered to be the parts manufacturer's certification of compliance with the original construction code that applied. Examples are tubes, pipes, heads, tubesheets, or subassemblies that are mechanically attached together. *Pressure-bearing replacement parts* preassembled by attachment welds must have the welding done in conformity with the original construction code. The supplier or manufacturer must certify in the form of bills of materials and drawings, with a statement of certification, that the material and fabrication comply. A commercial welded manway for installation in an exchanger shell or channel is an example.

When the ASME Code is the original construction code, *pressure parts fabricated by welding that require factory inspection,* and which require factory inspection by an AI, must be built by a manufacturer who has an appropriate ASME Certificate of Authorization and ASME Code symbol stamp. Such parts require inspection and stamping per the applicable ASME Code section. The manufacturer must supply a partial data report. When a code other than the ASME was used, the part must be manufactured by an organization certified as required by the original construction code.

When the organization fabricating the part cannot provide certification to the original construction code or equivalent, the part fabricator may obtain a National Board Certificate of Authorization. The manufacturer must document the replacement part on Form R-3 and apply the R stamp as described in NBIC, Appendix 2. Examples are replacement channels, replacement tube bundles, and replacement shells.

Authorization of repairs and alterations. You may not initiate a repair or alteration without the Inspector's prior authorization. The Inspector who makes the acceptance inspection should be the same one who authorized the work, but in all cases must be employed by the organization that employs the Inspector who authorized the work.

The NBIC requires the inspection and certification to be made by and Inspector employed by one of the following:

- A jurisdictional Authorized Inspection Agency.
- The Authorized Inspection Agency of the R certificate holder making the repair or alteration.
- The Authorized Inspection Agency which insures the heat exchanger.
- The Owner-User Inspection Organization. The company's organization and inspection procedures must be specifically approved by the jurisdiction, or when there is no jurisdiction, by the National Board. Inspectors employed by Owner-User Inspection Organizations may authorize and accept work only on pressure-retaining items owned or used by the company.

Welding. Welding must be performed in accordance with the requirements of the original code of construction used for the pressure-retaining item. The Welding Procedure Specification (WPS) for the welding must be qualified in accordance with the original construction code, but when this is not possible or practicable, may be qualified in accordance with Section IX of the ASME Code. An R certificate holder may use one or more of the Standard Welding Procedure Specifications shown in NBIC, Appendix A. Welders or welding operators must be qualified either as required by the original code of construction or ASME Code, Section IX. The R certificate holder must certify and maintain records of and make available to the Inspector results of procedure and welding performance qualifications.

The R certificate holder has to assign a unique identification mark to each qualified welder or welding operator and have a written procedure for identifying who made all welded joints. The mark may be stamped, using a low-stress stamp, adjacent to all welds made by the individual. Alternatively, the certificate holder may keep a record of welded joints and the welders or welding operators who made them.

If a welder or welding operator has not welded using a specific process within a period of 6 months or longer, the qualifications for that process expire. The qualification must be revoked if there is a specific reason to question a welder's or welding operator's ability to make welds that meet the specification. This does not affect qualifications not questioned.

The welding procedure must specify pre- and postweld heat treatment. If postweld heat treatment in accordance with the original construction code is inadvisable or impractical, you may use alternative methods or special welding methods acceptable to the Inspector. Part RD addresses the permitted alternatives.

Nondestructive examination. The same personnel requirements, methods, and acceptance criteria for nondestructive examination that applied to original construction apply to repairs and alterations. Where this is not possible or practical, you may use alternative methods acceptable to the Inspector and the jurisdiction where required. NDE personnel must be qualified as required by the original construction code. When not possible or practicable, they may be qualified in accordance with ASNT SNT-TC-1A.[4] You must follow the rules of the original construction code for calibration and its documentation of pressure gages and examination and test equipment.

Acceptance inspection and stamping. Before signing the report form, the Inspector must (1) review the drawings, (2) ensure that the welding was performed as required, (3) witness an applied pressure test, (4) make sure that the required NDE has been satisfactorily performed, and (5) be satisfied that all other things have been done to assure compliance with the NBIC.

The R certificate holder responsible for the work must stamp the exchanger or apply a nameplate as described in Appendix 2. It must be done only with the Inspector's knowledge and authorization. If the original stamping must be removed, subject to the jurisdiction's approval, the Inspector must witness making a facsimile, obliteration of the old stamping, and transferring the stamping to the new item. If stamping is on a nameplate, the Inspector has to witness its transfer to the new location. You must describe the relocation on the applicable NBIC form. *Do not restamp ASME Code items with the ASME Code Symbol.*

Registration of documentation. Organizations that stamp repairs or alterations with the R symbol may register them with the National Board. Some jurisdictions require such registration.

Additional requirements for repairs (RC-2000)

Following is a summary of NBIC's additional requirements for repairs.

Drawings. Before commencing repairs, prepare drawings with sufficient information for performing the work.

Authorization to perform repairs. Do not begin repairs without the Inspector's prior authorization. The Inspector must determine that the repair methods are acceptable. If the jurisdiction approves, the Inspector may give prior approval for routine repairs if the Inspector makes sure that the R certificate holder has acceptable procedures for doing them.

Routine repairs. Routine repairs are ones performed under the R certificate holder's quality system program, but in which the Inspector is not involved during in-process activity and stamping is waived. They have to be documented on a Form R-1, "Report of Welded Repair," with a statement on its line 9, "Remarks: Routine Repair." The R certificate holder must document in its quality system program how their control and implementation will be done. Before doing routine repair work, the holder should determine that routine repairs are acceptable to the jurisdiction where the work is to be done. These are acceptable routine repairs: (1) welded repair or replacement of a section of tube or pipe five (5) NPS inch diameter and under, and their attachments; (2) replacing or repairing non-load-bearing attachments where postweld heat treatment is not required; (3) weld buildup of wasted areas in shells and heads that does not exceed 100 in^2 (~645 cm^2) or 25 percent of nominal wall thickness or ½ in (~12.7 mm), whichever is less; and (4) corrosion resistance weld overlay not exceeding 100 in^2 (~645 cm^2).

Repair methods. The repair methods must be acceptable to the inspector. Some repair methods are discussed in Part RD.

These are some of NBIC's general suggestions for repair methods:

- Do not repair cracks in welds or base materials without providing a means for stopping crack propagation at the ends, for example, drilling holes at each end of the crack. Use nondestructive examinations to determine that the crack has not traveled.
- Before repairing cracks, find the causes and take the actions necessary to eliminate future cracking.
- Cracks in head knuckles and other regions of discontinuity, such as tubesheet skirts, are very serious threats to continued safe opera-

tion. Consider complete removal and replacement of tubesheets, and removal and replacement of affected plate areas. The welds of patches in knuckles should be ground smooth and radiographed.

- Repair minor cracks, isolated pits, and small plate imperfections only if the extent of the defect combined with continued deterioration is such that operating without a repair is hazardous. When repairing such defects, always remove the defect to completely sound metal.
- Before repairing metal wasted by erosion, such as that which results from wire drawing or poor inlet nozzle placement in channels, consider the following: (1) the cleaning and welding procedures to be used, (2) the most suitable kind of nondestructive examination, and (3) how to avoid impairing the structural integrity of the underlying pressure parts.
- Seal welding tubes that were originally only expanded is acceptable as a repair method. However, unless all welding surfaces are cleaned scrupulously before welding, anticipate further leaks.

Pressure-testing repairs. The R certificate holder is responsible for all activities relating to pressure-testing repairs. When a test pressure will exceed the set pressure of the pressure-relief device, prepare the device as its manufacturer recommends. This summarizes these activities, which apply to all repairs:

- Test to the higher of 80 percent of the MAWP stamped on the nameplate or operating pressure using water or other liquid medium.
- Test replacement parts used in repairs to the MAWP stamped on the pressure-retaining item being repaired.
- Test at a temperature not less than 60°F (~15.6°C) unless the owner provides information on the toughness characteristics of the material to indicate acceptability of a lower test temperature. During close examination, the metal temperature must not exceed 120°F (~49°C) unless the owner specifies a higher temperature and it is acceptable to the Inspector.
- Maintain test pressure for at least 10 min before the Inspector examines the exchanger.
- Maintain test pressure for as long as the Inspector requires to conduct the examination.
- If liquid test media would contaminate the process fluids, or when pressure testing is not practical, testing may be by these other methods: (1) pneumatic test, provided the R certificate holder has

the concurrence of the Inspector, the jurisdictional authority (where required), and the owner—precautionary requirements for pneumatic testing of the original construction code must be followed; and (2) alternatively, if the Inspector, jurisdictional authority (where required), and owner concur, NDE methods that verify the exchanger's integrity may be used.

Stamping repairs. Repairs must be stamped as required in NBIC, Appendix 2. If, after consideration of the type and extent of the necessary repairs, the jurisdiction accepts it and the Inspector concurs, nameplates and stamping may not be required for routine repairs.

Documenting and registering repairs. The R-1 certificate holder must document repairs performed in accordance with the NBIC on Form R-1, "Report of Welded Repair," as shown in Appendix 5. Use Form R-4, "Report of Supplementary Sheet," to record additional data when space is insufficient on Form R-1. The Inspector must indicate acceptance by signing the R-1. Form R-3 and the Manufacturer's Partial Data Reports for Replacement Parts must be attached to the R-1 and made a part of it.

The certificate holder must distribute legible copies of the completed R-1 with attachments to the owner or user, the Inspector, the jurisdiction (if required), and the Authorized Inspection Agency responsible for inservice inspection. The R-1 may also be registered with the National Board.

Repairs to exchangers built to ASME Code, Section VIII, Division 2. The user must prepare or have prepared a detailed plan covering the scope of the repair. A Professional Engineer (PE), registered in one or more states in the United States or provinces of Canada, experienced in pressure-vessel design, must review it. The PE's review and certification must ensure that the work involved in the repair is compatible with the user's design specification and the manufacturer's design report. After review and certification, the repair plan must be submitted for acceptance to the Authorized Inspection Agency (AIA) whose Inspector will make the acceptance inspection and sign the R-1.

Additional requirements for alterations (RC-3000)

Following is a summary of NBIC's additional requirements for alterations.

Design. The R certificate holder must establish controls to make sure that all necessary design information, applicable drawings, design calculations, specifications, and instruction are prepared, obtained, controlled, and interpreted to provide the basis for an alteration that accords with the original construction code. When the original construction standard requires a Manufacturer's Data Report, the holder must obtain a copy of the original to use in designing the alteration. If the report cannot be obtained, the holder must get agreements on the method of establishing the design basis for the alteration from the Inspector and jurisdiction.

Calculations. Before starting any physical work, an organization experienced in the design part of the construction standard used for the exchanger must complete a set of calculations, which must be made available for review by the Inspector who accepts the design.

Rerating. In the NBIC, *rerating* means increasing the MAWP or temperature, or decreasing the minimum temperature such that additional mechanical tests are required. It does not provide rules for derating and suggests that you contact the jurisdiction in which the exchanger is installed to find out whether any specific procedures must be followed. Rerate only after meeting to the satisfaction of the jurisdiction at the location of the installation the following requirements:

- Prepare revised calculations that verify the new service conditions in accordance with the certificate holder's quality control system.
- Establish rerating in accordance with the requirements of the construction standard to which the unit was built.
- Determine that current inspection records verify that the pressure-retaining item is satisfactory for the proposed service conditions.
- Make sure that the pressure-retaining item has been pressure-tested, as required, for the new service conditions.

Drawings. As appropriate, prepare drawings to describe the alteration that have enough information to satisfactorily perform it.

Pressure-testing alterations. A pressure test, conducted as dictated by the original construction code, is required when an alteration involves installing a replacement part and/or it will affect the design pressure, design temperature, or design rated capacity. The test may be performed at the point of manufacturer or installation. The R certificate holder is responsible for all activities relating to pressure-testing replacement parts. This summarizes some pressure-testing activities for alterations:

- When a test pressure will exceed the set pressure of the pressure-relief device, prepare the device as its manufacturer recommends.

- The pressure test of replacement parts and connecting welds must not exceed 1½ times the MAWP adjusted for temperature. (*Connecting welds* arc defined as ones attaching the replacement part to the pressure-retaining item.) When the original test pressure includes considering corrosion allowance, test pressure may be adjusted on the basis of remaining corrosion allowance.
- As an alternative to pressure-testing connecting welds at 1½ times the MAWP, after acceptance by the jurisdiction and Inspector, you may (1) pressure-test at the MAWP stamped on the pressure-retaining item or (2) subject the connecting welds to an NDE method that verifies the integrity of the repair.
- Test at a temperature not less than 60°F (~15.6°C) unless the owner provides information on the toughness characteristics of the material to indicate acceptability of a lower test temperature. During close examination, metal temperature must not exceed 120°F (~49°C) unless the owner specifies a higher temperature and it is acceptable to the Inspector.
- Maintain test pressure for at least 10 min before the Inspector examines the exchanger.
- Maintain test pressure for as long as the Inspector requires to conduct the examination.
- Except for the alternatives to pressure-testing connecting welds, if liquid test media would contaminate the process fluids, or when pressure-testing is not practical, testing may be by these other methods: (1) provided the R certificate holder has the concurrence of the Inspector, the jurisdictional authority (where required), and the owner, pneumatic test may be used and precautionary requirements for pneumatic testing of the original construction code must be followed; and (2) alternatively, if the Inspector, jurisdictional authority (where required), and owner concur, NDE methods that verify the exchanger's integrity may be used.

Stamping. Stamp in accordance with Appendix 2. Apply the nameplate or stamping next to the original manufacturer's stamping or nameplate.

Documenting and registering alterations. The R certificate holder performing the alteration must document repairs performed in accordance with the NBIC on Form R-2, "Report of Alteration," as shown in Appendix 5. Use Form R-4, "Report of Supplementary Sheet," to record additional data when space is insufficient on Form R-2. The Inspector must indicate acceptance by signing the R-2. Form R-3 and the Manufacturer's Partial Data Report for Replacement Parts must be attached to the R-2 and made a part of it.

The R certificate holder that certifies the "design change" must complete and sign the "Design Certification" section of the Form R-2. An Inspector must indicate acceptance of the design by signing the R-2's "Certificate of Design Change Review."

The certificate holder that performed the alteration must complete and sign the "Construction Certification" section of the R-2. An Inspector must indicate acceptance of the installation by signing the R-2's "Certificate of Inspection" section.

The certificate holder must distribute legible copies of the completed R-2 with attachments to the owner or user, the Inspector, the jurisdiction (if required), and the Authorized Inspection Agency responsible for in-service inspection. If the item is registered with the National Board, the holder must submit one original and one legible copy to the National Board. If the item is not registered with the National Board, one original and one copy of the R-2 may be registered with the National Board.

NBIC appendixes pertinent to heat exchangers

Appendixes listed above that apply to heat exchangers are 1, 2, 4, 5, 6, A, B, C, D, F, and G. In the following we summarize the most relevant parts of the appendixes.

Appendix 1. "Preparation of Technical Inquiries to the National Board Inspection Code Committee." Use the format described in this mandatory appendix to submit inquiries.

Appendix 2. "Stamping and Nameplate Information" details the format, location, and size of characters to use in stamping and has illustrations of required marking for repairs, alterations, rerating, and the marking of parts fabricated by welding with the use of National Board Form R-3.

Appendix 4. "Glossary of Terms" has been discussed above.

Appendix 5. "National Board Forms" has specimens of Form R-1, "Report of Welded Repair"; Form R-2, "Report of Alteration"; Form R-3, "Report of Parts Fabricated by Welding"; and Form R-4, "Report Supplementary Sheet." Each of these forms has a numbered circle wherever the preparer must enter data. The number is keyed to a "Guide for Completing National Board Form R Reports." The appendix follows the R form specimens with NR form specimens NR-1, "Report of Repair ❑ Modification ❑ or Installation of Replacements ❑ to nuclear components and systems in nuclear power plants" similarly keyed to a guide for filling in the required data. Also included in the appendix is a specimen of the boiler or pressure-vessel identification form referred to above, Form NB-7, "Pressure Vessels Report of Inspection"; Form NB-4, "Notice of New Business or Discontinuance" used by Authorized Insurance Inspection Agencies"; Form NB-5, "Boiler or Pressure Vessel Data Report First Internal Inspection"; and others not relevant to shell-and-tube heat exchangers.

Appendix 6. "Example of Repairs and Alterations" is a vary valuable set of examples of repairs and alterations.

Appendix A. "Standard Welding Procedures" can be a great time saver for Repair Organizations in that it has a series of ANSI/AWS Standard Welding Procedure Specifications (WPSs). The Repair Organization may use one or more of the listed standard WPSs as an alternative to WPSs that it has qualified. To use these WPSs, the Repair Organization must accept by certification contained in the Standard WPS full responsibility for application of the Standard WPSs in conformance with the application as stated in the Standard WPS.

Appendix B. "Recommended Preheat Temperatures" is a guide that suggests minimum preheat temperatures for various materials listed in the ASME Code by P numbers and/or the equivalent materials in other construction codes and standards.

Appendix F. "Pressure Differential between Safety or Safety Relief Valve Setting and Boiler or Pressure Vessel Operating Pressure" points out that if a safety valve or pressure-relief valve is subjected to pressure at or near its set pressure, it will weep or simmer. Consequently deposits may accumulate in the seat and disk area, which can eventually cause the valve to freeze closed, which would prevent the valve from opening at its set pressure. In such an event, unless pressure application is stopped, overpressure could exceed the exchanger's rupture pressure. Therefore, it is extremely important to maintain a pressure differential between the operating pressure and the valve set pressure to prevent weeping. Although the guide has caveats warning readers to consider the nature of the fluids being processed and the start-up, shutdown, and normal operating conditions, it recommends the following general set of differentials:

- For set pressures up to 70 psi (~4.83 bar), 5 psi (~0.34 bar)
- For set pressures between 70 and 1000 psi (~4.83 and 68.95 bar), 10 percent of the set pressure
- For set pressures above 1000 psi (~68.95 bar), 7 percent of set pressure.

The API Inspection Code (API 510)

Introduction to API 510

The API Inspection Code (API 510) is a code for maintenance inspection, rating, repair, and alteration of pressure vessels addressed to the customs and practices in the oil-refining, chemical, and petrochemical industries. It is very similar to the NBIC. In applying API 510, one should be thoroughly familiar with the recommended *Inspection Practices* of API 572. Although only a few of the states and provinces mention API 510 in their laws, rules, and regulations, those that require applying NBIC will usually accept API 510 except for provisions that conflict directly with NBIC or the jurisdiction's rules.

API 510's definitions parallel those of the NBIC. But there are enough subtle differences to make it necessary to repeat some of them here. Its statement of scope covers general applications and specific ones. Generally, API 510 is intended only for organizations which maintain or have access to an Authorized Inspection Agency. API 510 is generally restricted to organizations that maintain or have access to engineering and inspection personnel or those technically qualified to maintain, inspect, repair, alter, and/or rerate pressure vessels.

API 510 applies to vessels constructed in accordance with any of the several editions of the former *API-ASME Code for Unfired Pressure Vessels for Petroleum Liquids and Gases,* the several editions of Section VIII of the *ASME Boiler and Pressure Vessel Code,* to other recognized pressure-vessel codes, to nonstandard vessels, and to other vessels constructed noncode or approved as jurisdictional special. API 510 applies only to vessels after they have been placed in service, including items that are inspected by an Authorized Inspection Agency and/or repaired by a Repair Organization as defined by API 510. API 510 cannot be used when it conflicts with prevailing regulatory requirements.

API 510 has an alternative set of rules for pressure vessels in natural resource service such as drilling, production, gathering, transportation, lease processing, and treatment of liquid petroleum, natural gas, and associated salt water. The alternative rules are intended for services which may be regulated under safety, spill, emission, or transportation controls of various federal, state, and local authorities.

API 510 definitions

The following definitions paraphrase some of API-510's glossary terms. Not all terms in the glossary are included.

alteration A physical change in any component which has design implication which affects the pressure-containing capability of a pressure vessel beyond the scope of the items described in existing data reports. Alterations do not include comparable or duplicate replacements, adding any reinforced nozzle equal to or less than the size of existing reinforced nozzles, adding those not requiring reinforcement, or rerating. Note that, unlike NBIC, API 510 does not consider a nonphysical change to be an alteration.

principles of the ASME Code Because, in some circumstances, repairs and/or alterations cannot *conform* strictly with the ASME Code rules, and because API 510 requires application of its rules when they conflict with those of the ASME Code, API 510 applies the *principles* of the ASME Code. In practice the ASME Code rules are applied whenever practicable.

Authorized Inspection Agency API 510 defines the following as Authorized Inspection Agencies:

1. The inspection organization of the jurisdiction in which the exchanger is used.
2. The inspection organization of an insurance company which is licensed or registered to—and does—underwrite pressure-vessel insurance.
3. Owners or users of pressure vessels who maintain an inspection organization for activities relating only to their equipment and not for vessels intended for sale or resale. The owner-user agency shall designate inspectors with suitable experience.

Authorized Inspector (Inspector) An Authorized Inspector or an Inspector is an employee of an Authorized Inspection Agency who is qualified to perform inspections under this inspection code. Under this definition, Inspectors need not hold commissions from the National Board or from any jurisdiction, except when the jurisdiction's rules, laws, or regulations require them to hold such commissions.

jurisdiction A legally constituted governmental administration which may adopt rules relating to pressure vessels.

maximum allowable working pressure (MAWP) The maximum gage pressure permitted at the top of a pressure vessel in its operating position for a designated temperature based on calculations using the minimum (or average pitted) thickness for all critical vessel elements, exclusive of thickness designated for corrosion and loadings other than pressure. This is somewhat different than the ASME Code definition, because it applies to an existing, possibly corroded unit.

on-stream inspection The use of any of a number of nondestructive examination procedures to establish the suitability of a pressure vessel for continued operation. The vessel may, or may not, be in operation while the inspection is carried out. This is parallel with the NBIC external inspection.

repair The work necessary to restore a vessel to a condition suitable for safe operation at the design conditions. API 510 states that, if any restorative changes result in a change of design temperature or pressure, the requirements for rerating must also be satisfied. The API 510 definition also includes the addition or replacement of pressure or nonpressure parts which do not change the rating of the vessel.

Repair Organization The API 510 definition, which is somewhat different from that of the NBIC, states that a repair organization means one of the following:

1. The holder of a valid ASME Certificate of Authorization for the use of an appropriate ASME Code symbol stamp
2. Owners or users of pressure vessels who repair their own equipment in accordance with the API 510 code
3. A contractor whose qualifications are acceptable to the owner or user of pressure vessels and who makes repairs in accordance with the API 510 Code
4. One who is authorized by the legal jurisdiction

Rerating A change in either or both temperature ratings or the MAWP rating of a vessel. Rerating is not an alteration unless physical modification of the vessel is required. This is exactly opposite to the NBIC definition of an alteration in which there is no physical change! A rerating may consist of an increase, decrease, or a combination change to maximum allowable working temperature and pressure. Derating below original design conditions is a permissible means to provide for corrosion. Note that, in contrast, NBIC rerating provides only for an increase in maximum allowable temperature or pressure.

API 510 inspection practices

API 510 has substantial information about inspection practices, similar to that of the NBIC. However, they are refinery-operation-oriented. Its section on practices covers

1. Preparatory work
2. Modes of deterioration and failure
3. Corrosion rate determination
4. MAWP determination using measured thicknesses
5. Defect inspection
6. Parts inspection
7. Evaluating corrosion and minimum thickness

Inspection of pressure vessels in API 510

This section covers external inspection, inspection intervals, pressure test, safety and pressure-relief devices, and records. The text is similar to that of the NBIC.

Repairs, alterations, and rerating of pressure vessels under API 510

With the exceptions noted previously, this section is like the NBIC's. Under API 510, repairs, alterations, and rerating authorizations are as follows.

National Board authorization. The repair organization holds a valid Certificate of Authorization for the use of an R stamp issued by the National Board.

Holder of an ASME Code symbol stamp. The organization is in possession of a valid ASME Certificate of Authorization to apply an ASME Code stamp, provided the authorization is within the scope of the organization's quality control system.

Jurisdictional authorization. Units may be repaired by organizations working within a governmental jurisdiction, if the jurisdiction has issued authorization for the organization to perform such repairs.

Qualifications of owner-user Inspectors

API 510 educational and experience requirements for Owner-User Inspectors are summarized below. Suitable experience under API 510 rules consists of one of the following:

1. A degree in engineering plus 1 year of experience in the design, construction, repair, operation, or inspection of boilers or pressure vessels
2. A 2-year certificate in engineering or technology from a technical college plus 2 years of experience in the design construction, repair, operation, or inspection of boilers or pressure vessels
3. The equivalent of a high-school education plus 3 years of experience in the construction, repair, operation, or inspection of boilers and pressure vessels

An individual with a National Board or jurisdictional commission would probably be considered to have suitable experience. However, under API 510, *no formal instruction, training, or certification is required.*

How the National Board and API Codes Apply the ASME Code Rules

The NBIC applies the ASME Code design, materials, construction, and examination rules to repairs and alterations as written. When this is not practical, it requires the Inspector's concurrence for the deviation. It requires repairs and alterations to tubulars to be performed by an R-stamp holder. Its requirements for design calculations and stamping of alterations have been described above.

API 510 must allow for repairs and alteration to exchangers not built to the ASME Code. It recognizes that the ASME Code is written strictly for new construction but that most of its technical design, welding, examination, and materials requirements can be applied to the maintenance, inspection, rating, and alteration of operating heat exchangers. However, the ASME Code cannot always be followed for the operating units to which API 510 applies. This is because of its new construction coverage, such as revised or new-materials specifications, inspection requirements, certain heat treatments and pressure tests, and stamping and inspection requirements. Therefore, it states that "the engineer/inspector shall be guided by this inspection code in lieu of strict conformance with the ASME Code. To signify this the term *principles of the ASME Code* has been employed in this inspection code."

References

1. *National Board Inspection Code,* National Board of Boiler and Pressure Vessel Inspectors, Columbus, Ohio, 1995. A new edition is published every 3 years at about the same time as the new edition of the *ASME Boiler and Pressure Vessel Code.*
2. *Pressure Vessel Inspection Code, Maintenance Inspection, Rating, Repair, and Alteration,* 7th ed., API 510, The American Petroleum Institute, Washington, D.C., March 1992.
3. *Steels for Hydrogen Service at Elevated Temperatures and Pressures in Petroleum Refineries and Petrochemical Plants,* API 941, American Petroleum Institute, Washington, D.C.
4. *Recommended Practice for Nondestructive Testing Personnel Qualification and Certification,* ASNT SNT-TC-1A, American Society for Nondestructive Testing, Columbus, Ohio, 1984.

Chapter

7

Exchanger Cleaning

Introduction

In this chapter, we discuss cleaning tubular exchangers off and on line. Most exchangers are cleaned off line. However, on-line, continuous cleaning is gaining a strong position in suitable industrial applications. Off-line cleaning methods are mechanical, hydraulic, and chemical. Burning, which for many years was a strategy for removing coatings and deposits that other methods could not remove, is no longer acceptable because of the atmospheric pollutants it generates. Live steam jets have been used, but we consider them to be so unsafe that we will not discuss them. On-line cleaning may be either mechanical or chemical.

We begin by reviewing how the heat exchanger's configuration affects the choice of cleaning method(s) and discussing some of the tools available. Then we discuss on- and off-line cleaning systems and tools.

Methods suitable for cleaning fixed-tubesheet units

The tube side of a fixed-tubesheet unit can be cleaned by any of these methods, alone or in combination: (1) mechanical on-line, (2) mechanical off-line, (3) hydraulic, and (4) chemical. There is substantially no access to the shell side. Therefore, the only way to clean it is by circulating chemical solutions that will dissolve away foulants.

Methods suitable for cleaning straight-tube removable-bundle units

You can clean the tube side of straight-tube removable-bundle exchangers and the empty shell by any method or combination of methods. If the tubes are arranged on a square pitch or a sufficiently spread triangular pitch, you can clean the exterior of the removed bundle off line, mechanically and by hydroblasting. Some success has been achieved in hydroblasting the shell side of bundles with tubes on conventional triangular pitch, but the success rate depends on the kind of fouling and the tube diameter. The whole bundle can be immersed in an agitated chemical bath that forces solution through the bundle, or chemical cleaning solution may be circulated through the shell. Carefully consider the nature and concentration of the cleaning solution, the temperature at which the work is to be done, and postcleaning neutralization and flushing to remove all traces of the chemicals before returning the bundle to service.

Methods suitable for cleaning U-tube removable-bundle units

The ability to clean the tube side mechanically is limited by the smallest bend radius. Most U-tube units can be cleaned hydraulically with special jets and lances. Where the hydraulic equipment cannot traverse the bends, the alternative is to clean chemically. You can clean the shell side of the bundle by any of the methods used to clean straight-tube removable bundles.

The most suitable method or combination of methods depends on the kind of fouling, its tenacity, and the ways that have been found to be most successful in removing it. Foulant classifications and suggested off-line cleaning methods are tabulated in Table 7.1. Things to consider after selecting a method are listed in Table 7.2.

Off- versus On-Line Mechanical Cleaning

Off-line mechanical cleaning

Mechanical cleaning is limited by accessibility. Surfaces that are inaccessible for mechanical cleaning are the shell sides of fixed-tubesheet units, U-bends with radii smaller than 4 tube diameters, and shell sides of bundles in which the space between the tubes is less than 6.35 mm (~¼ in).

Mechanically cleaning tube exteriors. The following are used for mechanically cleaning tube exteriors off line: improvised scrapers such as flat, flexible steel strips and cutoffs from cross-flow baffles, sharpened to a scraping edge; rods and bars; and power-driven wire brushes. When you use power-driven equipment, the bristles and drive shaft must be small enough to fit between adjacent tubes and flexible enough to move around. Commercial equipment for off-line cleaning inside tubes is driven by compressed air, electricity, or water

TABLE 7.1 Foulant Classifications and Suggested Off-line Cleaning Methods

Type of coating	Examples	Origin	Suggested method*
Hard, tenaciously adhering	Barnacles	Organic	1, 1 + 2, 1 + 2 + 3
	Mineral-salt scales (deposits so firmly bonded that there is virtually no contact between surrounding fluid and the covered surfaces)	Precipitation	
	Baked-on varnishes and cokes	Organic	
Soft and loose	Powdery deposits	Settling	1, 2, or 3
	Sludges (amorphous deposits, seldom bonded to the supporting surfaces)	Precipitation	
	Slimes (gelatinous deposits that result from the application of microorganisms or the products of their life processes)	Biological	
Gumlike		Oil-refinery column bottoms	1 + 3

*Key: 1 = chemical; 2 = mechanical; 3 = hydraulic.

TABLE 7.2 Considerations for Off-line Cleaning Methods

Effects of process fluids and foulants on safety and health of the workers doing the cleaning
Protection requirements for passersby
How to safely collect and dispose of foulants and wash fluids and residues with minimum environmental pollution
Economics of doing the work with in-house maintenance personnel versus subcontracting to a cleaning contractor

power and may be provided with water flushing. The drill heads and brushes shown in Fig. 7.1 are used in combinations recommended by their manufacturers to suit the fouling. It is usually necessary to select the best types and combinations experimentally. Use of (1) spade shapes, fluted spade shapes, and twist drills for scales and (2) brushes after drilling and for powdery deposits and sludges is recommended.

Three general types of cleaners are commercially available, mostly for cleaning inside tubes: (1) cleaners with internal motors as shown in Fig. 7.2, (2) cleaners with external drives (also called "drill type" cleaners, pictured in Fig. 7.3), and (3) blowguns as illustrated in Fig. 7.4.

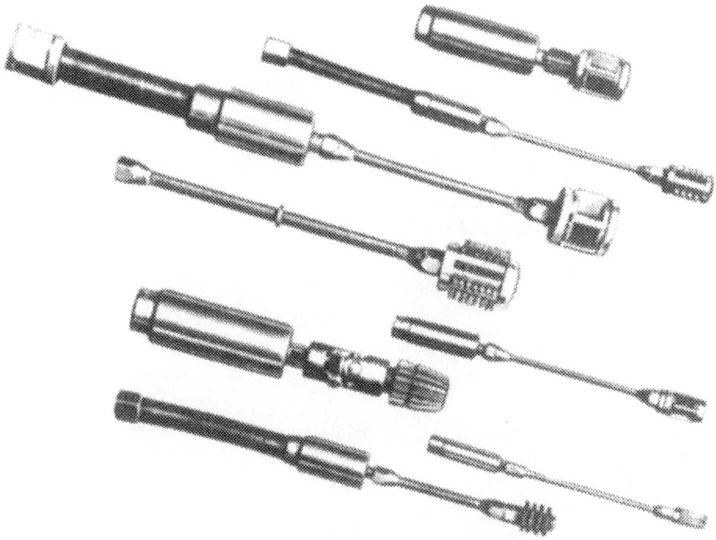

Figure 7.1 Drill heads and brushes. (*Elliott Company, Dayton, Ohio.*)

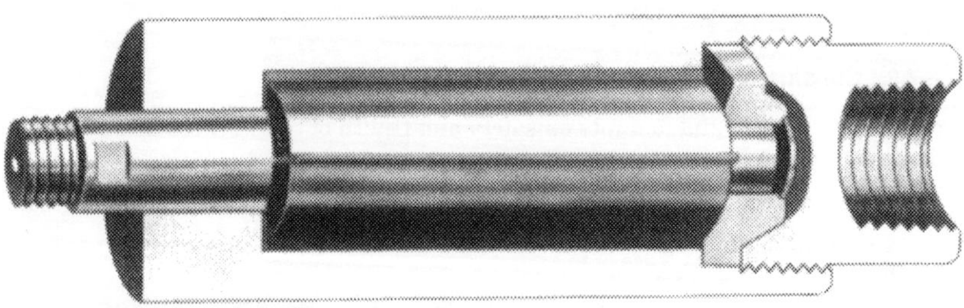

Figure 7.2 Internal cleaner motor. (*Elliott Company, Dayton, Ohio.*)

Figure 7.3 External (drill-type) cleaner. (*Indrisco, Springfield, Ohio.*)

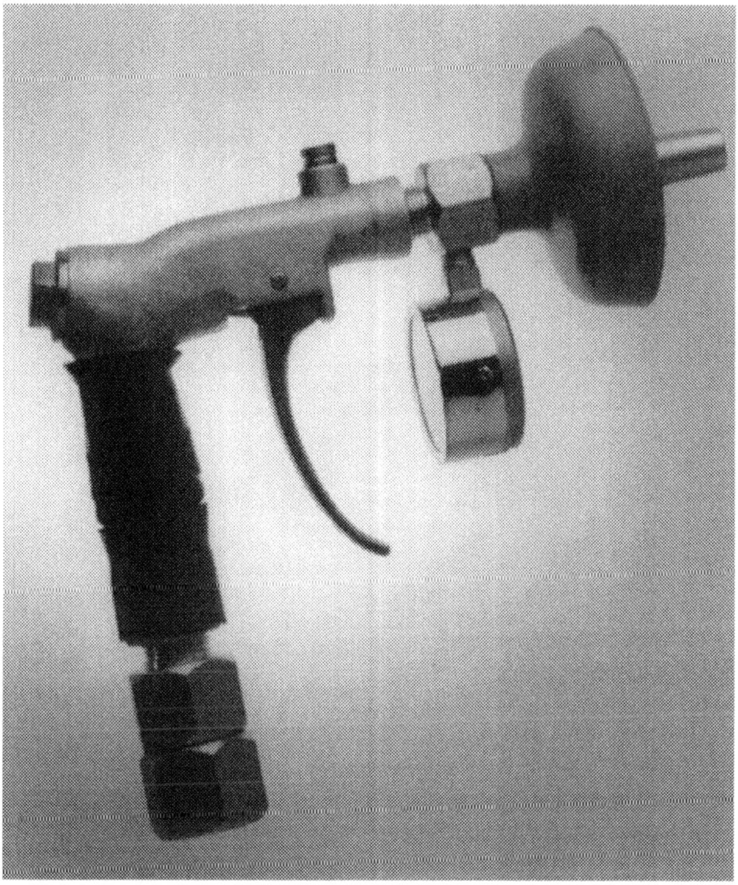

Figure 7.4 Water blowgun for projecting soft plugs through tubes. (*Conco, Verona, Pa.*)

Cleaners with internal motors, coupled to a cleaning head, enter the tubes and traverse their length as shown. Power is limited by the size of the housing that can fit into the tubes. External-drive cleaners are driven by motors or gearmotors external to the tubes.

Use blowguns and brushes or plastic tube cleaners on straight-tube units when the tubes are coated with soft easily removable deposits. Remove harder scales with Conco scraper type plugs illustrated in Figs. 7.5*c* and *d*. The plug is blown through the tubes by release of air or hydraulic pressure. The plugs are caught in a basket at the far side. The tube inside diameter must be constant throughout the tube length.

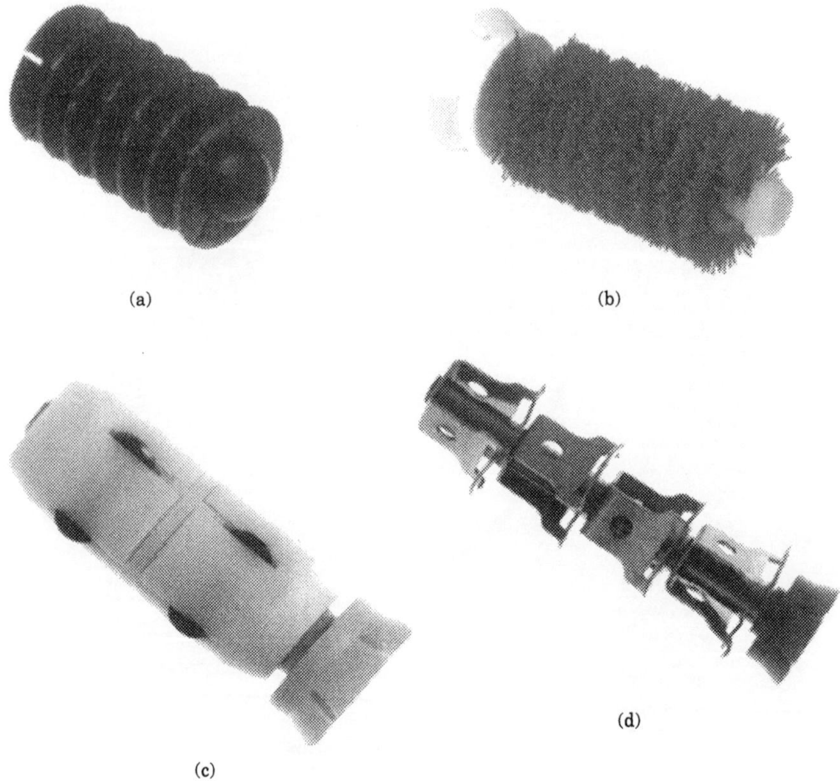

Figure 7.5 (*a*) Conco plastic tube cleaner; (*b*) brush cleaner; (*c*) calcium carbonate tube cleaner; (*d*) Conco stainless-steel tube cleaner. (*Conco, Verona, Pa.*)

Considerations for off-line mechanical cleaning. Mechanical cleaning with drills, brushes, and blowguns is primarily laborious and does not require high skill levels. When sufficient labor and supervision are available, it is generally economical for the plant maintenance force to clean mechanically. The investment is in the cost of the equipment and facilities for safe disposal of foulants removed from the unit. There is also some cost of safety engineering to protect personnel and property.

On-line mechanical cleaning

The simplest on-line cleaning system is to install piping and valves that allow periodic backflushing. Piping for backflushing is sized smaller than the regular inlet line. Provide at least enough backflushing flow to achieve a minimum velocity in the tubes of at least 1 m/s

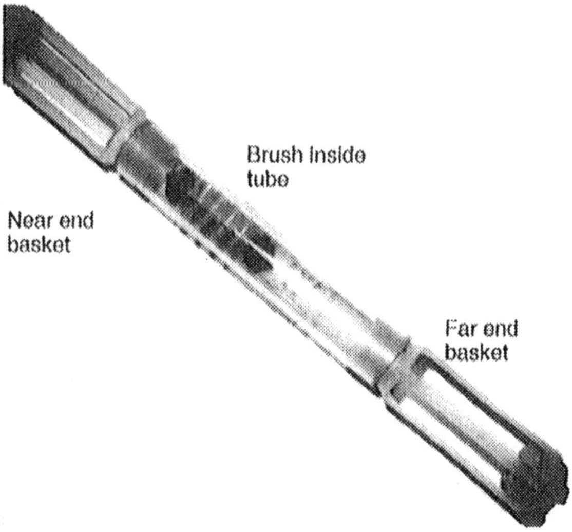

Figure 7.6 Demonstration specimen of brush and basket for on-line cleaning, showing baskets fitted into a plastic tube containing a brush. (*WSA Engineered Systems, West Allis, Wisc., photograph by MGT Inc.*)

(~3 ft/s) and preferably 1.5 m/s (~5 ft/s) to get the cleaning action needed to maintain the tubes clean.

This long-used system was improved on for straight-tube application by a system in which spirally wound brushes slightly larger in diameter than the tube interior are placed in the tubes and a basket is press-fitted into each end of the tube as shown in Fig. 7.6. The twisted wire central to the spiral brushes terminates in tapered nosepieces fore and aft. Normal flow sends the brush from the inlet-end basket to the outlet-end one. The spiral shape makes the brush rotate as it travels. Backflow also rotates the brush and sends it back to the inlet end. Each passage cleans away deposits. Using an automatic flow controller and diverter valves, you can set flow reversals on a timed cycle or manually reverse flow as needed to maintain cleanliness within the limits of the fouling resistances specified in the design.

An alternative to the basket-brush system, the recirculating sponge-ball, on-line system shown in Fig. 7.7, was developed for steam surface condenser cleaning. It is now also used in water-cooled process exchangers and to a lesser degree for maintaining cleanliness when other fluids circulate. This system consists of a device for injecting spongy balls into the inlet flow, dispersing them in the stream, collecting them after they pass through the exchanger, and recirculat-

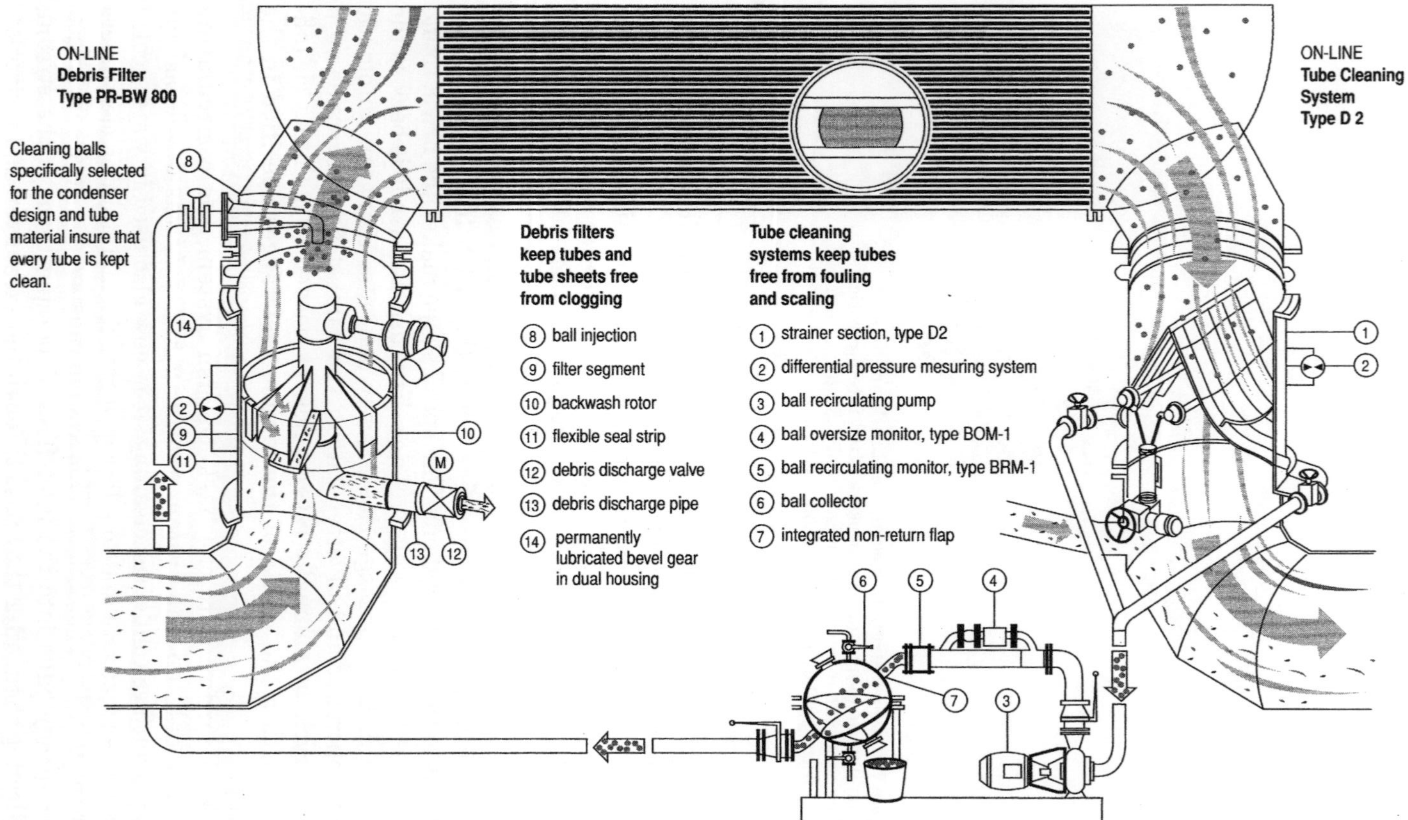

Figure 7.7 On-line ball cleaning system. (*Taprogge America Corp., Woodbury, N.Y.*)

ing and reinjecting them at the inlet. Circulation is through a distributor that directs approximately the same number of balls into each tube in a given period. The balls are slightly larger than the tube interior and are made of rubber or synthetic polymer. For maintenance cleaning, they are uncoated. However, when there is a sludge or scale accumulation, the system may be charged with balls coated with an abrasive.

Periodically, the balls are sized, and those that have become excessively worn are replaced (Fig. 7.8). Suppliers of these systems determine the initial ball charge based upon the number of tubes in the exchanger and circulation rate of fluid through the tubes. Typically the number of balls in the initial charge is about 10 to 15 percent of the number of tubes. The suppliers suggest that an average of 12 to 15 passes per hour is required to meet their normal cleanliness guarantees.

A primary concern with both systems is the need to minimize trash from entering the tubes. If your plant is located where cooling water may carry debris at various times during the year—for example, a river during spring floods, cooling towers during the fall season, and coastal areas during storm seasons—consider installing a debris filter ahead of these on-line systems.

Following is a summary of how the ball cleaning system works:

1. Balls of spongelike material, slightly larger than the tube I.D., are dispersed into the fluid before it enters the tubes. The balls must be compatible with the fluid circulated and be of approximately the same density.

Figure 7.8 Various types of balls used in on-line ball cleaning system. (*WSA Engineered Systems, West Allis, Wisc.*)

2. For the initial cleaning, the balls are coated with abrasive.
3. For regular continuous operation, the balls are bare.
4. The balls are collected at the discharge end of the unit and recirculated.
5. The system may be designed for continuous ball cleaning inside the tubes of a network of exchangers in which the same tube-side fluid is circulated.

Hydraulic and Chemical Cleaning

Hydraulic cleaning (hydroblasting)

In hydrocleaning,[1–3] powerful water jets are directed at fouled surfaces through nozzles or tips by lances and probes. The jets are not abrasive or directed at one spot long enough to erode it. However, abrasives can be added to the high-pressure water when needed to break up hard scales. The following materials and equipment are required for hydrocleaning:

1. Clean water.
2. High-pressure, high-volume pump and driver. Water pressures range up to 138,000 kPa (~20,000 psi). The usual upper limit is 69,000 kPa (~10,000 psi). Typical capacities and ranges are 35 m^3/h at 41,000 kPa [~150 gpm (gal/min) at 6000 psi] to 15 m^3/h at 69,900 kPa (~65 gpm at 10,000 psi).
3. Hydraulic hoses.
4. Lances or probes.
5. Control guns.
6. Nozzles.
7. Controls.
8. Safety shields, exclusion-zone ropes, and warning signs.

Manually and automatically operated equipment is available for cleaning inside and outside tubes. Figure 7.9 shows a typical hydroblast apparatus used to clean the outside of the tubes arranged to provide a cleaning lane of at least 6 mm (~¼ in). The bundle is placed on the rollers, and the nozzle manifold moves along the track, cleaning the bundle from end to end. Tightly spaced tubes and those on triangular pitch may not be capable of being cleaned externally.

Figure 7.10 shows a typical hydroblast apparatus using a rigid lance to clean inside the tubes. A dump foot gun is usually used for hands-free operation when using flex lances and hoses.

Figure 7.9 Hydroblast apparatus for outside of tubes. (*Butterworth Jetting Systems, Houston, Tex.*)

Figure 7.10 Rigid-lance hydroblaster. (*Butterworth Jetting Systems, Houston, Tex.*)

TABLE 7.3 Advantages Claimed for Hydrocleaning

Less time is required to organize personnel, equipment, prepare the unit, and perform the work than for other methods
The cost is less than for off-line chemical cleaning
There is less loss of metal than occurs as a result of the corrosive effects of chemical cleaning
There is no mechanical damage of the kind that can result from mechanical cleaning with drills and bits
Solidly plugged tubes may be bypassed in chemical cleaning, but hydrocleaned tubes are cleaned individually; with appropriate nozzles, solidly plugged tubes may be entered and cleaned
Disposing of foulants suspended in water and dilute water solutions is less hazardous than disposing of chemical cleaning solutions that contain foulants
The reaction between fouling products and water seldom produces noxious or poisonous gases; the vapors of chemicals used for cleaning and the products of their reaction with the foulants may be noxious or poisonous
It is not always possible to remove foulants chemically

Special training in safety and operation is required. It is recommended that contractors who specialize in the work be employed. Table 7.3 lists advantages claimed by proponents of hydraulic cleaning.

Considerations for hydrocleaning. The capital investment for hydrocleaning equipment is very substantial. Furthermore, workers who use the equipment must be very thoroughly trained. Despite the incorporation of "dead man" foot switches that shut down the high-pressure water when not depressed, the operation may be very hazardous. When flex lancing is used because there is insufficient space for rigid lancing, some means of preventing the stinger from backing out and causing injury and destruction must be provided.

Hydrocleaning organizations accumulate experience and keep current on safety requirements. Their personnel usually undergo regular training in maintaining and safely using the equipment. Furthermore, they have on hand a variety of nozzles and tips not generally available to operating companies' maintenance forces. Unless you can justify the capital investment and cost of training, it is desirable to employ the services of a reliable contractor.

Some exchangers may handle a product that fouls the tubes so badly and so quickly that cleaning is required daily or weekly. If you cannot avoid this problem with an exchanger by redesigning or changing the operation, it may be cost-effective to procure the equipment and train maintenance workers to do the hydrocleaning. Be very sure to study the hazards and safety requirements.

Chemical cleaning

The chemical cleaning[4] methods listed in Table 7.4 are used to preclean and pretreat new and retubed units,[5] remove fouling deposits and scales during operation (on-line cleaning), and clean units off line. When possible, it is efficient to do off-line chemical cleaning in place. Blind all lines leading to and from the unit beyond the chemical cleaning connections. Neutralize and thoroughly flush before removing the line blinds.

Table 7.5 lists reasons for precleaning and pretreating new units and pretreating newly cleaned equipment. Pretreatment must be

TABLE 7.4 Chemical Cleaning Methods

Water flushing
Solvent washing
Detergent washing
Inhibited-acid washing
Alkali washing
Bactericidal treatment
Combinations of the first six items listed

TABLE 7.5 Reasons to Preclean and Pretreat

1. Removing internal mill scale, loose weld spatter, and slag.
2. Removing dust, grit, and other pollutants, which foul the unit before start-up, collected on protective oily tube coatings.
3. Removing protective varnishes from pipe-shell interiors.
4. Removing foreign matter introduced in the shop, for example:
   - Weld rod stubs
   - Soapstone sticks
   - Marking crayons and chalks
   - Wiping rags
   - Articles of clothing
   - Remnants of food and food containers
5. Removing organic and inorganic substances deposited by hydrostatic test water.
6. Promoting formation of corrosion-resistant films, both on surfaces that depend for corrosion resistance on naturally forming films and on carbon steels.
7. Removing sensitizers that disrupt corrosion-resistant films, i.e., iron dust on austenitic tubes.

coordinated and compatible with the fluid system. Use either pumped recirculation or air agitation for dispersing cleaners and inhibitors.

On-line treatment for biological fouling associated with marine growth has traditionally been accomplished by chlorine injection into the cooling water. Recent environmental regulations have severely restricted the permissible chlorine levels in the discharge from water-cooling systems.

One solution to these limitations is to employ a *targeted chlorination system,*[6,7] which uses a manifold array of chlorine injection nozzles in the water box set close to the tubesheet. High concentrations of chlorine are pumped through individual nozzles on a timed sequential basis, killing marine growth in a very limited region of the tube layout. Total residual chlorine (TRC) easily meets the regulated discharge-level limits when it mixes with the total exit water flow. Tests have shown that this procedure expends significantly lower quantities of biocide. Other on-line targeted injection treatment chemicals have also been used to control biofouling.[8]

Considerations for chemical cleaning. In situ chemical cleaning requires handling cleaning agents and knowing their effects on the foulants and the materials of construction. It is generally cost-effective to implement the cleaning procedure and set up the pumps, valves, and other equipment with the plant workforce. However, it may be desirable to contract with a firm that produces cleaning agents for consulting on the most suitable agents to use and how to use them.

It is usually more cost-effective and desirable for disposing of waste chemicals to use contractors for chemically cleaning bundles removed from exchangers. Such cleaning consists of immersing the bundle in a bath of strong cleaning chemicals and subsequently flushing. Contractors who do this kind of work are equipped with appropriate tanks, heating systems, ventilation, and effluent waste-disposal facilities.

Deciding on Contract Cleaning Organization or Plant Maintenance Force

Whether to perform maintenance with the plant workforce or to hire a maintenance organization is discussed in Chaps. 8 and 12. Deciding whether to clean exchangers with the plant workforce or to employ the services of a contract cleaning organization requires answering some of the same questions asked in those chapters. Here, the decision depends on the required training, facilities for doing the work, scheduling requirements, availability and cost of plant maintenance labor, and safety considerations. The following covers questions to ask when doing the benefit-cost analysis used to make the decision:

- How much time is available to clean the exchanger? If a clean and tested spare bundle is available, you can clean and refurbish the bundle after the outage is complete and the plant back on line.
- If the exchanger has to be cleaned in place, can it be done with mechanical and/or high-pressure water-jet systems?
- Some mechanical systems require only a nominal capital investment and a short training period for safe and effective implementation.
- When chemical cleaning is required, it can be done either on or off site. We recommend that a contractor, regularly engaged in the business, be hired if at all possible. Their work crews are more experienced in the handling of the chemical than the plant force. Furthermore, they generally have arrangements in place for disposing of the used, foulant-laden chemicals. This may be the most expensive component of this cleaning option.
- Water jetting has the least environmental impact. Therefore, it is the preferred cleaning method if it can remove the foulant.
- A combination water jetting–mechanical device will usually remove all except the hardest scale from inside the tubes. Figure 7.11 shows such a device in use.

Figure 7.11 Rotary lancing machine in use to clean tube interiors. (*Stone Age, Durango, Colo.*)

References

1. "Halliburton Industrial Cleaning Services Industrial Catalog IC12029 (Revised)," Halliburton Services, Duncan, Okla. See also Industrial Technical Data Sheets IC12054, Troigon Service, and IC12047 HT-150, Industrial Cleaning Trailer.
2. "New Tool for Cleaning Tube Side of Heat Exchangers," Editors, *Maintenance News Lett.,* **5,** Oct./Nov. 1980.
3. D. Landry, "Turnaround Time to Clean Tube Bundles Cut by 40% Using Less Manpower," *Chem. Process.,* Aug. 1981.
4. J. P. Engle, "Cleaning Boiler Tubes Chemically," *Chem. Eng.,* Oct. 18, 1971.
5. G. W. Bardley and J. G. Smith, "Removal of Iron Oxide Deposits without Ferritic Ion Corrosion," *Mater. Protect. Perform.,* **11,** 48–51, Nov. 1973.
6. Condenser Biofouling Control—State of the Art Symposium," *EPRI Proceedings,* Nov. 1985, EPRI Report CS-4339.
7. C. Thompson, J. Carbonara, Y. Mussali, W. Chow, and B. Syett, "Advanced Targeted Chlorination Design," 1993 EPRI Condenser Conference.
8. M. Jakabowski, K. Small, and Y. Musalli, "Targeted Bromination at Brayton Point," 1993 EPRI Condenser Conference.

Chapter

8

Record Keeping; Preparing for Repairs and Alterations

Introduction

This chapter begins by describing files to set up and maintain to provide a complete history for each exchanger that enables evaluating the condition of the equipment. Then it discusses scheduling maintenance and repairs. The chapter closes with some recommendations for preparing to make repairs and alterations.

Organizing Inspection and Maintenance Records

Files need to be organized and faithfully maintained to optimize exchanger service life. The investment in setting up these files will be repaid many times during the lifetime of the equipment. Having ready access to records of previously performed repairs and alterations and to agreed-on repair/alteration procedures minimizes downtime and contributes to long, useful life.

Engineering, operating, and maintenance groups must agree on which records are necessary, how to maintain them, where and in what form to keep them, and the most efficient way to retrieve and refile. Cooperation among these groups is essential because (1) operating people must review the engineering files to understand what the exchanger is supposed to do and how it was designed to do it; (2) maintenance workers should be able to look at the operating logs and operators' notes about problems to determine what must be fixed or altered to restore a tubular to its required capacity; and (3) engineering groups must be able to research operating and maintenance records to assess how designs are performing, to troubleshoot the causes of problems, and to assess the results of repairs and maintenance.

Formats and systems

A variety of formats and systems are available for record keeping, from simple card files to computerized storage of scanned documents, data, and records.

Manual systems. A shipboard record-keeping scheme called the *Machinery Index,* which the U.S. Navy used for many years, exemplifies a manual system. Here, large paper forms were bound into a book. The forms identified each piece of machinery or equipment by its identifying tag number and listed nameplate data, drawings, catalogs, and other information and the location of these documents. As inspections were made and progressive or repair maintenance was done or alterations made, the hours of operation, description of the work, and notes about replacement parts, etc., were entered. Individuals who needed a complete history of a piece of machinery or who needed to examine drawings, parts lists, and so on could quickly find the necessary information by using the tag on the equipment as a key.

A similar civilian system uses the tag number or other item-identifying system as the key to filing and retrieving inspection and maintenance records. This also facilitates accounting and relating costs to the equipment. Typical practice has been to have a card filing system in place of bound sheet forms. Essential data for each exchanger (and other kind of equipment) are entered in a description box. Beneath the description are data entry lines that give a chronological record of inspections and maintenance. These data are tied to the maintenance job order, inventory tracking, and procurement systems. The card system is backed by a parallel set of similarly tagged and identified file folders, in which are stored copies of all the relevant documents for each equipment item. Manufacturer's installation, operating and maintenance, and repair manuals may be stored in the file folder if they are not too bulky. Otherwise they are appropriately tagged and stored on shelves. Tracings produced by the company's engineering department are usually numbered using a system that identifies the project number, job or work order number to which they apply, the specific item, and the responsible department.

Many heat exchanger manufacturers number their drawings in a way that keys them to their shop order number, which forms part of the serial number stamped on the nameplate. A typical drawing numbering system for a heat exchanger factory includes standard drawing sizes as part of the drawing number. Shops generally issue standard size copies of manually or computer-drawn drawings, with easily read print sizes to the people on the factory floor. This is because the factory workers must be able to read the prints under factory lighting conditions.

Standard drawing sizes in the United States are in inches: A = 8½ × 11, B = 11 × 17, C = 17 × 22, D = 22 × 34, E = 34 × 44, etc. This permits folding all drawings larger than A size to 8½ × 11 in, to fit into standard filing folders and mailing envelopes. Some standards for manual drafting require views and details to fit neatly into multiples of A-size space to enable issuing details neatly to the factory floor. This is not necessary for computer-generated drawings, since parts of a drawing showing the desired details can readily be printed out to a size that the plotter is capable of producing.

Although computer-generated drawing is becoming widespread, many drawings are still made on tracing paper. Many originals, and often prints, are filed in drawing file cabinets compartmented by the standard drawing sizes.

Many exchanger shops facilitate traceability by including the shop order number/serial number as part of all engineering and sales documents, specifications, purchase documents, manufacturer's data report forms, etc. If a buyer's purchase order covers several nearly identical exchangers, the factory usually adds alphabetically lettered suffixes to the shop order/serial number to identify the individual units. If a customer's order covers many diverse pieces of equipment, the exchanger shops generally issue separate shop orders for each size and type. However, they may also use any one of a variety of other schemes. Each shop also has a set of standard drawings for such things as weld details, davits for handling channels and channel covers, supports of various types and procedure specifications, NDE standards, and other material to which the job drawings refer.

A typical set of drawing numbers and titles for the previously discussed fictional Wonderful Heat Exchanger Company's fictional job no. 2301 might be as follows:

Drawing no.	Rev. no.	Title
J-2301-1-B	1	Setting plan for size 33-244 type AES horizontal oil cooler for Fictional Customer, Fictional Location
J-2301-2-D	3	Assembly of size 33-244 type AES horizontal oil cooler for Fictional Customer, Fictional Location
J-2301-3-D	2	Bundle details for size 33-244 type AES horizontal oil cooler for Fictional Customer, Fictional Location
J-2301-4-D	4	Shell, channel, return cover, and shell cover details for size 33-244 type AES horizontal oil cooler for Fictional Customer, Fictional Location
J-2301-5-C	2	Gasket and bolting details for size 33-244 type AES horizontal oil cooler for Fictional Customer, Fictional Location

The drawing number sequence and sizes vary with the type and size of exchanger and relevant details. Since drawings are abstractions intended to convey pertinent information and generate a mental picture in the viewer, it is desirable that they be to scale. However, rely only on the printed dimensions, since it is unlikely that drawings will be completely true to scale, even when computer-generated. Many manufacturers will use the acronym *NTS* (not to scale) liberally to discourage scaling drawings. Others may use standard blank, not-to-scale drawing forms for assemblies and details and simply fill in the dimensions and other necessary data needed by their customers and factory personnel.

If Wonderful were to build a set of several heat exchangers to the numbered drawings listed above, for example, a stack of three shells in series, the serial number for the first shell would be 2301-A; for the second shell, 2301-B; and for the third shell, 2301-C. They would probably issue separate drawings to show the differences between the three shells. If they received an order for a duplicate set of exchangers several years later, they would most likely copy the drawings either into a new computer file or by means of sepias and reissue them with new numbers. Other shops might just reissue the original drawings, revised to show that a second set of exchangers in series was built. In this alternative, the serial numbers of the second set would be 2301-D for the first new shell, 2301-E for the second shell, and 2301-F for the third.

Any one of a variety of other systems can prevail in different shops, but nearly all will tie serial numbers to the manufacturing drawings, exchangers, purchase orders, and other documents. However, National Board stamping begins with the National Board number of the first piece of equipment that a given shop registers with the National Board and proceeds serially.

Computerization. Modern computer data management and bar-coding technology and technology for copying, storage, and retrieval have made manual practices obsolescent, although card files and similar systems are still widespread. Eventually, manual filing will become obsolete.

Since powerful, relatively inexpensive desktop computers are widely available and local area networks (LANs) abound, we recommend keeping readily retrievable computerized records with backups. Depending on the circumstances, you can maintain the records on the hard drives of desktop personal computers (PCs) or in the storage facility of a network central processor where they are accessible to all PCs on the LAN. To make the system work properly, maintenance personnel must update the records as inspections are made or they perform work. Tied to the spare-parts inventory tracking system, a computerized system can expedite maintenance. Terminals may be located conveniently to give supervisors and mechanics ready access to the drawings, records, parts and tool inventories, etc. Hard copies can also be generated. There is an increasing number of commercially available programs to perform these functions. Because the needs of individual plants and organizations vary, most such programs can be custom tailored to reflect the specific needs of a facility.

Backing up files. The procedure in long time use for backing up files is to microfilm the original documents and to store the originals safe from damage or deterioration. The microfilms of the documents can be retrieved and printed out as required. With appropriate equipment, a complete up-to-date data file is thus available from backup. However, retrieving microfilm records and printing them out in useful form is time-consuming.

Scanning devices and computerization now permit storing all relevant maintenance files, drawings, photographs, tables, etc. on various media and transmitting them between far-apart locations electronically. You can even scan the National Board forms referred to in Chap. 6 into the computer for retrieval. With this system, search for and retrieval of maintenance documents is a matter of minutes instead of hours. Furthermore, drawings and photographs can be displayed graphically on monitors. Inexpensive color printers make printing out of enlarged photographs simple.

Documents to include in inspection and maintenance files

The following paragraphs suggest what to include in inspection and maintenance files.

Specification and procurement documents. Start by retrieving the final issue of the owner's or architect-engineer's original requirements specification, purchase specifications, and purchase documents. For process tubulars, they should include the complete information as shown on TEMA or API 660 heat exchanger specification sheets, including descriptions of stream flows and mechanical design and materials of construction data. For closed feedwater heaters, they should contain the HEI closed feedwater heater specification sheet.

Purchase specifications should have incorporated any of the owner's or architect-engineer's standard heat exchanger and pressure-vessel specifications and information related to performance guarantees. Purchase documents should have clearly stated what kind of inspections were to be performed and the document package that the manufacturer was required to provide. When feasible, include the inspection brief that was written for the exchanger(s).

These documents provide a roadmap to the path of exchanger procurement and production.

Manufacturer's original document package. Most of the information needed to compile and organize a complete document package is nearly always available in the manufacturer's files at the time of fabrication. Well-written purchase specifications require manufacturers to provide a data package to be delivered with the tubular. It should include the documents discussed in the following paragraphs.

Drawings. The document package should contain all the manufacturing drawings that will be required by the purchaser's maintenance force to perform any inspection, maintenance, repairs, and/or alterations that can reasonably be anticipated over the life of the equipment. These include setting plans, assembly, and detail drawings. Manufacturers that have developed design features they consider proprietary may require purchasers to execute confidentiality agreements to assure them that only the purchasers' maintenance forces will use the restricted information.

All the drawings should contain the information listed in Table 8.1 presented in checklist format. Heat exchangers with special features will require other items that should be included.

The setting plan drawing requires all the information listed in checklist format in Table 8.2. Assembly drawings should have all the information listed in the checklist in Table 8.3.

For most exchangers the accompanying detail drawings should show the information listed in the checklist in Table 8.4.

Approved procedure specifications. The package should include any approved procedures used in manufacturing and those recommended

TABLE 8.1 Information that Should be Shown on All Manufacturers' Drawings

- ☐ Manufacturer's name and address and shop location
- ☐ Drawing number and date; revision numbers and revision dates
- ☐ Reference drawing numbers and titles
- ☐ Owner's name and address, plant location, purchase order number, and item number
- ☐ Main centerlines of all views
- ☐ 0–180° and 90–270° centerlines of all end and plan views of bonnets, channels, tubesheets, baffles, support plates, and shells
- ☐ Boltholes identified as straddling or on centerlines
- ☐ Identification of working points

TABLE 8.2 Information that Should be Shown on Setting Plan Drawings

- ☐ All information listed in Table 8.1
- ☐ Service
- ☐ Size and type designations, operating position
- ☐ Gross and net effective surfaces
- ☐ Construction code, stamping requirements, registration data
- ☐ Design pressure and temperature, test conditions
- ☐ Nameplate stamping; nil-ductility temperature restrictions
- ☐ Overall dimensions
- ☐ Lifting lug and trunnion details
- ☐ Location of center of gravity full of water, dry and empty
- ☐ Clearances required for pulling tubes and removable bundles
- ☐ Working points coinciding with those on other drawings
- ☐ Axial distances from main working point to nozzles and supports and between adjacent nozzles and supports
- ☐ Radial distances to nozzle faces and supports from axial centerline and distances between faces
- ☐ Pitch angle for pitched units
- ☐ Pitch distance to pitched members
- ☐ Weights full of water, dry and empty, removable bundle or shell where applicable
- ☐ Axial, perpendicular and torsional nozzle loads (when provided by architect-engineer or owner)
- ☐ Support bolting pattern and expansion allowance
- ☐ Reactions at supports full of water, dry and empty and at operating conditions
- ☐ Shimming allowance for stacked units
- ☐ Flow direction arrows at nozzle centerlines

TABLE 8.3 Information that Should be Shown on Assembly Drawings

- ☐ All information listed in Tables 8.1 and 8.2
- ☐ Applicable manufacturer's standards (HEI, TEMA class, etc.)
- ☐ Materials of construction by applicable construction code specification number
- ☐ Corrosion-erosion allowances
- ☐ Nondestructive examinations to be performed
- ☐ Surface finishes and coatings
- ☐ Precautionary notes
- ☐ Sequence of assembly and disassembly

TABLE 8.4 Information That Should be Shown on Detail Drawings

- ☐ MATERIALS SPECIFICATION NUMBERS
- ☐ BUNDLE DETAILS
  - ☐ Tubesheet
    - ☐ Thickness
    - ☐ Outside diameter (O.D.)
    - ☐ Bolt circle diameter (BC); number, location, and diameter of boltholes
    - ☐ OTL or OTC
    - ☐ Drilling template and tolerances
    - ☐ Hole grooving and edge preparation
    - ☐ Machining details
    - ☐ Locations of pass-partition grooves
    - ☐ Details of attachments to the shell partitions of C- and N-type heads
  - ☐ Bundle handling and pulling details
    - ☐ Center of gravity and lift points
  - ☐ Details of attachments of C and N channels to tubesheets
  - ☐ Skirts and attachments of floating tubesheets
  - ☐ Skeleton
    - ☐ Cross-flow baffle and support dimensions and cuts
      - ☐ Hole drilling and finish
      - ☐ Cutoff lines, notches, ears, and slides or rails
      - ☐ Impingement protection details
    - ☐ Longitudinal baffles (with seals or welded in, shrouded zones and end plates)
    - ☐ Special baffle constructions, attachments, and sealing devices
    - ☐ Shrouding and separator details
    - ☐ U-tube schedule
      - ☐ Row number
      - ☐ Tubes in row
      - ☐ Bend radius
      - ☐ Straight length
    - ☐ Tube diameter, gage, and specification
- ☐ SHELL
  - ☐ Dimensions and transition section angles
  - ☐ Vapor belt, vapor dome, and expansion joint details
  - ☐ Impingement protection details

TABLE 8.4 Information That Should be Shown on Detail Drawings (*Continued*)

- ☐ Main body flange details
- ☐ Nozzle construction details
- ☐ Locations of weld seams and cut lines
- ☐ Details of weirs and flow-control devices
- ☐ Stuffing box, lantern ring, and gland details
- ☐ Entrainment separation and vortex control device details
- ☐ Supports and attachments details
- ☐ Nameplate, clip, and appurtenance details

☐ CHANNEL, BONNET, RETURN COVER, AND SHELL COVER DETAILS

- ☐ Dimensions and transition section angles
- ☐ Erosion protection details
- ☐ Flange and closure details
- ☐ Bolts and gaskets
- ☐ Nozzle details
- ☐ Locations of weld seams
- ☐ Pass-partition attachment details
- ☐ Diaphragm and pass-partition cover construction

by the manufacturer for maintenance and repairs. The common ones are:

- Welding procedures
- Repair-welding procedures used during construction and recommended for maintenance
- Tube-to-tubesheet joining procedures
- Tube-to-tubesheet joint repair procedures used during construction and recommended for maintenance
- Heat-treating procedures
- Nondestructive examination procedures
- Preassembly cleaning procedures
- Testing procedures used during construction and those recommended to be performed after repairs or alterations
- Posttest cleaning procedures
- Procedures for preparation for surface coating of exterior

- Coating procedures
- Procedure for preparation for shipment
- Procedure for draining and drying after factory testing
- Procedure recommended for draining and drying after in-service testing
- Recommended procedures for inerting during shutdowns for maintenance and for interrupted production
- Procedures for skidding, crating, boxing, and use of shipping shells
- Procedure for meeting special shipping requirements

Manufacturer's data report. Manufacturer's data reports, discussed in Chap. 6, provide a quick rundown of design and test conditions, materials of construction, sizes, nozzles, and other information about the construction.

Nameplate data. Nameplate rubbings are valuable because installation placement frequently makes access to the nameplates difficult.

Certified mill test reports and location map of all pressure-bearing materials. Mill test reports are especially important for these parts:

- Tubes, including locations of heats if exchanger is tubed with more than one heat. This is usually the situation when a unit has more than 120 tubes. (See the heat map in Fig. 1.3.)
- Tubesheets.
- Channel, shell, return-cover, and shell-cover flanges.
- Channel barrels.
- Shells.
- Welding consumables.

Heat-treatment charts and records. Heat treatment charts will generally be in the form of circular or strip charts. It is important to have records of temperatures at various locations in the furnace.

Manufacturer's records of nondestructive examinations and tests. These records should include detailed descriptions of the procedures used and the results. The package should include the following:

- Visual examinations
- Radiographs (when dual-radiograph technique is specified)
- Fluid penetrant examinations
- Magnetic particle examinations

- Eddy-current tests
- Ultrasonic scans
- Gas and bubble tests
- Helium leak mass spectrograph examinations

Manufacturer's manual for installation, operation inspection, and maintenance. The manual should include the following items:

- Exchanger data sheet or feedwater heater data sheet showing the design basis; alternative data sheets predicting performance for alternate modes of operation; data sheets or supplements showing effects of overloads, bypassing shells in multishell units, and similar information.
- Schematic installation sketch.
- Recommendations for safe operation. The manual should include discussion of and warnings about the hazards of operating pressure vessels and equipment that handles dangerous substances. In addition, it should have warnings about the effects of exceeding the exchanger's design basis.
- Schematic of liquid level for units in which control of liquid level is required for successful operation. The text should fully discuss liquid-level control, including how to verify it and the consequences of improper level.
- Recommended inspection and testing schedules.
- Original tube thickness measurements and recommendations for frequency of making thickness measurements.
- Recommended frequency of performing hydrostatic and other leak tests.
- Recommendations for inspecting shell, channel, and bundle for erosion and corrosion; these recommendations should include suggestions for making measurements both on and off line.
- Recommendations for tubesheet, tube, and tube-to-tubesheet joint inspections and repair and tube plugging procedures. By far, the largest number of exchanger failures occur in the tubes, tubesheets, and tube-to-tubesheet joints.
- Procedures for examining and repairing ligaments.
- Procedures for locating leaking tubes.
- Procedures for locating the axial position of tube leaks. These may include borescopy, using the MGT Inc. axial tube leak position finder (see Fig. 4.1), or other means.

- Procedures for plugging tubes; recommendations for insurance plugging.
- Sketches of and information about the kinds of plugs to use.
- Description of original manufacturer's tube plugging map and recommendation for maintaining tube plugging map during life of unit.

Reports of failures and failure analyses

The actions taken to restore production when an exchanger fails affect its subsequent maintenance and life. At one extreme, restoring production is so urgent that the failed part is replaced, bypassed, or plugged off as quickly as possible and the unit put back on line. No effort is made to determine the cause(s) of the problem. At the other extreme, production, maintenance, and engineering cooperate in analyzing the failure, determining how to effect repairs, and the appropriate course of action to avoid future outages. In either case, write reports to assist future troubleshooting and maintenance. Arrange these reports chronologically together with the records of welded repairs and alterations.

Records of welded repairs and alterations

The forms used for recording welded repairs and alterations made to the NBIC have been discussed previously. The National Board keeps R-1 and R-2 forms showing repairs and alterations attached to the original manufacturer's data report form on file. You can obtain copies from the National Board by mail or telecopier. However, to avoid losing time during an unplanned outage, it is better to have copies on hand.

For repairs and alterations to the API Inspection code, it is a practical necessity to keep the records of repairs and alterations in the maintenance files. Retrieving copies from the jurisdiction or insurance agency responsible for on-site inspection may be very time-consuming.

Deciding Whether to Use In-Plant Maintenance Force or to Contract a Repair or Alteration Organization

Deciding whether to do maintenance with the in-plant workforce or to employ the services of a contract repair organization depends on the answers to these questions:

1. Is the plant workforce qualified to do work that must conform with jurisdictional, inspection code, and inspection agency requirements? Most owners will have to hire a repair or alteration organization that meets the jurisdictional and code requirements if the work to be done consists of repairs or alterations to the pressure parts in accordance with the NBIC or API inspection code and requires the on-site AIs and jurisdictional approval. Depending on its authorizations, an owner or user that has an Owner-User Inspection Agency in place and employs an Owner-User Inspector may be able to do the work. Even when the in-house maintenance organization satisfies the jurisdiction's, the inspection code's, and the AI's requirements, the workers may not be qualified to do the work.
2. Is the plant workforce trained, and does it have the necessary skills to perform procedures not governed by the jurisdiction, inspection code, or inspection agency responsible for inspecting the exchanger? A typical procedure that is often poorly done because of insufficient training and practice is rerolling tubes in expanded tube-to-tubesheet joints.
3. Can the plant maintenance people identify the kinds and extent of problems they may encounter while doing the work? Examples are (*a*) recognizing that parts of a tube bundle cage should not be reused because of baffle-tube support deterioration or excessively corroded tie rods and spacers when retubing a bundle and (*b*) perceiving that the surfaces under washed away tube-to-tubesheet welds may also be wormholed.
4. If an unanticipated problem is encountered, can the plant crew handle it? If it is planned to use in-house staff to retube a bundle, but it is then discovered that replacement tie-rods, spacers, baffles, and tube supports are required, there may be a substantial delay in the work if replacement materials are not on hand or plant facilities are not equipped to produce the parts.
5. Is the equipment required for the proposed maintenance procedure on hand, or must it be rented or purchased? If it is not on hand, how long will it take to get?
6. Are there enough workers and supervisors to perform the work in the allotted time?
7. What are the safety considerations?
8. Will it be necessary or desirable to hire a heat exchanger specialist to consult about the repairs or alterations?
9. What is the availability of the maintenance contractors specialized in doing the work?

10. Does the contractor have regular, full-time employees, with skill and experience in heat exchanger maintenance work, or are workers hired for the job from the local labor pool?
11. Can the contractor complete the work in the time required?
12. Are there any labor union contract restrictions?
13. How do the costs of doing the work with the maintenance staff compare with contracting it out?

Materials Documentation

Most original purchase specifications require the manufacturer to furnish mill test reports for all parts of the pressure envelope. Equally important for maintenance are those for non-pressure-bearing parts such as zone enclosures, alloy baffles, and supports. Make sure to tie the mill test reports to the bill of materials. When retubing, repairing, or altering an exchanger, make sure to document all materials used. Maps of the locations of these materials are very helpful in future maintenance and troubleshooting problems attributable to materials.

Tube heat maps

A sketch of a tube heat map was shown in Fig. 1.3. Make a similar tube heat map that shows the location of the heat or heats of tubing that replaced the removed tubes for partially retubed exchangers.

Maps of locations of critical construction materials

When there is a hurry to get on with a repair or alteration, it is very helpful to be able to tie heat numbers on the various parts of the exchanger to the mill test reports in the document package. However, trying to find the heat numbers may be tedious and time-consuming. The work will be expedited if the procurement specification requires the manufacturer to provide a sketch that shows where each heat of material listed on the mill test reports has been used.

Weld seam maps

Most tubulars are fully insulated. When searching for a leak or when a repair or alteration is to be performed, it is very important to know where the shell and channel seams are. If the specifications have not required the manufacturer to produce a weld seam map, mark up a drawing, or make a sketch for the maintenance file that will show this information. In procuring new exchangers or replacement shells or channels, require the manufacturer to show the seam locations. It is also helpful to have the manufacturer note the welding procedure number(s) of the approved welding procedure(s) used to make the seams.

Documentation of Procedures

Procedures were discussed in various contexts in earlier chapters. Whether maintenance staff or a contract maintenance organization makes repairs and alterations, be sure that approved procedures, qualified as required by the NBIC or API Code, are on hand and available to the people who do the work. The following paragraphs discuss some important ones.

Welding procedures

All welding procedures, used in maintenance, whether for repairs or alterations to an inspection code, should follow the requirements of ASME Code Section IX, the NBIC, or the original construction code. They should state the kind of welding that the procedure covers, the thickness range, electrical characteristics, position in which the work is done, parent metal P numbers, and electrode classifications. The procedures should indicate the joint preparation, speeds of welding, bead sizes, electrode sizes, gas cup sizes, shielding gas compositions and flows, and all the other variables involved in welding. It is important to understand that welding procedures list two kinds of variables: essential and nonessential. *Nonessential* variables may be changed without requiring requalification; *essential* ones may not be.

If you plan to use a new procedure, for example, for manual metal arc weld repairing tube-to-tubesheet welds made by gas tungsten arc welding, it will be necessary to make and test specimens in order to qualify the procedure and the welder who performs the welding.

Note that in order to seal-weld tubes to tubesheets, under the ASME Code Committee's interpretation of the Code rules, the procedures, and the welders who use them, must be qualified.

Tube-to-tubesheet joining procedures

Except when the tubes are welded to the tubesheet, the ASME Code does not now require qualifying tube-to-tubesheet joining procedures. Nonmandatory Appendix A has some requirements for testing joint strength when the tubes are intended to stay the tubesheets. The reliability of an exchanger depends on many connections of tubes to tubesheets. Therefore, it is very important to have carefully thought out, repeatable procedures for making and repairing such joints.

Tube-joining procedures should incorporate, or include by reference, subsidiary procedures for cleaning and drying the tube ends and tube holes. For tube rolling and hydroexpanding, torque or pressure settings should be tied to tube yield stress. This means that if the tube manufacturer supplies more than one heat of tubing, you have to adjust the settings in accordance with the variation in yield stress from heat to heat. Because torque controls may drift, the procedure should require regular dynamometer checking (torque analyzing) and adjustment. Be sure that the procedures are available to the workers doing the joining or joint repairing.

Explosive expanding procedures require establishing the appropriate charge size for each heat of tubing.

Heat-treating procedures

Depending on the material of construction and its thickness, cutting into and rewelding the pressure envelope may require heat treating. It may not be possible to remove the unit from its site for the necessary heat treatment. On-site heat treatment may be performed using electric blankets. It is very important to review the procedure to be used, especially its description of how the temperature is to be measured, controlled, and recorded, and how the structure is to be cooled. The heat treatment must restore ductility to the pressure-retaining shell and channel parts. At the same time it must not adversely affect the properties of the tube metal.

Scheduling

Exchangers are part of a network of interdependent equipment components. Consequently, from the heat exchanger's perspective it is rarely possible to inspect off line, do needed maintenance and cleaning, or make required repairs at the most optimal times without affecting production capability while the plant is on line. The plant demand and equipment maintenance requirements always set the plant outage schedule for inspection, cleaning, and repairs and alterations. This means that some maintenance that can be deferred without hazard to life, health, and property may be put off longer than they should be. This holds true even when the need for cleaning, repairs, or alterations makes production inefficient or limits output.

When production rates are based on demand forecasts, output is ordinarily below maximum system capacity. Here, you can boost production to store sufficient inventory to accommodate a preplanned shutdown. However, if the forecast is too low or there is a surge in demand that outstrips capacity, production may continue beyond the point where deferring necessary work causes serious difficulty later on. It may not be possible to predict the nature, extent, and ultimate cost of doing so. Although the cost of purchasing product for resale can be estimated immediately, the unpredictability of the costs of deferring the work impedes the benefit-cost analysis for determining the desirability of immediate or deferred maintenance. Consequently, the decision is usually to continue operating rather than charging off immediate shutdown and maintenance costs.

Power stations cannot inventory their output for later sale. However, power demand forecasting is very advanced. Therefore, fossil stations generally can plan maintenance outages for times of low demand, or when surplus power can be economically imported. Refueling and safety requirements drive outage planning in nuclear stations.

Scheduling cleaning

The functions of cleaning exchangers are to permit thorough inspection and to remove foulants and scale that have reduced their performance. Plant designers try to match deterioration and fouling rates of system components so that no single one limits output. However, corrosion, erosion, and fouling rates vary with the fluids flowing, and their concentrations, flow rates, and temperatures. Furthermore, since temperatures and concentrations in a given fluid nearly always change as the fluid passes through the system, deterioration rates and fouling buildup often varies with the location of a piece of equipment.

In a tubular heat exchanger, the rates of deterioration and fouling may vary across the whole unit. An example is a multipass unit with hot fluid entering the tubes that suffers more corrosion in the inlet pass than in the outlet one.

Cleaning the tube and shell sides of heat exchangers requires considering fouling.[1] The following suggests a basis for scheduling cleaning in which fouling is deposited from cooling-tower water (Fig. 8.1):

1. Achieve steady-state operation.
2. Measure the four terminal temperatures and pressures and the flow rate on each side.
3. Calculate the clean overall coefficient U_c and actual shell- and tube-side pressures drop.
4. Repeat the measurements and calculations at regular intervals. Measurements and calculations may be made manually, or automatically using a fouling monitor setup.
5. Calculate the fouling rate from Eq. (8.1):

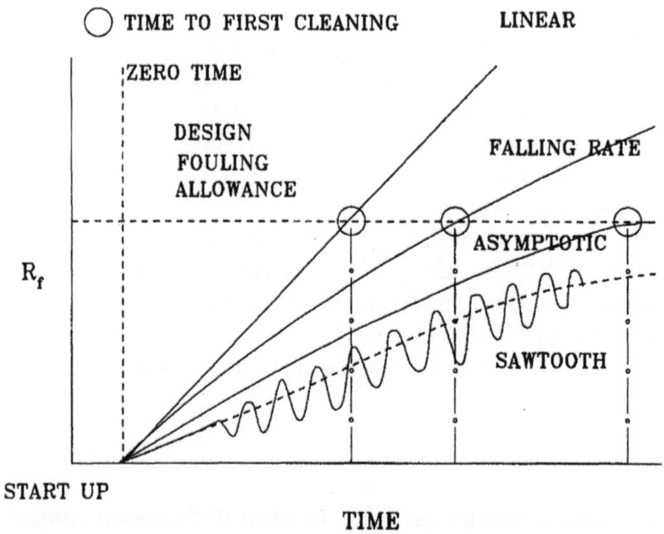

Figure 8.1 Variation of apparent differential fouling with time for industrial cooling waters.

$$R_f = \frac{1}{U_o} - \frac{1}{U_c} \tag{8.1}$$

where R_f is the fouling resistance U_o is the actual operating overall coefficient, and U_c is the clean overall coefficient.

6. Plot the fouling resistance against time.
7. If there is a time delay before the onset of measurable fouling, relocate the origin of the plot to set time zero at the onset point.
8. Extrapolate the plot as a linear falling rate or asymptotic rate to the point of intersection with the design fouling allowance.
9. Plan to clean at the time shown by the intersection point.
10. Continue to monitor to refine the extrapolation.
11. After cleaning, repeat the process to establish a cycle.
12. Alternatively, plot the rate of pressure-drop buildup.

Figure 8.1 is a plot of fouling resistance with time for three kinds of industrial cooling water fouling. In linear fouling, resistance to heat flow increases linearly with time. This indicates that the shear force of the water on the foulant never becomes great enough to peel off accreted layers. In falling-rate fouling, shear forces due to flow make it increasingly difficult for layers to build up. Asymptotic fouling is the desired type. Here the rate of shear approaches the rate of accumulation at the design fouling allowance. The fourth curve is a generalization for any of the types when acid cleaning fluid is injected into cooling-tower water to remove deposits.

An alternative, but similar, method used in refineries makes use of a performance index I_p (Fig. 8.2). Equations (8.2) and (8.3) define I_p:

$$I_p = \frac{U_{so}}{U_{co}} = \frac{\text{observed service overall coefficient}}{\text{observed clean overall coefficient}} 100 \qquad (8.2)$$

$$I_p = \frac{1}{1 + R_f U_{co}} 100 \qquad (8.3)$$

Scheduling repairs and alterations

To set a schedule, the scheduler needs some idea of the work to be done, who will do it, and how long it will take. Furthermore, a logical sequence of doing the work must be prepared. The first step is to make an on-line external inspection, including nonintrusive thickness measurements by ultrasonic scanning; the second is to examine the

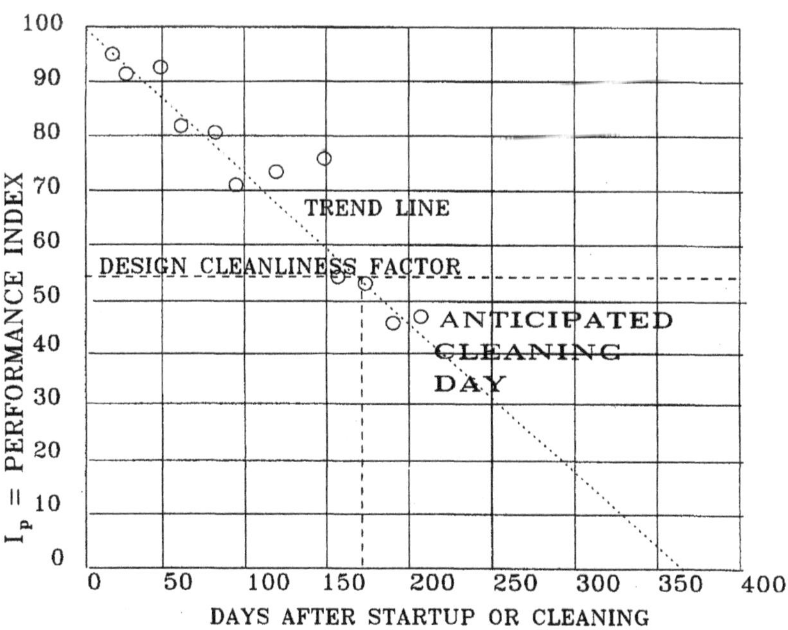

Figure 8.2 Performance index–time–trend-line method for setting cleaning schedules.

operating logs; the third is to review the maintenance record subsequent to the last scheduled outage.

The on-line inspection will reveal readily apparent leaks, suspicious distortions, and regions of the pressure envelope that may require restoration or repairs by cutting and patching. Be sure to examine multicorrugation expansion joints for squirming. Look for any bulges or cracks in insulation, wet spots or discolorations under the unit, and similar indications of trouble. Ask operating personnel about any unusual sounds or smells coming from the exchanger. Enhance the ability to listen by using an industrial stethoscope. Assist the sense of smell by using a hydrocarbon detector.

Rattling sounds (chattering) may indicate broken tubes or excessive vibration. An increase in odors indicates the onset of a leak. Operating logs may reveal overloads or other excursions that could have caused damage. Use the throughputs and terminal temperatures to calculate whether performance has deteriorated.

If there are broken tubes, it will be necessary to stabilize them in the region of the break and plug the ends. If tubes are vibrating, it will be necessary to determine the mechanism, find out the extent and character of the damage, and take the necessary actions to isolate the damaged ones and stabilize the others. Prepare the necessary materials and procedures before the shutdown.

When planning a maintenance outage for feedwater heaters, compare the actual terminal temperature differences (TTD) recorded with the ones the manufacturer predicted for the flows specified at the time of procurement. Also compare the observed TTD for the present flow with the oldest recorded TTD for the present flow to determine if there is any deterioration. If the unit is constructed with a drains subcooler, compare the recorded drains subcooler approach (DCA) with that specified by the manufacturer for the design flow and with the oldest recorded DCA for the present flow. Substantial deviations indicate trouble spots to investigate and prepare for.

If the maintenance record reveals that the unit had been shut down for plugging and restored to service between the previous and present scheduled outages, examine the plug map carefully. Properly recorded plug map information can provide clues to the causes of tube failures. Based upon these clues, prepare ahead for corrective and restorative measures.

Scheduling inspection, repairs, and alterations during prescheduled outages. Inspect off-line during prescheduled outages to monitor suspected conditions and to examine parts of the exchanger not accessible during operation. The inspection agenda may have many items. Depending on the conditions that the inspection discloses, repairs may be required.

When critical-path scheduling software is available, adapt it for scheduling maintenance and repairs. Some software may also be adapted for planning personnel and machine deployment and subcontract work.

List the items to be inspected and decide on the sequence to follow. For each part it is decided to inspect, assume that every kind of repair that might be required will be performed. As inspection proceeds, the critical path will change to reflect the time required to delete or add maintenance procedures. This permits rapidly deciding when to defer a repair procedure that can be put off to a future outage. Table 8.5 in checklist

TABLE 8.5 Checklist of Inspection Items to Include in the Critical Path

- ☐ Channels and channel covers: look for the indications indented to the right
    - ☐ Pitting
    - ☐ Erosion
    - ☐ Graphitization
    - ☐ Linear corrosion—especially at the interior welds of nozzles and the welds of pass partitions to channel barrels and tubesheets
    - ☐ Crevice corrosion under gaskets and between the channel covers and channel flanges and channel flanges and tubesheets
    - ☐ Lamellar tearing of tubesheets in the region of pass-partition-to-tubesheet welds
    - ☐ Cracking of pass-partition-to-channel and -tubesheet welds
    - ☐ Corner cracking of frames to which removable pass-partition covers are bolted in D-type heads
    - ☐ Condition of studs in pass-partition bolting frames of D-type heads
    - ☐ Condition of pass-partition covers and gaskets in D-type heads
    - ☐ Condition of parts of special high-pressure closures
- ☐ Tubesheet faces: these are the locations of considerable maintenance activity; some things to examine for possible trouble are indented
    - ☐ Ligaments: eroded ligaments may require restoration by welding; cracked ligaments may indicate conditions potentially life-threatening to the exchanger; allow time to chase the cracks to their end, make appropriate repairs; allow for the possibility that there will be several fluid penetrant inspections and repairs as the fluid penetrant reveals weld flaws and ligament cracks not visible to the unassisted eye
    - ☐ Tube ends: whether the tube-to-tubesheet joints are expanded only, welded only, or welded and expanded, tube-end "feathering" or raggedness indicates a problem resulting from excess turbulence or entrance velocity; allow time in the schedule to dress up the tubesheet face and tube ends; also allow some period to make up or procure and install a wasting plate or to procure and install protective tube-end ferrules
    - ☐ Tube-to-tubesheet welds: these welds are subject to washing away, pitting, uncovering of subsurface corrosion, and cracking; if there have been previous indications of these problems, be sure to allow time in the schedule to clean the surfaces thoroughly and to make repairs

TABLE 8.5 Checklist of Inspection Items to Include in the Critical Path (*Continued*)

- ☐ Tube interiors: if tube corrosion has been predicted, it is prudent to measure remaining thickness and search for pitting; a good, but time-consuming sequence for examining tubes is (1) borescope the interior to see what it looks like, (2) clean the interior (preferably by hydroblasting), (3) locate discontinuities by ultrasonic axial scanning, and (4) examine the tubes in which the ultrasonic axial scanning indicated discontinuities by Internal Rotary Inspection Scanning (IRIS); alternatively, tube condition can be assessed by eddy-current examination
- ☐ Bundles of removable-bundle exchangers: the time required to pull, inspect, and reinstall a bundle depends on the type and size of the bundle and the location of the heat exchanger in the system; use records of previous extractions and reinsertions to estimate how much time to allow in the schedule; look at the items to the right
    - ☐ Baffles or supports for edge thinning, tube-hole enlargement and general erosion and corrosion
    - ☐ Fretting and wear of tubes and baffles or supports where the tubes penetrate
    - ☐ Tube cracking just aft of the inner face(s) of the tubesheet(s)
    - ☐ Tubes in the vicinity of the tubesheets and inner faces of the tubesheets for deadspot corrosion
- ☐ Shells and return covers: full inspection for pitting, line corrosion, crevice corrosion in the regions or bundle baffles or supports, etc., requires scheduling prior cleaning; the regions to examine are listed to the right
    - ☐ Crevice corrosion where the tubes penetrate the baffles or tube supports
    - ☐ Tube erosion in way of the shell inlet and outlet connections
    - ☐ Internal welds where connections penetrate
    - ☐ Sites where hot fluid enters
    - ☐ Locations where cross-flow baffles and tube supports rest on or near the shell
    - ☐ The places between the inlet-end tubesheet and the first baffle or support, and between the last baffle or support and the outlet-end tubesheet or bundle end

form lists some typical items to schedule for inspection and inclusion in the critical path.

Maintenance supervisors can estimate the personnel and time required for inspection and repairs. Add reasonable contingencies to their estimates for dealing with unforeseeable problems.

For work to be contracted, prepare inquiry specifications that describe the equipment and work to be performed. Allow sufficient time for the contractors to examine the site and to make their proposals. Add time to that quoted by contractors depending on how well they are known and their past records of performance.

Reference

1. *Fouling in Heat Exchange Equipment,* HTD-Vol 17, American Society of Mechanical Engineers, 1981.

Chapter

9

Maintaining Tubesheets, Bonnets, Channels, and Covers

Introduction

This chapter deals with maintaining closures, channels and bonnets, and tubesheets. It covers bolted and non-bolted closures, repairing gasket surfaces and field remachining flanges, correcting excessive flat-cover deflection, strategies for dealing with chronically leaking flanged, gasketed joints, and replacing diaphragm seals in channels and manways. It deals with maintaining pass partitions and pass-partition covers and cover flanges. Its channel coverage includes replacing channels and covers. The chapter closes with a discussion of wormholing and repairs to wormholed tubesheets.

Flange Design Basics: Bolting and Unbolting Closures

The following summary of how flanges are designed under the present ASME Code rules is intended to provide some insight into the considerations for bolting and unbolting.* A discussion and references

*The rules are found in Section VIII, Division 1, Appendix 2 and Division 2, Appendix 3. The discussion applies generally to flange design rules of other pressure-vessel codes which are similar to those of the ASME Code. Although the rules cover flanges with ring gaskets, they do not prohibit flanges with other than ring gaskets such as ones using full-face gaskets or that use means other than bolting for fixing or clamping the flange to provide effective restraint against flange deflection. However, such flanges must be designed in accordance with good engineering practice, and the design method must be acceptable to the Authorized Inspector. Rules for clamp connections are found in Appendix 24 of Section VIII, Division 1.

to the original work on which the rules are based may be found in *Modern Flange Design,* 7th ed., Bulletin 502, available from Taylor Forge International, Memphis, Tenn.*

The ASME Code rules for flange design do not apply to nongasketed flanges and to ones in which there is edge contact. For such flanges and for pairs of identical flat-faced flanges separated by a metal spacer or with edge contact, it is convenient to use the ASME Code nonmandatory Appendix Y of Section VIII, Division 1.[1,2,3,4,5,6]

Design basis

Most heat exchanger flanges are designed as elastic structures, using an analysis built on the work of Waters, Taylor, and colleagues.[7,8,9] Other methods, such as two- and three-dimensional finite element analysis, are used for high pressures and large diameters, and odd-shaped flanges with discontinuities.

The limits in the scopes of the TEMA Standards and ASME Code rules are 3000 psi (~20,700 kPa) and a product of diameter times pressure of approximately 60,000 lb/in (~10,508,000 kPa•mm). In addition, the TEMA Standards apply to exchangers 60 in (~1524 mm) or less inside diameter, with the intent to limit shell wall to 2 in (~50.8 mm) and stud diameter to 3 in (~76.2 mm). However, they state that the criteria contained in the standards can be applied to larger diameters and refer the user to Section 10, "Recommended Good Practice," for units outside the scope.

Stand-alone flanges

Flanges are designed as stand-alone structures for these heat exchanger configurations: (1) channel flanges for channel-to-cover connections and channels bolted to a tubesheet welded to the shell and extended as a flange, (2) shell flanges for channels welded to a tubesheet extended as a flange for bolting to the shell, and (3) channel and shell flanges of removable-bundle stationary tubesheets extended as a flange when the channel and shell flanges are fastened to the tubesheet by one set of stud bolts threaded into tapped holes. In these cases, the moments applied to the tubesheet by bolting up and pressure must be considered in the tubesheet design.

*This edition was released circa 1980 by the Taylor-Bonney Division of Gulf + Western Industries. Subsequently, Hackney of Dallas, Texas purchased the product line and machinery for producing forged flanges of 24 in diameter and smaller. The current organization, Taylor Forge International, produces larger-diameter ring and hub flanges. They are located in Memphis, Tenn.

Paired flanges

Flanges of removable-bundle exchangers with stationary tubesheets sandwiched between removable bonnets or channels and the shell, and in which the bolting is not threaded into the tubesheet, must be designed as a pair. The flange design must take into account the differences in gasket dimensions and properties and flange configurations on each side of the stationary tubesheet. When the flanges and/or gaskets are not the same, calculate the bolt loads for the most severe conditions of operating and/or gasket seating loads applied to each side simultaneously. After determining the loads for the most severe loading conditions, calculate each flange, following the rules of the applicable division.

Available bolting

In two- and multipass tube-side units, the bolting must be adequate to yield not only the ring of gasket material inside the bolt circle but also the gasket webs that seal the pass compartments from one another. Since only one set of bolts is available to yield both gaskets of paired flanges during bolting up, and to maintain the joints on both sides tight under pressure during operation, the match of the shell-side gasket with the tube-side one must consider tube-side gasket rib compression.

Effects of misdesign

Insufficient gasket compression, deflection of the tubesheet when the tube-side pressure is higher than that in the shell, flange rotation, and noncollinearity of the shell-side and tube-side gasket rings may lead to interpass leakage.[10]

Materials requirements

The materials used for flanges must meet ASME Code general requirements for materials for pressure-containing parts. Some specific additional requirements are as follows:

1. Flanges made from ferritic steel must be given a normalizing or full-annealing heat treatment when the thickness of the flange section exceeds 3 in (76.2 mm).
2. Material on which welding is to be performed must be proved to be of good weldable quality.
3. Welding must not be performed on steel that has a carbon content greater than 0.35%.

4. All welding on flange connections must comply with the division's requirements for postweld heat treatment.
5. Fabricated hubbed flanges may be machined from a hot-rolled or forged billet. The axis of the finished flange must be parallel with the long axis of the original billet, but need not be concentric with the finished flange.
6. Hubbed flanges, not machined from a hot-rolled or forged billet, may not be machined from plate or bar stock material unless the material has been formed into a ring with the original plate surfaces parallel to the axis of the finished flange. The original plate surface need not be present in the finished flange. The joints in the ring must be welded butt joints.
7. Bolts, studs, nuts, and washers must meet the ASME Code divisions' requirements (see Fig. 9.1). Bolts and studs smaller than ½ in (12.7 mm) are not recommended for heat exchanger construction. If smaller bolts or studs are used, ferrous bolting material must be alloy steel. Precautions must be taken to avoid overstressing small-diameter bolts.

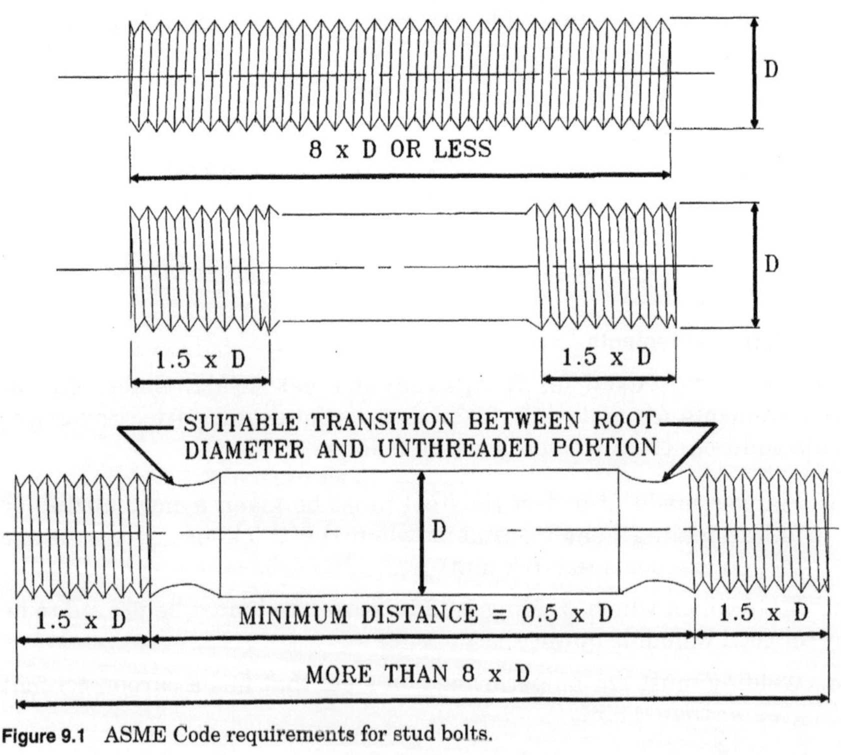

Figure 9.1 ASME Code requirements for stud bolts.

Regarding requirement 6, the thickness to be used to determine postweld heat treatment and radiography requirements for the rolled structure is the lesser of the flange thickness or width (outside radius minus inside radius). The back of the flange and the outer surface of the hub must be examined either by the magnetic particle or liquid penetrant method.

Circular flange design

Circular flange design involves the following steps:

1. Selecting gasket material suitable for the fluid environment
2. Selecting a desirable gasket confinement
3. Picking the proper gasket type and width for the operating pressure and temperature
4. Choosing a flange facing appropriate for the selected gasket and service
5. Determining the bolting required: (*a*) to yield the gasket and (*b*) to sustain sufficient load on the gasket to maintain a seal when the hydrostatic end force due to pressure unloads the gasket
6. Optimizing the hub proportions, flange width, and thickness within the stress levels permitted in the various parts of the flange

The following paragraphs discuss various aspects of these steps.

Assumptions and design considerations

The ASME rules assume rigidity and that the boltholes do not materially affect strength.* The design method of the ASME Code considers only two load systems, illustrated schematically in Fig. 9.2: (1) the load required to seat the gasket and (2) the loads that the internal pressure applies. This figure (where W is the flange design bolt load for the operating conditions or gasket seating, as may apply) shows not only the loads but also lever arms on which they act to cause the flange to bend. Calculations are made for each load system to determine the required bolting and the proportions and thicknesses that will withstand both sets of loads without exceeding permitted radial and tangential stress levels in the flange ring and longitudinal stress in the hub (when present).

*Flanges designed in accordance with the ASME Code rules, especially those with large diameters or subject to unusual operating conditions, may not be rigid enough to maintain the tightness required to prevent unacceptable leakage. Readers who encounter such conditions should avail themselves of the method of checking flange rigidity described in ASME Code, Section VIII, Division 2, Appendix M.

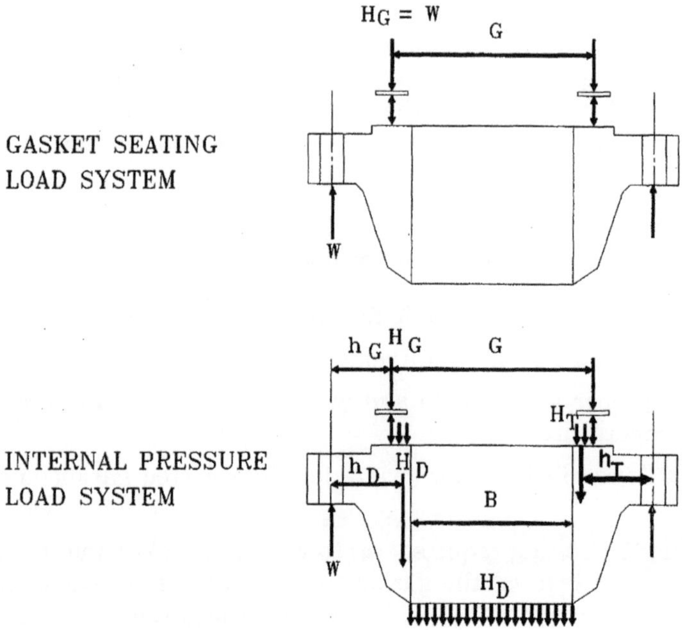

Figure 9.2 Flange load systems.

Bolt loads

When a channel or bonnet is bolted to a tubesheet, only one set of loads (those imposed on the tube side) must be investigated. However, when the tubesheet is clamped between a pair of flanges (such as a channel and a shell flange), the tube-side flange, gasket, or both, may not be the same as the shell-side flange, gasket, or flange and gasket. The loads must be found for the most severe condition of operating or gasket seating applied simultaneously to both sides. This condition may be

1. Gasket seating at atmospheric temperature on one flange together with operating pressure and temperature on the other
2. Gasket seating on each flange at the same time
3. Operating conditions on each flange at the same time

After the most-severe-condition loads have been found, calculations to determine the required bolting, proportions, and thicknesses must be made for each flange.

Requisites for flange bolting

There are two requisites for the flange bolting, which are derived from the load systems:

1. At atmospheric temperature and pressure, the bolting must be able to supply enough initial load ($W = W_{m2}$) to seat the gasket.
2. At the design pressure and temperature, the bolting must be able to supply force ($W = W_{m1}$) sufficient to resist the hydrostatic end force that tends to part the joint, and to maintain enough compression on the gasket and joint-contact surface to keep the joint tight.

Initial tightness depends on the gasket properties and flange design. The load applied by bolting up must be adequate to yield the ring of gasket that closes the connection and make it flow into surface irregularities. It must also be able to deform the gasket ribs under the pass partitions to prevent pass-partition bypassing.

Operating tightness depends on the gasket's resiliency. (It must continue to seal when applied internal pressure tends to separate the mating parts.) At high operating pressures, when W_{m1} is greater than W_{m2}, it determines the amount of bolting required. At low or moderate pressures, the load required to seat the gasket W_{m2}, will govern.

Bolting-up load system

The total cross-sectional area of the bolts required for seating A_{m2} is a function of the gasket material, the effective contact area to be seated, and the allowable bolt stress at atmospheric temperature S_a. For this condition, $A_{m2} = W_{m2}/S_a$. The total cross-sectional area of the bolts required to meet the second condition A_{m1} is a function of the gasket material, the effective gasket contact area to be kept tight under pressure, the design pressure, and the allowable bolt stress at the design temperature S_b. For the second condition, $A_{m1} = W_{m1}/S_b$. The minimum cross-sectional bolt area A_m required for the connection is the greater of A_{m1} and A_{m2}.

The load W_{m1} is substituted for W in the design of the flange when considering operating conditions; W_{m2} is modified to take account of the operating conditions when they govern A_m, as well as the amount of bolting A_b actually provided.

The adjusted load W, for the atmospheric case, is the average of A_m and A_b multiplied by the allowable stress at atmospheric temperature S_a. This provides a margin against abusing the flange by overbolting. When additional safety against mishandling is desired, or when the flange must be able to withstand the full available bolt load, W may be adjusted to A_bS_a.

The load W required to seat the gasket is

$$W_{m2} = \pi bGy \tag{9.1}$$

Here in consistent units:

b = effective gasket seating width; this differs from the actual gasket width dimensions
G = gasket load diameter
y = gasket design seating stress
W_{m2} = bolt load required for seating the gasket, i.e., to make it fill the flange and tubesheet irregularities, thereby effecting a seal

On bolting up, the gasket exerts the load H_G over the surface with which it is in contact. As illustrated, at equilibrium the gasket load and bolt load are equal. Note that the effective gasket width actually changes with flange deflection. However, this is not considered because of the assumption of rigidity.

Pressure load system

The second load system has the four forces shown on the lower sketch of Fig. 9.2:

1. Hydrostatic end force on the area inside the flange, labeled H_D in Fig 9.2 and shown in Eq. (9.2). This is the product of the internal pressure and projected area of the shell cover or channel. It acts on the flange ring through the hub (or in the case of a ring flange through the channel barrel or shell). If the hub is tapered, its line of action is midhub.

$$H_D = \frac{\pi}{4} B^2 P \qquad (9.2)$$

where, (in consistent units) B is the corroded I.D. of the flange and P is the design pressure.

2. Pressure force that acts on the exposed part of the flange face, labeled H_T. This is the difference between the total hydrostatic end force H and H_D, the hydrostatic end force on the area inside the flange. The total hydrostatic end force H is the force that would act on the circle of diameter G when pressure P is applied:

$$H = \frac{\pi}{4} G^2 P \qquad (9.3)$$

Force H_T is expressed as follows:

$$H_T = H - H_D \qquad (9.4)$$

If the gasket covered the raised face shown in the figure, then H_T would be zero, but leakage is assumed to be possible as far as G. H_T is assumed to act on the circle midway between the I.D. of the flange B and the gasket line of action G.

3. The total joint compression load H_p is given as

$$H_p = 2\pi bGmP \tag{9.5}$$

where (in consistent units) m is the ratio of compressive gasket stress to design pressure required to maintain the joint tight under pressure load. For example, when m is 3, the compressive gasket stress at pressure P must be at least $3P$ for the joint to be tight. If the gasket were theoretically perfect, leakage would not start until the fluid pressure just exceeded the gasket stress; for this condition, the value of $m = 1$.

The load W ($= W_{m1}$), supplied by the bolts, is in equilibrium with $H + H_p$ (Fig. 9.3). This equilibrium condition is expressed as

$$W_{m1} = H + H_p = \frac{\pi}{4}G^2P + 2\pi GmP \tag{9.6}$$

Gaskets and gasket factors

Most gasket manufacturers will suggest suitable materials. The gasket type, width, and flange facing and confinement are related.

The suggested (nonmandatory) values for gaskets tabulated in the ASME Code are based on a 1941 work of Rossheim and Markl that suggested rules for finding b and presented tables of values of m and y for commonly used gasket materials.[11] Considering that the value of m required to make a tight seal varies inversely with initial bolt load and that at zero initial pressure m is very high, they developed an empirical formula for y that approximates the m value required to produce a bolt load that would seal the joint at zero internal pressure.

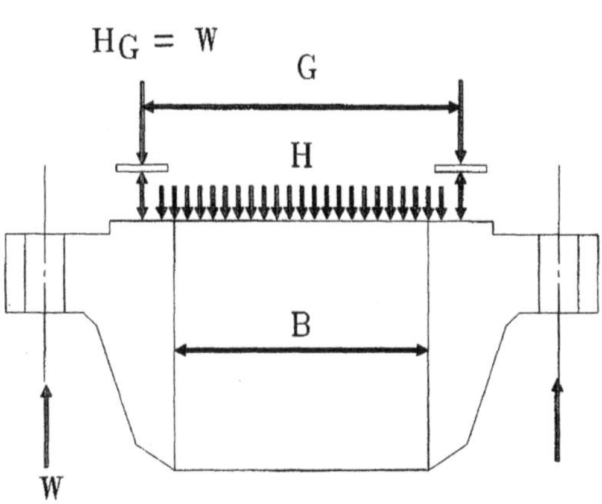

Figure 9.3 Balance of loads due to pressure.

In Eq. (9.6), the first term on the right-hand side is the total hydrostatic pressure load acting on the effective gasket diameter G. The second term is the gasket reaction over an annular area $2b$ wide on the same G. Theoretically, to achieve a seal, the gasket must resist the load $2b\pi GmP$ when the internal pressure equals P. However, the gasket resists the total load shown in Eq. (9.6). Applying pressure P relieves the load $(\pi/4)G^2P$, leaving $2b\pi GmP$ as the theoretical load to seal tightly at pressure P. Factor m in the ASME Code tables has a safety margin over the theoretical m.

ASME Code values for m and y are *not* mandatory because they are not founded on an acceptable theory or backed by sufficient experimental investigation. The published values are based on experience and some experimentation.

Because of the many variables that must be examined, testing is time-consuming.[12] Much has been done to try to establish the interrelationships between gasket dimensions, gasket material properties, leakage, and tightness criteria and to reexamine the method and parameters in current use.[13–21]

What constitutes a tight joint depends on its service. A leak rate on the order of 0.001 cm^3/s might be acceptable for water service; one of 0.0001 cm^3/s (~6 drops/h) might be acceptable for other liquids. For dense gases an acceptable rate might be 0.00001 cm^3/s and for lethal substances, 0.000001 cm^3/s. The exchanger user must set the criterion that applies for the given service. The ASME Code's tabulated values for m and y should be modified to reflect the applicable criteria.

A proposed procedure that considers leak criteria and tightness points out that y and m are not independent and suggests establishing "Code-like factors" for m and y for various leak rates and contained fluids.[22] These are based on full gasket width (N as defined in the ASME Code), minimum gasket design seating stress, and actual seating stress.

Flange thickness and proportions

Calculations to determine optimal flange thicknesses and proportions are done iteratively. For a trial set of flange dimensions, moments are determined for each load system as the products of the various loads and their distances from the bolt circle to their lines of action. Because of the assumption of flange rigidity, the shortening of lever arms due to rotation is neglected in calculating the moments. Longitudinal hub stress, radial flange stress, and tangential flange stress produced by the moments are calculated. (For hubless flanges or where the designer chooses to neglect the strengthening effect of the hub, only the tangential flange stress is determined.) These stresses must not exceed the code's allowable limits.

The stress calculations require computing various factors that involve the assumed thicknesses and shape, and kind of flange. One set of factors concerns the ratio of flange O.D. to I.D.; another is related to the hub proportions. Other factors that depend on these sets are used in the actual stress calculations. The code has graphs of the dependent versus independent factors for flanges joined integrally to the channel barrel or shell (integral-type flanges) and those attached in such a way that the barrel or shell does not contribute to their support (loose-type flanges). It also provides equations for calculating the factors suitable for computer programming. The moment and stress calculation methods are fully detailed in the ASME Code and therefore are not repeated here.

Bolt spacing

When a design meets the ASME Code rules, but the bolt spacing is so wide that it may leak, consider adjusting it by increasing the flange thickness, increasing the number of bolts, or both. A theoretical analysis of bolt spacing indicates proper load distribution when the bolt spacing is $(2d + t)$, where d is the bolt diameter, in (mm) and t the flange thickness, in (mm).[23] According to this analysis, if the bolt spacing exceeds this value, multiply the total moment calculated for each load system by

$$\sqrt{\frac{\text{Bolt spacing}}{(2d + t)}}$$

The TEMA Standards suggest a maximum spacing of

$$B_{\max} = 2d_B + \frac{6t}{m + 0.5} \tag{9.7}$$

When the distance exceeds $B_{\max}$, the standards recommend that the total flange moment determined by ASME Code design methods be multiplied by a correction factor of

$$\sqrt{\frac{B}{B_{\max}}}$$

In the TEMA calculation

B = bolt spacing, centerline to centerline, in

d_B = nominal bolt diameter, in

t = flange thickness, in

m = gasket factor used in ASME Code flange calculations (explained above)

Flanges with full-face gaskets

The ASME Code does not now have rules for designing flanges with full-face soft gaskets, but does not prohibit their use. An analysis proposed by Rossheim in 1943, which has been used successfully, is outlined in *Modern Flange Design,* Bulletin 502, edition VII. It makes these assumptions:

1. Rigidity at the bolt circle before pressurizing.
2. Inner flange edge unrestrained.
3. Uniform gasket pressure on the annular surfaces on each side of the bolt circle.
4. Ring effect and boltholes cause no significant loss of strength.
5. A countermoment exists between the bolt circle and the flange O.D.

Required tightness

As with tube-to-tubesheet joints, required tightness depends on the service and much has yet to be done to relate service conditions, leak rates, and the suitable ASME Code–like factors. Most flanged joints designed by the ASME Code's rules are tight when tested and in use. The ones listed below may leak if the rules have been applied without further consideration. When investigating flange leakage, always review the design of flanges with these characteristics:

- Very large diameter
- High design pressures
- High design temperatures
- Severe temperature gradients
- Differences between bolt material and flange material thermal expansion coefficients
- Unusual gasket arrangements
- Services in which there may be short-term temperature or pressure excursions beyond the design conditions

Bolt tightening

The objectives of bolt tightening are to

1. Provide the force required to seal the gasket and maintain tightness during operation
2. Stress all the bolts equally
3. Apply the bolt stresses in such a way that gasket compression and flange stress are not distributed unequally during bolting up

Allowable and actual bolt stresses. The maximum allowable bolt stresses tabulated in the ASME Code rules are used to determine the minimum required amount of bolting.* However, the design value and the bolt stress that might actually be present, or that might be needed for conditions other than design pressure, may exceed the allowables.

Initial tightening prestresses the bolts. The stress that it develops in the bolts must be adequate to resist conditions that tend to produce a leaking joint but not so much that it yields the bolts or flange assembly, thereby producing enough relaxation to result in leakage.

The usual hydrostatic test is 1.5 times the design pressure, multiplied by the ratio of atmospheric to design temperature allowable stresses. Consequently, the initial bolt stress must be more than that allowed for the design. Otherwise, the bolt strain developed during the test could let the joint separate. The ensuing gasket decompression could be so great that the joint would leak. Just increasing the bolt stress for hydrotesting proportionately does not consider that the safety factor required for hydrostatic testing need not be as high as for operating conditions. On the other hand, it does not consider that a stress-strain analysis may show that 150 percent of the tabulated allowable design stress is not sufficient. Proper analysis of how much initial bolt stress is needed during prestress and application of test pressure requires considering changes in bolt elongation, flange deflection, and gasket load.

There is a conservative margin between the ASME Code design stress values for bolting and the bolt yield stresses. Therefore, it is permissible to exceed the design bolt stress, if you guard against excessive flange distortion and gasket crushing.

The margin between the code's yield and allowable stresses is considerably less for flange and tubesheet materials. For this reason, where permanently tight bolted joints are desired, always use the lower allowable stresses for materials where the code provides two sets of allowable stresses.

---

*For a full discussion of bolting and flange rigidity design considerations for bolted flanged connections, see Appendix S of Section VIII, Division 1 and Article 3-5 of Appendix 3 of Section VIII, Division 2 of the ASME Code.

Bolting up. Manually tightening bolts is most often done on bolt sizes smaller than 25 mm (~1 in). The mechanic's feel for the right amount of tightness is developed over a period of time. The probable bolt stress developed manually with standard wrenches when bolting up is complete is

$$S = \frac{7{,}880{,}000}{\sqrt{d}}\text{kPa} = \frac{45000}{\sqrt{d}}\text{psi} \tag{9.8}$$

where (in consistent units) S is the bolt stress and d is the bolt diameter.

Using hand wrenches, it is easy to overstress small-diameter bolts and hard to stress large-diameter bolts enough to achieve tightness. Most shops and maintenance departments use air-driven impact wrenches as a convenience for nut tightening on small studs and as a necessity on large ones.

Torquing. Instead of relying on mechanic's feel, it may be desirable to use a calibrated torque wrench for bolting up. Various publications have equations that relate bolt stress to torque. Table 9.1 lists the suggested values for SA-193 Gr. B-7 stud bolts. At room temperature, the ASME Code's allowable stress is about 25,000 psi (~172,370 kPa) for these bolts. Good results will be achieved by using the torque

TABLE 9.1 Suggested Bolt Torques, ft•b, for Various Bolt Stress Values

				Values of torque in ft•lb for bolt stress, psi		
Nominal diameter, in	Threads per inch	Root diameter, in	Root area, in^2	7,500	15,000	30,000
½	13	0.400	0.126	30	40	60
⅝	11	0.507	0.202	60	90	120
¾	10	0.620	0.302	100	150	200
⅞	9	0.731	0.419	160	240	320
1	8	0.838	0.551	245	368	490
1⅛	8	0.963	0.728	355	533	710
1¼	8	1.088	0.929	500	750	1,000
1⅜	8	1.213	1.155	680	1,010	1,360
1½	8	1.338	1.405	800	1,200	1,600
1⅝	8	1.463	1.680	1,100	1,650	2,200
1¾	8	1.588	1.980	1,500	2,250	3,000
1⅞	8	1.713	2.304	2,000	3,000	4,000
2	8	1.838	2.652	2,200	3,300	4,400
2¼	8	2.088	3.423	3,180	4,700	6,360
2½	8	2.338	4.202	4,400	6,600	8,800
2¾	8	2.588	5.259	5,920	8,800	10,840
3	8	6,324	2.838	7,720	11,580	15,440

value shown for about 30,000-psi (~206,840-kPa) stress, which is 20 percent above the allowable. Note that the specified minimum yield stress for SA-193 B-7 bolts of 2½ in (63.5-mm) diameter and smaller is approximately 105,000 psi (~724,000 kPa) and for larger diameters is 95,000 psi (~655,500 kPa). This allows a substantial margin for bolting-up purposes.

Whether bolting up manually or torquing, the bolt-nut assembly must be properly lubricated, the threads must be properly cut for the required fit class, and foreign matter must be excluded from the mating surfaces. Otherwise the assumed or calculated load delivered to the bolts will be greater than the actual load.

Recommendations for manual bolting up and torquing flanged joints are as follows. Tighten the bolts manually in several rounds. Usually four passes in which the feel of the resistance to tightening seems to increase by about 25 percent is adequate for manual bolting. Torquing should similarly be done in three to four rounds in equal increments of torque. The sequence in which bolts should be tightened can be set in either of the following ways:

1. Divide the flange into four quadrants—A, B, C, and D—and number the bolts in each quadrant from 1 to *n*. For each round of wrenching, bolt up in a sequence that starts with bolt 1 in quadrant A, proceeds to bolt 1 in quadrant C, then to bolt 1 in quadrant B and to bolt 1 in quadrant D. Then in the same sequence of quadrants, bolt 2, bolt 3, and so on.
2. Divide the flange into four quadrants: A, B, C and D. Number the bolts sequentially, beginning with the first bolt in quadrant A. Make a four-column table headed A, C, B, D. Under each column heading list the bolt numbers in that quadrant. The bolting sequence then goes left to right across the columns. Figure 9.4 illustrates how this system works. The sequence for the 24-bolt flange is 1, 13, 19, 7, 2, 14, 20, 8, 3, 15, 21, 9, 4, 16, 22, 10, 5, 17, 23, 11, 6, 18, 24, 12. For the 20-bolt flange it is 1, 11, 16, 6, 2, 12, 17, 7, 3, 13, 18, 8, 4, 14, 19, 9, 5, 15, 20, 10.

Controlled bolt elongation

The only way to truly determine bolt stress (within the limits of the assumptions about material properties) is to measure the bolt elongation. The bolts may be heated to establish a precalculated elongation and the nuts run down to contact with the flanges. Alternatively, elongation during bolting may be measured with an extensiometer or micrometer. These are time-consuming procedures and would not ordinarily be done.

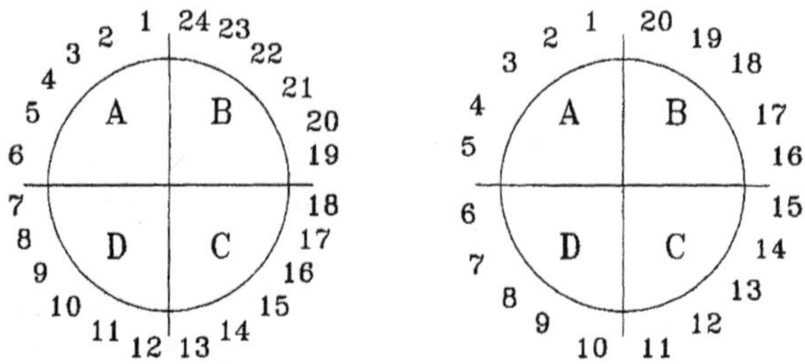

A	C	B	D
1	13	19	7
2	14	20	8
3	15	21	9
4	16	22	10
5	17	23	11
6	18	24	12

A	C	B	D
1	11	16	6
2	12	17	7
3	13	18	8
4	14	19	9
5	15	20	10

Figure 9.4 Determining bolting-up sequence.

However, inexpensive portable ultrasonic devices (e.g., the "Stress-Mike," a product of Strestel Inc., Scottsvalley, Calif.) are now available that measure the elongation and read out the bolt stress directly. The more costly bolt tensioning method eliminates false readings.

Bolt tensioning

In *bolt tensioning,* a hydraulic bolt tensioner stretches the bolts and loads them to a predetermined stress level. The relationships are given in Eq. (9.9). You can verify the bolt stresses by ultrasonically measuring the elongation and converting it to stress. Bolt tensioners require stud bolts longer than those tightened by torque wrenches. The ends of groups of four or multiples of four bolts, distributed equally in the four quadrants, are simultaneously gripped by hydraulic rams and stretched within their elastic limits a predetermined amount. With the studs in this condition, the nuts are run down hand tight to the flange, following which the grips are released. The stress in the studs is calculated from the elongation using

$$S = Ee \tag{9.9}$$

where, in consistent units, S = bolt stress
E = Young's modulus
e = elongation

Unless the tensioner can grip all the bolts at the same time, add bolt stress in several increments, using a sequence that will distribute the loads uniformly.

Unbolting

A common practice in unbolting flanged joints is to use a nut runner to loosen and remove the bolts one after the other around the bolt circle. Avoid this procedure because it can plastically deform the mating parts and studs. The reason is as follows. When the unit is depressurized, if all the bolts are stressed equally and the joint was tight or nearly tight before shutdown, the gasket reaction and consequent flange bending moments are distributed in the more or less uniform pattern (described in Ref. 12). If one bolt is completely removed, and the mating parts are stiff enough to prevent gasket relaxation between the bolts on either side of the empty hole, the stress in the remaining bolts must increase to absorb the gasket reaction. As bolts are removed, the bolt stress in the remaining ones may increase enough to yield the bolts. The bolted-together parts may warp in the vicinity of the tight bolts opposite the completely unbolted region.

In some refinery operations, a standard maintenance procedure is to remachine tubesheets and flanges after each disassembly. This reflects the observation that the flanges are frequently out-of-flat when the unit is taken apart. The probable principal cause of the distortion is the failure to use a suitable unbolting procedure. The following are recommendations for unbolting:

1. Unbolt in the sequence described under bolting up.
2. Release the bolt loads in several passes that unload the bolts in approximately equal increments.

Opening and Closing Quick-Opening and Nonbolted Closures

There are too many kinds of quick-opening and nonbolted closures to discuss here. However, some general safety rules for depressurizing, cooling down, draining and flushing, and venting apply to them all.

Depressurizing and venting

Quick-opening channel closures should be designed so that they cannot be opened when they are under pressure. Be aware that designs that appear to satisfy this condition may be unsafe. An example is a modified breech-lock closure, sometimes called a *wedge-type closure.* The name derives from the closures used on the breeches of large naval weapons. In these guns, the rear end of the gun barrel (breech), into which bag powder and the projectile are loaded, is machined with an interrupted female thread. The cover or plug is machined with a male interrupted thread which fits into the interruptions in the female end. The interruptions in the breach are machined down to just below the depth of the female thread root. The interruptions in the plug are machined down to just below the depth of the male thread root. The width of threaded parts of the plug are sufficiently smaller than the widths of the unthreaded parts of the breech to permit the plug to be inserted. The plug is inserted and rotated to make a tight, strong seal. The thread configuration makes it necessary to exert some effort to turn the male plug in the female breach so as to get a tight metal-to-metal seal.

Some quick-opening closures use modifications of the breech lock in which wedges and tapered slots take the place of the male and female machined parts. When you rotate the cover, the screwing action forces the mating parts to compress a gasket. In some designs the wedges force the cover into tight contact with a metal sealing element that the hydrostatic pressure forces into the space between the cover and the barrel. These closures are almost always used on high-pressure channels. As a result, the covers are thick and heavy. Therefore, they are fitted with handwheel and lever systems or hydraulic-ram openers to effect the rotation required to relieve the locking action of the wedges.

After a gun is fired, there is no residual pressure in the breech. Therefore, opening the gun to reload is not dangerous. However, the mechanical device used to rotate the channel cover may be able to exert enough effort to overcome the friction between the wedges and slots, even when there is a small amount of pressure in the channel. The results can be disastrous. To get an understanding of the problem, consider the following.

A forged cover for a 1371.6-mm (54-in)-I.D. channel constructed in accordance with ASME Code rules for flat heads to withstand 6895 kPa (1000 psi) and having 3.2-mm (⅛-in) corrosion allowance would be about 184.2 mm (7¼ in) thick and would have a mass of about 2130 kg (~4700 lb). The force that a mere 35 kPa (~5 psi) would apply to the surface would be approximately 51,700 N (~11,450 lb), sufficient to accelerate the cover at about 0.75 m/s^2 (~2.45 ft/s^2). At this rate of acceleration, a worker standing 0.6 m (~2 ft) in front of the cover would be struck with this force in just 0.64 seconds.

Swing-bolted closures are not as subject to the same hazards as non-bolted quick-opening ones. However, a common bad practice for quick access is to bolt up only every other bolt This can overstress the bolts and bend the cover between the swing bolts. Knocking the swing bolts back for a quick opening when there is some pressure residing in the channel can have results similar to that described above.

Because of the hazard of failing to depressurize a quick-opening cover, make sure that *both sides* of the exchanger are at atmospheric pressure, with an open vent before opening. Do not count on there being no leak between the shell and tube sides. It is not always possible to know in advance if there is a leak of noxious, hazardous, or lethal material between the sides, or if pressurized fluid has leaked from the shell into the channel.

Good practice is to purge the tube side with inert gas whenever the process fluid is flammable, explosive, noxious, or lethal. When analyzing the purge effluent shows that the danger of fire or explosion has been eliminated, circulate clean, breathable air until the oxygen content in the channel is in the range of 18 to 20 percent. Purging and flushing with air will assist in cooling the structure to permit safe and comfortable entry.

Cooling down

Most of the time, the reason for opening a closure is to gain access as quickly as possible. Most people cannot keep their flesh in continuous contact with a surface that has a temperature much above 60°C (~140°F). Surface and atmospheric temperatures of 49°C (~120°F) and higher are extremely uncomfortable to work in. In emergency situations, it is feasible to work for limited times in temperatures as high as 49°C (~120°F), if there is a constant flow of breathable, room-temperature, fresh air.

If the situation permits, it is desirable to flush the tube side with cooling water before draining and blowing out. But be careful not to shock a hot structure. This applies especially to closed feedwater heaters and vertical waste-heat boilers and catalytic reactors in which the top tubesheets may be operating at high temperatures. Sudden impingement of cold water on the hot tubesheet can cause so severe a shock as to destroy it. Whenever feasible, have a thermal analysis performed to determine a suitable cooldown rate. In any event, do not cool down more rapidly than at the manufacturer's recommended rate.

The authors know of an instance in which spray cooling the tubesheets of a pair of supercritical feedwater heaters caused multiple leaks in its welded and hydroexpanded tube-to-tubesheet joints.

Repairing Corroded or Damaged Gasket Surfaces

The gasket surfaces of alloy-faced flanges, covers, and tubesheets are subject to various kinds of corrosion and damage:

- Applied linings or facings may crack as a result of differential thermal expansion and high gasket loading.
- Integrally clad surfaces produced by explosive or roll cladding may laminate, bend, and crack at unjoined regions.

- Fusion-welded surfaces may have had subsurface porosity, inclusions, and lack of penetration into the parent metal.

Following are suggestions for making repairs.

Preparing the surfaces of cracked applied flange linings

The best surface preparation is to completely remove the applied lining to make way for a fusion-weld facing or replacement by a new surface ring. Grind away any corrosion products on the base-metal surface under the ring. Be sure to get to the root of any corrosion, using visual inspection and fluid penetrant examination to verify this. It is also necessary to completely remove the attachment welds. Figure 9.5 illustrates the ways to apply a lining. The preferred way permits filet welding the facing to the flange, filet welding the flange to the barrel, and filling in the space after these welds have been inspected.

Preparing laminated regions of explosive or roll-clad flange surfaces for repairs. If the parent and cladding metal are compatible for fusion welding, gouge out the damaged region to just past the boundaries of the laminated area. When the metals are not compatible, prepare for a replacement.

Preparing pitted and corroded flange surfaces for weld repairs. Clean away with scrapers and wire brushes all visible corrosion. Grind or machine the surface until the roots of all visible pits and cracks have been removed. Fluid-penetrant-examine to verify that there are no pits or cracks not visible to the unaided eye. It is especially important to expose the bottom of the damaged region. Clean the surface with a safe, volatile solvent until no visible foreign matter remains.

Recommended welding processes

Build up the surface with weld metal until there is sufficient crown to machine to the original thickness. Use manual metal arc welding ("stick" welding) because it is the process most forgiving of impurities.

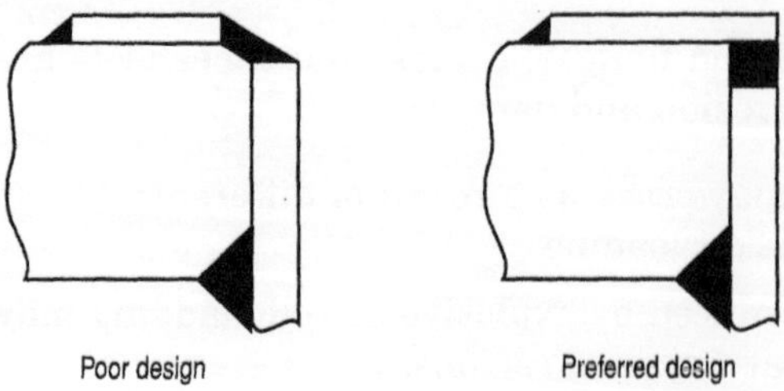

Figure 9.5 Applying linings to flange faces.

Whenever possible, the welding should be uninterrupted until it is complete. Use extreme care to remove any slag between laying down beads.

It is common practice to use gas tungsten arc welding [GTAW; also called TIG, which is the acronym for *tungsten inert gas*] for repair welding because it leaves no slag and is assumed to produce a denser deposit. However, the surfaces to be welded must be very clean and free of foreign matter and impurities. Unless cleanliness and freedom from contamination is certain, this process is likely to produce unsatisfactory welds.

Restoring surface finishes

After inspecting the weld deposit, restore the gasket surface by grinding or machining. Cover surfaces can be readily remachined and gasket grooves remilled in the maintenance shop. Channel-flange and tubesheet surfaces must be remachined on site. This may be done successfully by manual surface grinders for low-pressure joints. High-pressure ones and those with milled gasket retention grooves require the use of portable tools. These are available either for purchase or from contract field machining organizations; a typical portable machine tool supplier is Climax Portable Machine Tools.* A typical supplier of contract field machining and maintenance services is New Castle Engineering, Inc.†

Field Remachining Warped Flanges

If there is sufficient excess metal in a warped flange, its flatness can be restored by field remachining. The reduced thickness will make it more subject to bending. Therefore, it is important to find out why it warped.

Determining the cause of flange warping

The ASME Code rules for designing the bolted connections are concerned only with not exceeding the allowable stresses but do not consider elastic deflection. Without exceeding the allowable stresses, the parts may bend so much that there is contact at the edges. For this reason, consider using plain faces only for relatively low pressures and with the more easily compressed gaskets.

When the parts bend, the gasket is loaded differently than the design assumes. The outer edge may be crushed and the inner edge

---

*Climax address is PO Box 230, Newberg, OR 97132, tel. (800) 338-8311.

†New Castle address is PO Box 568, New Castle, IN 47362, tel. (317) 529-7473.

hardly deformed. A joint that should have been tight may leak. Further tightening the bolts usually makes the problem worse. Here, if the mating parts remain elastic, the deflection further reduces the amount of gasket sealing area. However, the additional bolt loads may plastically deform the flanges. Better ways to deal with elastic flange bending are (1) replace the gaskets with ones more easily compressed, (2) replace the gaskets with ones that have limit rings, and (3) install a peripheral ring metal spacer about half the gasket thickness between the flange edges to limit the deflection. Each of these schemes requires evaluating the flange design. If you contemplate installing a peripheral ring spacer, investigate the flange and bolt stresses for the contact condition. Limit rings may not relieve the problem but may prevent gross crushing under various load conditions.

Measuring the degree of flange distortion

When the parts are available for inspection, lay a straight-edge across the gasket surface. Measure the out-of-flatness and cupping at the quarter and eighth points to determine how much material must be removed to restore a flat surface perpendicular to the barrel centerline.

Correcting Excessive Deflection in Flat Covers

Flat covers may be in full compliance with ASME Code rules and TEMA standards, yet deflect elastically enough to cause trouble. The symptoms of excessive deflection are

- Erosion-corrosion at pass-partition locations
- Shifting of gasket compression outward
- Persistent recurrence of leakage after achieving tightness by retightening

There is only a modest effect on the thermal efficiency of bypassing around pass partitions. However, if there are substantial pass-to-pass differences in pressure, the erosion caused by the leakage will cause increased bypassing and deterioration of the pass partitions and cover.

Measuring cover deflection

The way to determine how much the cover is deflecting under pressure is to lay a straight-edge across the cover after bolting up but before applying pressure. Then apply operating pressure and remeasure. Make the measurements at the eighth and quarter points.

Correcting excessive cover deflection

Plastic cover deflection can be removed by reflattening in a press brake. This is not a technique to be applied by amateurs because the cover must be bent so that when it springs back, it is at just the right level. After flatness is restored, stiffen the cover to prevent future deflection.

Ways to stiffen covers after flattening

The simplest way to stiffen the cover is to back it with a disk to provide what is essentially a compound circular beam. It can be made to the full diameter and drilled for the cover bolts. To do this will require replacement bolts long enough to accommodate the backer. Alternatively, fillet-weld to the cover a disk small enough to fit inside the bolt circle without interfering with the cover bolts.

The cover can be stiffened by adding ribs to the back side to provide additional moment of inertia. These may simply be a cross in which the members lie on perpendicular major diameters, or you may weld a central pipe section to the cover and attach stiffening ribs to it and the cover. The rib ends may lie inside the circle tangent to the inner edges of the bolt circle to avoid interfering with bolting and unbolting. The ribs may be tapered with the high end near the center.

Note that any material welded to a pressure part must be acceptable under the rules applied by the inspection code. Furthermore, the Inspector may require you to produce calculations showing the original and reduced deflection.

Strategies for Dealing with Chronically Leaking Gasketed Flanged Closures

One of the nastiest maintenance problems is dealing with flanged joints that leak in an otherwise satisfactorily operating exchanger. The seepage may be at the joint of the tubesheet to the channel, the tubesheet to the shell, the channel cover to the channel barrel, the shell cover to the shell, or, worst of all, the return cover to the floating tubesheet. In the latter case, the stream mixing may be intolerable. It is a bad idea simply to cuss and retighten the bolts because this may contribute to the problem. A better way is to try to find the cause of the leakage and eliminate it.

Finding the causes of and fixing flanged joint leakage

These are some of the things to do to find out why a gasketed joint is leaking and some ways to fix the problem:

1. Examine the gaskets.
2. Measure the flange distortion.
3. Check for insufficient bolting.
4. Check for excessive bolting loads.
5. Examine for unequal bolt loading.
6. Investigate for differential thermal expansion between bolts and flanges.
7. Investigate the thermal and/or pressure cycling on gasket tightness.
8. Examine the gasket selection.

By carefully examining the gasket when you disassemble a flanged joint, you can see how it has been compressed. When the mating parts are plane and rigid, compression is uniform across the gasket ring and circumference. If the inner or outer edge of the gasket ring is compressed more than the other edge, it is safe to assume that the mating parts were either distorted before or during bolting up or are distorting under pressure. If there is more compression near the boltholes than midway between adjacent boltholes, the mating parts are scalloping.

Measure for flange distortion by inserting feeler gages between the flanges in the bolted-up assembly at the locations of the bolts and midway between. A regular variation indicates the kind of deflection called *scalloping*. To find out why there is more gasket compression at one edge than the other, measure the out-of-flatness of the disassembled parts as previously described. Tubesheets with expanded tube-to-tubesheet joints often take a bow as a result of tube rolling. If the parts are flat within the usual tolerances, they are probably bending elastically under pressure. In fixed-tubesheet exchangers not fitted with shell expansion joints, differential thermal expansion between the tubes and shell may cause the tubesheets to bow.

If tubesheet bowing not resulting from differential thermal expansion between tubes and shell is the cause of leakage, the cure is to remachine the tubesheet gasket surfaces to flatness. Before taking this step, review the tubesheet thickness calculations to make sure that sufficient metal remains after the remachining. If there is insufficient tubesheet thickness, it may be necessary to build up the surface to be remachined with weld.

To try to cure leakage caused by elastic deflections, measure the bolt stresses with an ultrasonic instrument. If the stresses at operating conditions are much above the allowables used in design, carefully backing off the nuts may allow the structure to rearrange the application of the load to the gasket and seal the leak. This procedure is exactly opposite the first approach of most people trying to seal a leak. Therefore, most people are reluctant to try it.

If the flange design conforms with ASME Code rules and there is no excessive bolt stress but the flange deflects so much on bolting up that the joint leaks, you can fix the problem by means of a backing flange. If the ring portion of the flange is recalculated, using about a 20 percent lower design stress, the total thickness of the ring part will increase by about 10 percent. Therefore, supply a backing ring one-tenth the thickness of the flange ring with the same I.D., O.D., and bolthole drilling. If it is necessary to split the backing ring, adjust the backing-ring thickness in accordance with the ASME Code rules for split flanges. A different kind of fix is to install a split ring spacer at the flange edges as discussed under remachining warped flanges.

If it is possible to determine that the cause of tubesheet bowing is differential expansion between the tubes and shell of a fixed-tubesheet exchanger, it may be feasible to stop the channel-to-tubesheet joint leakage by replacing the gasket with a wider, softer one. The replacement must be compatible with the original flange design.

Review the flange design bolting calculations to compare the bolting required to seat the gasket with the bolting required to resist pressure. If gasket seating was the determining load and the exchanger is multipass, check to see if gasket rib compression was considered. If it was not, there is probably insufficient bolting to seat the gasket properly.

One possible fix is to replace the gasket with a more easily compressed one. If this is not acceptable, short of replacing the flanges with new ones designed with sufficient bolting, little can be done to stop the leaks. As a temporary measure, it may be possible to install C-clamps between the bolts to provide additional sealing force. These should not become part of a permanent solution.

Although the ASME Code flange design rules make some allowance for more bolting than is required, this may not be enough to prevent overbolting. This can occur in three ways: (1) considerably more bolt cross-sectional area is provided than is needed to seat the gasket or resist the pressure, and in tightening the joint more bolt load is applied than is required to seat the gasket; (2) because of the wide spread between the allowable stress used in designing the flange and the bolt yield stress, excessive bolting-up stress may have been applied without yielding the bolts, or the bolts may have been excessively tightened because of a policy of regularly tightening all flanged joints; and (3) the thermal coefficient of expansion of the mating parts may be greater than that of the bolts, effectively further tightening the bolts as the exchanger comes to operating temperature. Whatever the cause, evidence of excessive bolt loads is crushed gaskets and flange edges that make contact with each other because of excessive deflection. The fixes for this problem are

- If the flange is rigid and strong enough to withstand the bolt load that caused the gasket crushing, replace the gasket with one that has a higher yield strength.
- Alternatively, for rigid, sufficiently strong flanges, replace the gasket with one that has a limit ring.
- If, after unbolting, the flanges are within acceptable flatness tolerances, replace the gasket and rebolt. Monitor the bolt load on rebolting to make sure that the design load is not exceeded by more than 20 percent during rebolting.
- Instead of regularly tightening bolts on flanged joints, install spring washers under the nuts. See below for dealing with differential expansion problems.

If the bolt stress is not measured during bolting up, the bolts may be unequally loaded. One can surmise that this has happened when the flanges are flat and the gasket has been uniformly compressed except in the region of one or two bolts. If the gasket has not been damaged, the fix for this situation is to rebolt in the pattern described above, making sure that after the final round all bolts are equally stressed.

Mechanical designers of bolted connections should consider differential thermal expansion between the flanges and tubesheets on one hand and the bolts on the other hand. This requires the thermal designer to estimate the flange, tubesheet, and bolt metal temperatures at the most severe case of start-up, operation, shutdown, or upset. The estimates may not be on target. Therefore, when a flanged closure begins to leak at the operating temperature but is tight under bolting-up conditions, consider measuring the bolt, flange, and tubesheet metal temperatures when the exchanger reaches operating temperature.

If the bolts expand more than the mating parts, the gasket has probably unloaded, leading to the observed leakage. The quick common strategy is to retighten the bolts to correct for the unloading. The danger in doing this is that when the unit cools to assembly temperature during a shutdown, the gasket and bolts will undergo additional loading as a result of thermal deflection opposite that which caused the unloading. For this situation, instead of retightening, install spring washers under the nuts to maintain the load on the studs constant.

A successful, less common, alternative is to replace the studs with longer ones, installing under the nuts thick washers of a material that has a higher thermal coefficient of expansion than the studs. These fixes may be combined.

More difficult to deal with is the problem when the mating parts expand more than the studs. This situation can occur when the flanges and tubesheets are austenitic stainless steel and the bolts are SA-193 Gr. B-7 or have a similar thermal coefficient of expansion. The approximate ratio of stainless-steel thermal coefficient of expansion to that of the alloy bolting is 1.25 to 1. This means that the stainless parts will expand about 25 percent more than will the alloy bolts. Since bolting up takes place at room temperature, the restrained expansion resulting from the change in temperature as the structure reaches its operating temperature further loads the bolts. The increased bolt load may or may not yield the bolts, depending upon the margin between allowable bolt stress and bolt yield stress. However, the increased load can cause gasket overloading and crushing and permanently warp the gasket contact surface or surfaces out of true plane.

The options for solving this problem are limited. They include changing the bolt material to one that has a thermal coefficient of expansion closer to that of the high-alloy parts, replacing the gasket with one that has a limit ring to limit the compression, and installing a flange-edge spacer that will limit the compression that the bolts can apply to the gasket. All of these strategies require an analysis of the mechanical design of the mating parts.

Applying loads to gaskets cyclically may have a ratcheting effect. The gaskets may become progressively harder with no evidence of crushing. Nonmetal gaskets may become brittle. Metal ones may crack. The fixes for this situation are (1) estimate or determine from the records the time to gasket failure and institute a gasket replacement program; (2) install spring washers to maintain constant loading on the gaskets; or (3) replace the gaskets with spiral wound ones, which are less likely to fail under cyclical loads. The third fix requires examining the flange design.

Many leaks are the result of improper gasket selection. As stated above, the m and y values tabulated in the ASME Code are not mandatory because there is no sound theoretical basis for the way they were established. Although flanged connections designed with the listed values are usually tight, some are not. Therefore, when there is a chronically leaking connection, examine the values used in design. When there is room for higher ones, consider substituting a gasket with the higher values.

Leaks may also stem from failure of a gasket as a result of contact with the process stream. Evidence is progressive deterioration from the inner edge outward. Replace the gasket with a more suitable one.

Instituting a regular bolt-tightening regime

If bolt stress decreases after initial tightening because of gasket creep or relaxation, restore tightness by retightening the bolts. At high service temperatures, gasket creep may also allow leakage. This may also be corrected by retightening once or every so often. However, in designing for high temperature, consider the relaxation properties of the materials, especially where creep is the controlling factor in design.

Regular bolt tightening may reestablish design bolt stresses in flanged connections that are initially tight but leak after a period of operation. The effects of gasket unloading and creep may be mitigated by installing spring washers. Spring washers under the nuts may even eliminate any need for retightening.

Precautions

Before routinely retightening bolts, make sure to establish the bolt stresses to be maintained and the permissible bolt-to-bolt variations. Measure the existing bolt stresses with an ultrasonic micrometer. If the stress levels and variations are in the desired range, do not do any retightening. If ultrasonic stress measuring equipment is not available, establish the desirable torque values to maintain. Do not exceed these values.

Do not retighten by serially tightening bolt after bolt around the flange. This is a sure way to start a joint leaking that has previously been tight and possibly to cause permanent damage. Follow the bolt tightening sequence described above.

When not to retighten closure bolts

Some closures depend for tightness on the hydrostatic end force created by the internal pressure. Boiler-type manways are typical. These consist of an oval or obround frame welded into the head or shell that it penetrates and an internal cover. The inner face of the manway frame is machined to a suitable gasket surface. The manway cover consists of a dished oval or obround plate of the same overall diametral dimensions as the frame. It has a flat rim, and its top surface is finished to receive a gasket. The convex (outer) surface of the cover is fitted to receive two stud bolts set on its major axis and straddling the minor axis. The distance between the studs is about one-half the length of the major diameter of the formed part.

The narrow width of the cover and the gasket are inserted through the frame and rotated to capture the gasket between the inner surface of the frame and the cover rim. Clamps that bear against the outer face of the frame are set on the studs, and nuts are tightened against the clamps. This provides some initial seating force to the internal gasket. When internal pressure is applied inside, it provides the required gasket-seating force.

After the cover is subjected to pressure, the compression of the gasket relieves the clamp bolt loads and the nuts may become loose, or at best only handtight. Therefore, these studs are usually double nutted. These studs are not required to be tight to maintain gasket tightness. It is difficult to convince people who are accustomed to having all bolts tight, not to tighten them when they notice that they are not snug. However, retightening these studs after the cover is under pressure may create just enough additional load to distort the cover and permit leakage past the gasket. The leakage will progressively worsen because of wire-drawing damage to the gasket surface and gasket.

The authors designed a temporary repair of such a boiler-type manway for a vessel containing hot pressurized water in which the maintenance force had overtightened the dog bolts. For this purpose, a split clamp was fastened beneath the dogs to hold the manway cover in place. The dogs were removed, and the manway frame was drilled and tapped to convert it to a stud pad to which was fitted a suitable gasket and a bolted flat cover thick enough to resist the internal pressure. The general scheme and the bolting information are illustrated in Figs. 9.6 and 9.7.

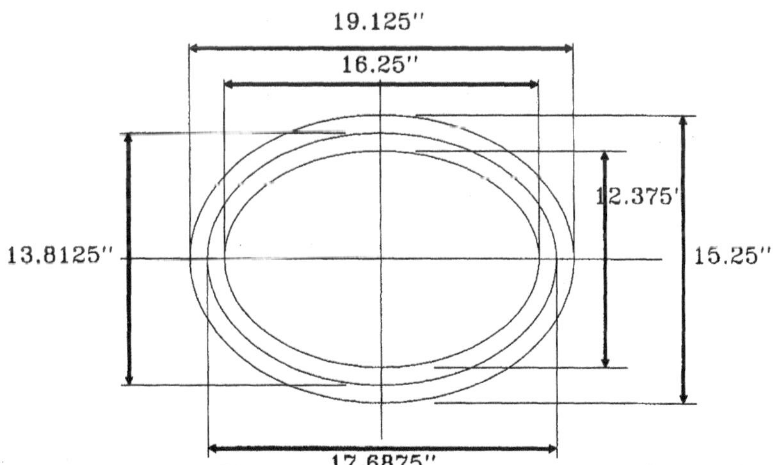

Equivalent A = 17.591 y, psi m
Equivalent B = 14.842 Comp Asbestos 1600 2
Equivalent C = 16.279

Assume Gasket is Full-Faced
Assume 20 5/8" SA-193-B7 Studs
Bolt spacing = 2.552"

INSIDE DIMENSIONS CORRODED

Figure 9.6 Dimensions and assumptions for recalculating boiler type manway frame as a stud pad.

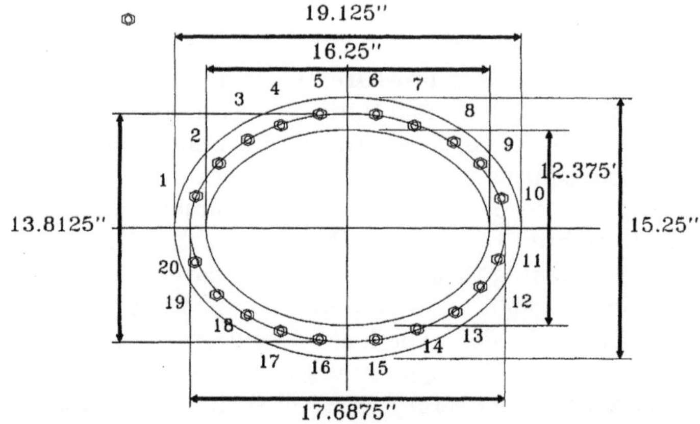

Bolting Sequence: 1,11,6,16,2,12,7,17,3,13,8,18,4,14,9,19,5,10,20

Apply torque in 4 increments following the recommended sequence in each round.

Recommended torque: 60 ft–lb
Note: This value stresses the bolts 20% more than the allowable values tabulated in the ASME Code, but is well below the 105,000 psi yield stress.

Figure 9.7 Recommended bolting sequence and bolting torque.

Opening and Resealing Diaphragm Closures

Diaphragm closures eliminate gasket sealing problems by substituting for the gasket a metal disk welded to a channel or manway flange. The sole function of the flange bolts is to resist the hydrostatic end load that the internal pressure applies to the diaphragm. The diaphragm is sealed to the flange by a fillet weld around its periphery. The manufacturer will have calculated the weld size required to contain the pressure and prevent leakage. Here are some simple procedures and precautions for maintaining diaphragm seals.

Cooling, depressurizing, venting, draining, and purging

Take the same precautions for venting, draining, and purging when opening a diaphragm-sealed joint as for any other. However, because it is necessary to grind, burn, or arc away the filet weld, be sure that any flammable or explosive atmosphere inside the exchanger has been purged and the unit inerted.

Backup diaphragm

Diaphragms can be reused many times if they have not been excessively corroded and if the edge weld removal has not created a condition that will produce unsound filet welds. You cannot know in advance if it can be used again. Therefore, it is prudent to have a replacement diaphragm on hand before opening a diaphragm-sealed closure.

Removing the diaphragm

If the closure is a studded one, after the cover has been removed, carefully remove and inspect the studs. If the studs have been bent or the threads been too badly damaged to rechase them, procure replacements. Protect the tapped holes from damage. The most frequent protections consist of painting the interior of the holes with a weld-spatter-resistant coating and stuffing wicking into the holes to prevent intrusion of foreign matter. Provide more positive protection by making up loosely threaded or push-fit brass stud-hole plugs (Fig. 9.8).

Preparing the surface

Examine the surface before replacing a diaphragm. Grind or file out any pits or grooves, down to sound metal. Build up with weld and metal-grind the surface smooth. Clean away all foreign matter with a volatile solvent safe for use in the area and on the metal. Tack the diaphragm into place. Make the closing weld in accordance with the manufacturer's recommendation. Do not exceed the manufacturer's recommended weld size. Use the recommended welding process (usually manual metal arc). After rewelding the diaphragm into place, remove the stud-hole protectors. Wire-brush out the stud holes, blowing out any foreign matter with an air nozzle. Use an appropriate

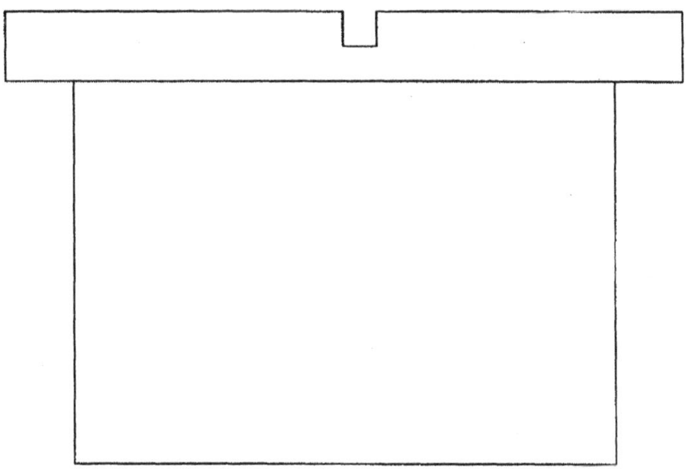

Figure 9.8 Stud-hole protector plug.

thread lubricant and reinstall the studs. Because most covers are heavy and clumsy to handle, slip protective sleeves over the studs before hanging the cover. After the cover is in position, remove the stud sleeves. Release the cover suspension and bolt up as described above.

How to Correct Pass-Partition Bending

Pass partitions sometimes bend even though they meet the minimum thicknesses specified in the manufacturer's standards. The bending may be the result of a greater pressure difference between the passes on either side of the partition than was assumed in the design. In addition, a pass-partition plate with two gasketed edges may not be clamped as tightly as is assumed by the method used in designing the thickness. Bending may also result when the welds joining the pass partition to the barrel or tubesheet fail as a result of restrained thermal deflection.

Dealing with the stresses imposed by differential thermal deflection is beyond the scope of this work. The following suggests repairs for bent partitions and ways to prevent subsequent failures.

Straightening bent pass partitions

If a pass partition is not so badly bent that the material is torn or crimped, it can be straightened. When the corrosion resistance of a material is not adversely affected by heating, you can heat it with an acetylene torch and jack or hammer it back into position. Austenitic and other metals must be straightened by cold jacking. Small hand-operated portable jacks are generally adequate. The springback of the metal makes it necessary to jack the plate past the neutral position. When the jacking force is released the plate relaxes to the desired location. Because you cannot easily estimate how far to deflect the plate, the work requires several jackings and relaxings.

Ribbing

When the pass partition has been restored to its design position, you may prevent future deformation by welding a rib across the direction of the bend. The rib moment of inertia in the direction of the bending should be adequate to resist the load applied by the pressure differential. A disadvantage is that ribbing causes increased turbulence and pressure drop.

Lateral stiffening columns

An alternative to ribbing is to install round bars as columns to stiffen the plate. These should be fully welded at the attachment points. It is not a good idea to install a stiff column across a channel barrel when there is a possibility of restrained thermal expansion. In this case install a column on only one side of the pass-partition plate.

Replacing bent pass partitions

The best permanent solution to pass-partition bending is to cut out the bent pass partitions and replace them with thicker ones. If you follow this course, you will probably need to taper the pass-partition edges to fit the gasket ribs at the tubesheet and channel cover.

Removing and Replacing Feedwater Heater Pass-Partition Covers

Manway-access feedwater heaters have built-in pass partitions. If the pass partition is simply a solid plate welded to the head and channel barrel, access to the side away from the manway can be gained only by burning windows into the pass-partition cover. The edges of the windows should be about 50 mm (~2 in) wide. In addition, a central 50-mm (~2-in)-wide rib should be left. To seal the pass partition, lapped cover plates may be welded to the lips. Alternatively, studs may be welded to the lips and covers gasketed and bolted in place. In either case, the corners of the rectangular windows should be burned to a generous radius and ground smooth to avoid stress concentration that could cause the corners to tear.

When disassembling bolted-on pass-partition covers, inspect the corners of the bolting flanges for cracking and tearing. Arc and grind out any cracks to their roots and reweld. After rewelding, grind the welds smooth.

Restoring Damaged Feedwater Heater Pass Partitions

Channel pass partitions of full-opening feedwater heaters that are welded to the tubesheet and channel barrel fail because of erosion, erosion-corrosion, and thermal stress cracking at the welds. A redesign and major alteration to the construction is necessary to overcome the latter problem. However, you can deal with the erosion and erosion-corrosion problem by burning or arcing out the damaged region, cleaning the edges of the hole, and patching with replacement plate. Although such patches are frequently applied by lapping the patch plate over the hole, it is far better to fit a plate into the hole and make a butt-welded repair. This will avoid creating additional turbulence that hastens subsequent failure.

Replacing Channels and Covers

Moderately corroded or eroded channel covers can be restored by machining down to the bottom of the wasted region, replacing the lost metal with weld metal, and remachining. Replace ones that are too far gone with identical ones; however, if a coarse-grained steel was used on the original, substitute a fine-grained one. In the authors' opinion, coarse-grained steels should be phased out of use in pressure vessels. If erosion or corrosion has progressed more rapidly than was originally estimated, consider replacing the cover with an integral alloy metal facing resistant to the flowing fluid. The cladding may be laid down fusion-weld-metal, integral-roll-clad, or explosion bonded.

Purchase a new channel or bonnet when the existing one is so badly eroded or corroded that it cannot be restored by replacing lost metal with weld deposits. Provide the manufacturer with a template of the bolting flange on the existing channel to make sure that the replacement fits. Be sure to specify for ASME Code–stamped exchangers that the manufacturer is to furnish a partial data report form, U-2, or the appropriate National Board form. Instruct the manufacturer to show on the data report that the channel is a replacement. Advise the manufacturer of the serial number, manufacturer, and National Board number of the heat exchanger and have them noted on the U-2.

Repairing Eroded and Wormholed Tubesheets

As noted in Chap. 5, the rate of leakage through a tube-to-tubesheet joint is a function of the square of the pressure difference between the tube and shell sides, the hole shape and size, and an inverse function of the viscosity of the leaking fluid. When the pressure on one side of the tubesheet is considerably higher than that on the other, leaks will occur through even the smallest leak paths. The consequent erosion, which looks like the holes bored by worms, is called "wormholing" or *wire drawing*. Wormholing occurs in the following places: (1) tube-to-tubesheet welds, (2) the surfaces of the tube and tube hole, and (3) into the ligaments between holes. If the tubes are more resistant to abrasion than is the tubesheet, the ligaments can be severely damaged. The problem can be so severe that the tubesheet is rendered unsafe.

When an inspection discloses washed-away tube-to-tubesheet welds, or if there has been leakage from the high-pressure side to the low-pressure one, through the joint, there is a good probability of wire-drawing damage to the ligament. One of the reasons for full-depth expanding tubes into holes in which the primary joint is made by welding is to reduce the prospect of leakage that causes wormholing. However, this is not a guarantee against it.

Always examine eroded tube-to-tubesheet welds to determine whether cracks or porosity have been exposed. Remove weld metal until you get to the root of the crack or porosity. If the crack has not progressed into the tube and ligament, it is feasible to make a suitable reweld. Repair welding requires cleaning as thoroughly as can be achieved. However, removing all foreign matter from the region of the crack is very unlikely. Therefore, do not consider any welding process other than manual metal arc (stick). Use the smallest-diameter electrode available and weld in multiple passes. Inspect between passes for porosity and cracking.

If the wormholing has progressed into the ligaments, grind out the area—if at all possible to the farthest extent of the damage. This will require closing off the tubes in the damaged area. To repair the wormholed region properly, remove most of the tube embedded in the tubesheet. Leave enough tube insertion to anchor and stabilize the tube. Grind out the wormholed area as far as possible. Press-fit into the holes round weldable-metal bars, setting their front ends just below the ground-out region. Then build up the tubesheet with layers of weld metal until the original thickness has been restored.

References

1. R. W. Schneider, "Flat Face Flanges with Metal-to-Metal Contact beyond the Bolt Circle," *J. Eng. Power, Trans. ASME, Series A* **90**(1), 82–88, Jan. 1968.
2. E. O. Waters and R. W. Schneider, "Axisymmetric, Non-identical, Flat Face Flanges with Metal-to-metal Contact beyond the Bolt Circle," *J. Eng. Industry, Trans. ASME, Series B* **91**(3), 615–622, Aug. 1969.
3. E. O. Waters, *Derivation of Code Formulas for Part B Flanges,* Welding Research Council Bulletin 166, Oct. 1971.
4. R. W. Schneider and E. O. Waters, "Some Considerations Regarding the Analysis of Part B, Code Flanges—1974," ASME paper 75-PVP-48, 1975.
5. R. W. Schneider and E. O. Waters, "The Background of ASME Code Case 1828: A Simplified Method of Analyzing Part B Flanges," *J. Pressure Vessel Technol., Trans ASME* **100**(2), 215–219, May 1978.
6. R. W. Schneider and E. O. Waters, "The Application of ASME Code Case 1828," *J. Pressure Vessel Technol., Trans. ASME* **101**(1), 87–94.6, Feb. 1979.
7. E. O. Waters and J. H. Taylor, "The Strength of Pipe Flanges," *Mech. Eng.* **49,** 531–542, May 1927.
8. E. O. Waters, D. B. Westrom, D. B. Rossheim, and F. S. G. Williams, "Formulas for Stresses in Bolted Flanged Connections," *Trans. ASME* **59,** 161–169, 1937.
9. E. O. Waters, D. B. Rossheim, D. B. Westrom, and F. S. G. Williams, *Development of General Formulas for Bolted Flanges,* Taylor Forge and Pipe Works, Chicago, 1949.
10. K. P. Singh, "Study of Bolted Joint Integrity and Inter-Tube-Pass Leakage in U-Tube Heat Exchangers, Part I: Analysis; Part II Applications," *J. Eng. Power, Trans. ASME* 1978.
11. D. B. Rossheim and A. R. C. Markl, "Gasket-Loading Constants," *Mech. Eng.* 647–648, Sept. 1943.
12. H. D. Raut and G. F. Leon, *Report of Gasket Factor Tests,* WRC Bulletin 233, Dec. 1977.
13. H. D. Raut, A. Bazergui, and L. Marchand, *Gasket Leakage Behavior Trends,* Welding Research Council Bulletin 271, Oct. 1981.

14. J. R. Payne and A. Bazergui, "More Progress in Gasket Testing—the PVRC Program," *1981 Proc-Refining Dept.* (American Petroleum Institute) **60,** 271–290, 1981.
15. J. H. Bickford, "That Initial Preload—What Happens to It?" *Mech. Eng.* 57–61, Oct. 1983.
16. A. Bazergui and J. R. Payne, *Progress in Gasket Testing—Milestone Results,* Welding Research Council Bulletin 292, Feb. 1984.
17. G. F. Leon and J. R Payne, "Current Research on the Behavior of Bolted Joints," paper 4.3, presented at 1984 National Design Engineering Show & ASME Conference, Chicago, March 1984.
18. J. R. Payne, A. Bazergui, and G. F. Leon, "A New Look at Gasket Factors," *Proc. 10th International Conference on Fluid Sealing, VHRA,* Innsbruck, Austria, April 3–5, 1984, paper H1, pp. 345–363.
19. A. Bazergui, L. Marchand, and J. R. Payne, "Effect of Fluid on Sealing Behaviour of Gaskets," *Proc. 10th International Conference on Fluid Sealing, BHRA,* Innsbruck, Austria, April 3–5, 1984, paper H2, pp. 365–368.
20. Design Division Problem XIII, "Re-evaluation of Gasket Factors Used in Flange Design," *Long-Range Plan for Pressure Vessel Research,* 7th ed., Welding Research Council Bulletin 298, Sept. 1984.
21. A. Bazergui, L. Marchand, and H. D. Raut, *Development of a Production Test Procedure for Gaskets,* Welding Research Council Bulletin 309, Nov. 1985.
22. J. Payne, A. Bazergui, and G. F. Leon, "New Gasket Factors—a Proposed Procedure," *Proc. 1985 Pressure Vessel and Piping Conference,* Vol. 98-2, New Orleans, June 1985.
23. I. Roberts, "Gaskets and Bolted Joints," *J. Appl. Mech., Trans. ASME,* **17,** 169–179, June 1950.

Chapter

10

Plugging, Ferruling, and Sleeving

Introduction

This chapter opens with a review of failures in tubes and tube-to-tubesheet attachments. It discusses establishing the axial positions and radial orientations of tube failures before plugging and recommends some ways to locate the damage.

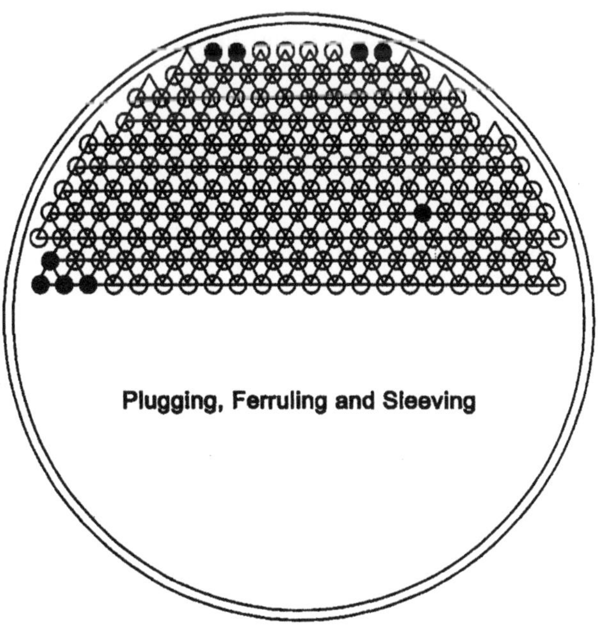

It suggests using records of past tube and tube-to-tubesheet joint examinations, plugging, ferruling, and sleeving to explore patterns of failures and mechanisms. (See Chap. 8 for recording methods.) Then it addresses reasons for plugging, illustrates several kinds of plugs and reviews their service suitability, and describes some plugging techniques and procedures. An examination of the uses and installation of ferrules, sleeves, tubesheet shields, and flow deflector sketches common to all heat-exchanger-using industries follows.

The service, operating circumstances, pressure and temperature conditions, and nature of the fluids flowing, strongly affect which device to choose and how to install it. Operation and maintenance of process exchangers (loosely defined as those used in the chemical and petrochemical, oil-refining, pharmaceutical production, paper manufacture, and mining and associated industries) differ in many ways from that which prevails in power generation. Within power stations there are different criteria for steam surface condensers, closed feedwater heaters, and auxiliary heat exchangers. Therefore, we augment the general discussion with information specific to such units.

Failures in Tubes and Tube-to-Tubesheet Joints

To establish a basis for responding to failures in tubes and tube-to-tubesheet joints,* and to take actions that will prevent recurrences, it is necessary to investigate the causes of the deterioration. In doing so, it is important to keep in mind that the tube and shell sides are interrelated. Before undertaking a fix on one side, consider how it will affect the other.

The following sequence of inquiring into these kinds of failures applies to installed units that must continue operating and ones undergoing postmortem examinations:

1. Peruse operating records. Compare recorded process stream compositions or steam properties, flows, pressures, and terminal temperatures with the design-point conditions specified on the original heat exchanger, feedwater heater, or surface condenser specification sheet.
2. Examine inspection and maintenance records for failure trends. These are indicated by plug maps or output of software for trending the results of tube inspections.
3. Look for failure patterns in the tube field and along the tubes.

*See Yokell's discussion of the symptoms and causes of troubles in heat exchangers in a 1983 article.[1]

4. Obtain specimens of failed tubes from damaged and intact regions for analysis. If feasible, trepan specimens from tube-to-tubesheet welds that were found to be leaking. By doing an autopsy on a unit, you may gain insight into the causes of its demise by subjecting boat sections cut out of the tubesheets to analysis. Perform metallurgical and chemical analyses of the specimens to determine the character of the deteriorations.
5. Methodically go through the list of suspected failure causes and mechanisms to find those most likely to apply.

Regarding item 1, *perusing operating records,* look for a history of pressure-relief valve trips and pressure and temperature excursions that exceeded the nameplate MAWP stamping. Determine whether the equipment was operated beyond its design overload capability. See if it ran within the manufacturer's permissible design overload range for unusually long periods. Find out if the unit was designed for substantially continuous base-load conditions but currently operates cyclically or in peaking load service conditions. If such a change was made, establish when. Determine if rapid deterioration began at the time of the change. Check records of start-ups and shutdowns to determine if the manufacturer's recommended procedures were followed or disregarded. Examine rates of heating to operating temperatures and cooling to temperatures suitable for maintenance access. See if start-up vents were opened to purge large quantities of noncondensibles and whether operating vents were maintained open to prevent noncondensible buildup.

Examine level-control records of equipment that requires maintaining a minimum liquid level for proper operation—for example, waste-heat steam generators, and closed feedwater heaters. If the unit is a process exchanger, check the recorded terminal temperatures against the design heat exchanger specification sheet. If it is a feedwater heater, look at the recorded terminal temperature differences (TTDs) and flows.* See if the recorded data are close to the values for the design-point condition shown on the feedwater heater specification sheet. Similarly, compare records of drain subcooler approach temperatures (DCAs) with the design-point condition DCA.† If the exchanger, or string of which it is a part, is fitted with an inerting system for

*The HEI *Standards for Closed Feedwater Heaters* define *terminal temperature difference* as the difference between saturation temperature corresponding to the entering extraction steam and the outlet feedwater temperature in degrees Fahrenheit. This value could be either positive or negative.

†The HEI *Standards for Closed Feedwater Heaters* define the *drain subcooler approach* as the temperature difference between the drains leaving the shell side of the heater and the entering feedwater on the tube side.

protection during shutdowns, check whether it was inerted when taken out of operation. After completing such checks, look into the possibility that the operating mode or (more likely) a change to the operating mode might be the primary cause of deterioration.

General failure patterns

The following associates some tube and tube-to tubesheet joint failure patterns with possible causes.

Failures in peripheral tubes midway between or at supports and baffles. Failures midway between or at supports and/or baffles almost always mean shell-flow-induced tube vibration contributed to or produced the damage. Some causes of tube vibration are shell fluid flow well above the maximum flow for which the unit was designed, excessive tube support spans and enlargement of baffle or tube-support holes resulting from corrosion and/or erosion, and vibration.

Be aware that the TEMA- and HEI-specified maximum support spans and minimum baffle and support plate thicknesses are not based on vibration considerations. Only a vibration analysis can determine span distances appropriate for the tube material, diameter, gage and pitch pattern, fluid flows, composition, and state of the shell fluid.

Paragraph RCB-4.2 of the TEMA Standards specifies that when the maximum unsupported tube length is 36 in (~914.4 mm) or less, or for tubes larger than 1¼ in (~31.75 mm), standard tube holes are to be 1/32 in (~0.79 mm) over the O.D. of the tubes. Paragraph 5.8.1.1 of the HEI *Standard for Power Plant Heat Exchangers* (HEI PPS) has the identical requirement when the unsupported length is 30 in (~762 mm) or less. Where the unsupported length exceeds 36 in (~914.4 mm) for tubes 1¼ in (~31.75 mm) diameter and smaller, standard tube holes are to be 1/64 in (~0.4 mm) over the O.D. of the tubes.

The TEMA Standards state that for pulsating conditions, tube holes may be smaller than standard. They allow overtolerance of 0.010 in (~0.254 mm) except that 4 percent of the holes are allowed an overtolerance of 0.015 in (~0.381 mm).

The HEI *Closed Feedwater Heater Standards* (HEI CFHS) Paragraph 3.7.2.1 require baffle and support holes to be drilled 1/64 in (~0.4 mm) greater than the nominal tube O.D., except that holes in subcooler baffles may be 1/32 in (~0.79 mm) greater.

Tube-hole clearance is considered in vibration analysis. (See the TEMA Standards Section 6, HEI PPS Paragraph 5.14 and HEI CFHS Paragraph 3.7.2.2 for further information and recommendations for dealing with flow-induced vibration.). In general, the tighter the clearance, the less likely it is that the tube will endure vibration damage. The clearances allowed by the standards are based on what is practical to construct. If the baffle or support holes are made larger than the standards permit, tube life will probably be reduced by fretting at support and baffle penetrations and wear at points of interspan collisions.

Some ways to fix insufficient impingement protection and vibration problems are discussed in Chap. 11.

Tube damage clustered about shell inlet centerlines. Tube damage clustered about shell inlet centerlines can indicate the following: (1) impingement protection was improperly designed for the *actual* flow, (2) impingement protection has broken loose from its attachments or has shattered, (3) two-phase flow has worn a hole into the impingement plate or worn the plate away, or (4) the impingement protection consists of an unsuitable perforated plate (strainer plate).

To look into the possibility of improper impingement plate design, calculate the value of ρV^2 of the measured volume of fluid flowing through the shell inlet and into the bundle. Here, ρ is the fluid density and V the average lineal velocity in the connection. Match the calculated values with the limits in the TEMA Standards Paragraph RCB-4.6. Using as-built construction drawings, check the impingement plate design to see if it meets the size and escape-area requirements of the standards. (See the TEMA Standards Paragraph RGP-RCB-4.62 and Figs. RGP-RCB 4.6211 through 4.6214 and RGP-4.6221 and 4.6222.)

Impingement damage may be self-limiting if the tubes were stabilized with close-fitting rods when plugged. Outer damaged tubes secured this way protect the inner ones. A history of several failures occurring in the inlet region after start-up of a new exchanger followed by a decline in the rate and number of failures with continued operation indicates that this has happened.

A broken-off plate may lodge in the bundle at a different location and batter the tubes to destruction. Pieces of impingement plate may pummel the tubes and snag in the tube nest with similar results.

Holes and wearing away of impingement plates may occur when one phase is a dispersed gas or a suspended abrasive solid. Deal with this phenomenon by changing the material of the impingement plate, increasing its thickness and changing flat plates to wrap-around ones.

Tube damage centered on shell outlet connection(s). Vortexing exit fluids are the usual cause of damage to tubes in the vicinity of shell outlets. After plugging or replacing the perforated tubes, install a cross-shaped antivortex baffle in the shell outlet to prevent recurrences.

Failures in innermost U-tube rows. Cold bending tubes into hairpins (U-bends) is more of an art than a science. The amount of springback after bending depends on the material properties and diameter-to-thickness ratio. It cannot be precisely controlled by the common methods used to bend tubes for process exchangers, feedwater heaters, and power-plant auxiliary exchangers. Different, more costly techniques are used to bend nuclear steam generator U-tubes because bending tolerances for such U-tubes are more stringent.*

Stiffness of the bent part increases with decreasing bend radius. Consequently, the innermost U-tubes have the least flexibility. As a result, the centerlines of U-tube legs may not be concentric with the centerlines of the baffle or support holes into which they are loaded. Depending on the degree of mismatch, the tubes may bind in the baffle-support holes during loading as they progress toward the tubesheet.

Each factory has its own way of dealing with this problem. Some factories apply lubricant to the tubes. Others chill the tubes to shrink them. Sometimes, frustrated workers hammer the U-bends to drive the tubes through the bundle. This method often damages the tubes and leads to early deterioration.

U-tubes may also fail because poor rolling technique twists the tubes, creating residual tensile stresses in the straight legs and U-bend, which promotes stress-corrosion cracking (SCC) in the vicinity of the tubesheets and the U-bends. SCC requires a material sensitive to such corrosion, an environment that causes it, and a tensile stress that approaches the yield stress. SCC may also occur at lower stress levels.

Failures in outer U-tube rows. Failure patterns in straight legs of U-tubes are similar to those in straight tubes. However, the larger-radius U-bends may suffer vibration damage if they are not adequately supported to maintain shell-flow-induced vibration within acceptable limits. Vibration effects may be intensified if the rows of U-bends are too close to permit vibrating tubes to move freely without rubbing on or colliding with each other.

When the shell inlet is beyond the U-bend end of a bundle, and not protected by a plate to deflect the incoming stream, erosion of the outer bends is possible. If the tubular is a closed feedwater heater that receives drains from a higher-pressure one into the space behind the bundle, some of the drains flash. The resulting two-phase flow will erode the U-bends unless they are protected.

---

*U-tubes for recirculating nuclear steam generators are bent to closer tolerances. However, the added cost of bending equipment and methods cannot be justified for non-nuclear-service tubulars.

Failures distributed randomly throughout the tube field. When tube failures occur at random throughout the bundle, the most likely causes are general corrosion and erosion. Only metallurgical analysis can determine the mechanisms. These may be pitting corrosion, intergranular corrosion, galvanic corrosion, corrosion from formation of oxygen-concentration cells, exfoliation of the surfaces of nonferrous tubes from which copper is leached, or combinations of these mechanisms. Corrosion may also reduce the ability of the tubes and tube-to-tubesheet joints to resist fatigue. Following is a summary of some corrosion mechanisms.

General corrosion. General tube corrosion in exchangers is almost always associated with erosion of the surfaces. When the fluid does not contain suspended solids or has suspended solids softer than the tubes, attrition of the surfaces is slow. When suspended solids are harder than the metal surface, the tube is abraded.

Pitting. Pitting develops in highly localized areas on tube surfaces.

Galvanic corrosion. Galvanic corrosion is associated with flow of current from a less noble metal (anode) to a more noble one (cathode).

Oxygen cell corrosion. Oxygen-cell corrosion results from a difference in the amount of oxygen in solution at one point on a surface relative to another.

Damage at the channel-side tubesheet face. High velocities of fluid entering through radial channel connections may erode protruding tube ends of tubes expanded into tube holes and the fillet welds of some types of welded tube-to-tubesheet joints. If the inlet fluid is at high pressure, tube end and weld erosion-corrosion may open a leak path between the tube and the hole. Fluid traversing the path may further erode and corrode the tube and tubesheet, opening additional paths that have the appearance of wormholes. The wormholing (also called *wire drawing*) attacks the softest metal. When the tubesheet is alloy-clad carbon steel, the attack on the softer carbon steel may be so severe that it creates lateral cavities behind the cladding.

Free-stream distribution of fluid exiting a nozzle is conical. The included angle depends on the fluid's properties. In shallow channels and bonnets constructed with axial inlets, the plane of the tubesheet intersects the projection of the conical flow in a circle smaller than the circle tangent to the outer tubes. Therefore, more fluid flows through tubes bounded by the smaller circle than those outside it. Evidence of such maldistribution is tube-end erosion-corrosion in a circular pattern centered on the centerline of the inlet nozzle.

Friction of fluid entering tubes may wash away welds and erode tube ends (feathering). The fluid regime does not stabilize for a distance of 10 to 20 tube diameters after fluid enters the tubes. The contraction effects may enhance tube-end erosion-corrosion. Expanding the tube into the tubesheet may compound this outcome because the flow area becomes slightly smaller as the stream emerges from the expanded region to the unexpanded region.

Bubble testing welded tube-to-tubesheet joints with soap solutions or other homemade bubble formers may cause SCC in sensitive tube and weld materials subjected to tensile stress because of high chloride ion concentrations. Although austenitic stainless steel generally does not suffer SCC at below about 50°C (~120°F), long-time exposure in this range may induce it.

Chapter 5 states that solutions used for bubble testing tube-to-tubesheet joints should be commercially prepared bubble formers that meet the requirements of Section V of the ASME Code with chloride ion content not exceeding 15 parts per million (ppm). Soap solutions, detergents, and bubble formers that do not meet these constraints promote SCC in austenitic tube ends when they run down the tubes during testing tube-to-tubesheet joints. Nevertheless, some organizations do "air and soapy water" tests with "soapy water" made from soap flakes or detergents intended for household or industrial use.

The tube end and the joint that anchors it to the tubesheet provide reaction to the bending, torsional, and tensile or compressive loads applied to the body of the tube. Therefore, this region is failure-prone. Tubes and joints subjected to shell-flow-induced vibration may fail at this site. Near the tube end, failures may result from work hardening, SCC, and fatigue. SCC can develop both inside and outside the tube end and for some distance beyond the inner face of the tubesheet.

There is a transition from the expanded diameter to the original diameter near the backface of tubes expanded into the holes. Stress in the tube varies from compressive inside the tube hole to tensile in and beyond the transition zone. This sensitizes the tubes to SCC in the transition zone and behind the tubesheet. The tensile stress level in the transition zone may approach the tube yield stress. If the tubes have been bulged by rolling behind the tubesheet rear face, the residual tensile stress may exceed the tube yield. In thin-walled tubes, an error in expanding technique, in which the tubes are inadvertently expanded beyond the rear tubesheet face, may lead to SCC failures.

Excessive columnar loads on the tubes created by forward step rolling may contribute to tube breakage in the region of the transition zone. The ends of tubes welded to the tubesheet without subsequent expansion may have high tensile stress levels resulting from the shrinkage of the welds joining the tubes to the tubesheets.

Crevice corrosion failures. When tubes are joined to the tubesheet only by frontface, fillet welding, a deficiency of oxygen in the crevice between the tube and the hole may cause acidity changes in the entrapped fluid, buildup of ions in the crevice, or depletion of an inhibitor in the shell-side fluid. Attack on the surfaces within the holes is of greater intensity and may be far more severe than on surrounding surfaces.

Tube-to-tubesheet joint weld failures. Three reasons tube-to-tubesheet welds fail are (1) corrosion in the heat-affected zone, (2) improper thermocouple placement during welding, and (3) localized stress.

Complex carbides may precipitate in the heat-affected zones behind the welds of austenitic tubes to tubesheets. An atmosphere that the austenitic stainless steel ordinarily resists may corrode this zone.

The emplacement of thermocouples to control temperature for pre- or postweld heat treatment is critical to avoiding distortion and undesired metallurgical changes. Improper placement may be an unknown cause of failures due to the distortion and altered metal structure.

Frontface fusion welds probably fail more from localized stress than from axial loading.[2] Here is a general description of the mechanism:

1. The tubesheets flex under load, intensifying the stress in the ligaments.
2. Residual stress due to weld cooling approaches the yield stress of the weld metal. Added to operational stress, residual stress raises the energy level to that required for crack propagation.
3. Tubes not expanded into the tubesheet after welding undergo free relative deformation between tubes and tubesheets due to expansion and contraction. The welds must also carry all the loads due to pressure on the tubesheet. For fixed-tubesheet exchangers, this includes loads resulting from differential expansion between the tubes and the shell. Stresses produced by loads on such welds resulting from harmonic vibration of the tubes add to the other stresses in the welds.

Hidden and undetected flaws. The analysis of tube-to-tubesheet weld failures should consider the possibility of undetected flaws. Some common ones are hidden gas pockets, blow holes, and weld cracking.

Failure may occur because of large gas pockets that are either open to the surface or covered by a thin membrane. The porosity may have been caused by lubricants used in prerolling, or the tubes may not have been sufficiently cleansed of drawing lubricants, oxides, or protective coatings. Near-surgical cleanliness is required to prevent porosity.

Incomplete deoxidization of surfaces to be welded may cause blow holes in welds. Good practice is to machine the oxide coating off the tube ends and abrasively clean the tube exteriors and hole interiors just before welding. Blow holes may also be caused by an inadequate welding technique that permits introducing nonmetallic inclusions, or burning through at the tube wall.

Buried porosity, weld-shrinkage, fatigue, and corrosion-induced fatigue cause cracks in the weld that propagate into the tube and tubesheet. Weld shrinkage stresses can initiate cracking in fissures of crack-sensitive materials. Fillet welds have exposed roots in which cracks can be caused by improper weld composition, heat-affected-zone hardening, hot cracking in all-metal combinations, and aftercooling cracking in carbon steel.

Frontface fillet welds of closed feedwater heaters may fail by propagation of a leak path from the root notch at the steam side through the tube to the feedwater side. The failures are attributable to fatigue or brittle fracture.

Crack failure paths are at the fusion line of weld-metal–base-metal junctions, and along the weld grain boundaries. The boundary between weld metal and parent metal may be the weakest link, especially with different tube and tubesheet thermal expansion coefficients. Tube-wall cracks propagate normal to the wall; cracks in fissure-sensitive materials propagate along the weld grain boundaries.

Failure patterns in steam and vapor generators

In most steam and vapor generators, heat recovered from a hot fluid flowing through the tubes is used to vaporize feed in the shell. The heat source may be oil, heated in a furnace installed specifically for generating the steam or vapor, or waste heat from a process stream. Steam or vapor may be generated in the shell, or the feed may be superheated and flashed into an external drum. Shells in which steam is generated may be fitted with dry pipes to improve exiting steam quality.

These exchangers may be positioned horizontally or vertically. Horizontal units may be configured as fixed-tubesheet, floating-head, or U-tube units. Virtually all vertical process steam and vapor generators are single-pass fixed-tubesheet designs. Many nuclear steam generators are vertical, channel-head-down U-tube units. Straight-tube horizontal units may be designed for single- or multipass tube-side flow. U-tube horizontal designs are usually arranged for two-pass tube-side flow, with the plane of the bends horizontal or vertical.

Tube and tube-to-tubesheet joint failures in waste-heat steam and vapor generators are most often associated with inadequate control of feed flow and level.

Tube and tube-to-tubesheet joint failures in horizontal steam and vapor generators. Horizontal steam and vapor generators are designed to operate with liquid maintained at a level below which all the tubes are submerged and above which is a vapor space. In such operation all the tubes in a single-pass unit are at approximately the same average metal temperature. The metal temperature is determined by the hot fluid inlet temperature, the feed inlet temperature, the film coefficients of heat transfer, and the degree of fouling on each side. (There may be modest differences in tube metal temperature between tubes near the bottom and those near the top due to hydrostatic head and incomplete mixing of the feed.)

The tube metal temperature of the tubes in multipass generators declines from pass to pass as the hot fluid surrenders heat to the feed. The plane of the bends in U-tube equipment affects the influence of low liquid level on the structure.

If the level falls below the upper tube rows, the metal temperature of tubes in the vapor space will lie between the shell-side steam or vapor temperature and the temperature of fluid entering the exposed (dry) tubes. This is because the film coefficient of heat transfer of dry steam or vapor is low relative to that of the heating medium. The submerged (wet) tubes will come to a temperature between the temperature of the pool of feed and a temperature somewhere between the hot fluid entrance and exit temperatures. Therefore, free expansion of the dry-tube lengths will be greater than that of the wet ones.

Failure patterns in horizontal straight-tube designs. Because of differential expansion, exposed tubes in straight-tube units are compressed and submerged ones stretched. The magnitudes of the resulting compressive and tensile forces depend on the metal temperatures, mean thermal coefficient of expansion between construction temperature and metal temperatures, and the dry-/wet-tube cross-sectional area ratio.

The dry-tube cross-sectional area is considerably less than that of the submerged ones. Therefore, they endure greater unit loads, causing damage where the dry tubes penetrate tube supports and where they are joined to the tubesheets. This is because the compressive forces may be large enough to bow or collapse the exposed tubes, break their tube-to-tubesheet joints, and bend the bundle into a banana shape. The bending causes the tubes to bind in the tube supports, where they become crimped and abraded. Fluctuating levels that alternately expose and submerge the tubes enhance the abrasion and may cause fatigue failures.

Failure patterns in horizontal U-tube designs. The pattern of failures resulting from low level in horizontal U-tube configurations depends on whether the plane of the bends is horizontal or vertical. If the plane is vertical and the inlet pass is at the bottom, all the inlet legs will be at a metal temperature between that of the pool of liquid being vaporized and a temperature somewhere between the hot fluid inlet and outlet temperatures. Because the thermal coefficient of heat transfer in the vapor is low relative to that of the fluid in the tubes, the exposed leg temperature will be about the same as that of the hot fluid exiting the inlet leg.

The outer bends are the ones with the greatest bend radii and therefore are the most flexible. This protects the structure somewhat from the effects of the difference between hot- and cold-leg expansion and upper-leg exposure. Usually, the larger bends can deflect substantial amounts under the imposed loads without harm. Damage patterns in such equipment may consist of fatigue failures at the U-bend and tube and support wear at support plate holes.

When the plane of the bends is horizontal, all the flooded tubes are at approximately the same temperature: between the bulk temperature of the feed pool in the shell and a temperature between the hot fluid inlet and outlet temperatures. Exposed tubes which are not cooled by feed will be at approximately the hot fluid inlet temperature. The effects of the differential expansion would be inconsequential if the axial tube movement were to be unconstrained. However, most of these units are built with vertically cut semisupports held at fixed distances from each other by tie-rods and spacers. Tie-rod and spacer assemblies in the hot vapor space expand more than do ones in the feed. This tilts the tube supports, which constrains the free movement of the hot tubes.

Failure patterns consisting of breaks in upper tube legs at supports nearest the U-bends and leaks in the tube-to-tubesheet joints indicate inadequate level control in such generators. The latter is striking since all the broken joints are in the same plane.

Failure patterns in vertical fixed-tubesheet steam and vapor generators. In vertical fixed-tubesheet units vaporizing in the shell, hot liquid heating media are usually routed up and hot-gas media down through the tubes. Hot gas may be the product of catalytic combustion in a bed above the top tubesheet, or from reactions in catalyst-filled tubes. The fluid to be vaporized enters the shell near the bottom tubesheet. The shell may be fitted with an expansion joint to accommodate differences in thermal expansion between the shell and tubes. Its design deflection depends on the shell and tube thermal coefficients of expansion and calculated average metal temperatures at start-up, shutdown, normal operating, and anticipated upset conditions.

The failure pattern to look for in vertical fixed-tubesheet generators depends on the heating medium, how the unit was started and shut down, the constancy of feed flow, and the maintenance of level position. To start operation correctly, the shell should be filled to an appropriate level before hot fluid enters the tubes. But if hot fluid flow precedes running feed into the shell, the tubes will be heated to the hot fluid temperature while the shell remains at approximately ambient. If a shell expansion joint has not been provided, tubes may buckle and tube-to-tubesheet joints break. If one not designed to accommodate the differential expansion at this condition is installed, it may be damaged beyond repair by starting this way.

When cold feed is turned on, it splashes on the underside of the hot top tubesheet and may initiate ligament cracks and tube-to-tubesheet joint failures. This creates a generalized pattern of cracked ligaments and broken tube-to-tubesheet welds. The top tubesheet may also be bent out-of-flat.

Damage patterns in once-through nuclear steam generators. Identifying damage and making repairs to tubes and tube-to-tubesheet joints in once-through nuclear steam generators is done remotely. Probes designed for the purpose are used to inspect a minimum of 5 percent of the tubes and their joints to the tubesheets at each refueling outage (~12- to 24-month intervals). This amounts to about 750 tubes. When defects are found, additional (and as many as *all*) the tubes and joints are inspected. It is imperative that primary-side hot, radioactive water not leak into the secondary (steam) side. When instrumentation discloses a leak, under current rules, the plant operator has a maximum of 72 h to fix it. Consequently, midcycle planned shutdowns may be done to avoid unplanned forced outages.

The tube failure mechanisms in these generators are well known.* They consist of (1) fretting due to flow-induced vibration, possibly aggravated by erosion-corrosion; (2) pitting as discussed previously; (3) wastage or localized corrosion near the tubesheet (currently controlled by appropriate water treatment and water blowdown rates); (4) denting in which growth of corrosion products deforms the tubes where they pass through carbon steel supports; and (5) SCC.

Caustic intergranular attack (IGA) and SCC may occur in crevices between incompletely sealed expansions of tubes into tubesheet holes and where incompletely removed sludge leaves a residue on the secondary side of bottom tubesheets. Because the skin temperature of the hot side is higher than that of the cold side, caustic may concentrate in the crevices, promoting the IGA.

SCC may also attack where the expanded-in section of the tubes emerges from the tubesheet because of high residual stress in the

*For U-tube steam generators, Foster Miller Inc. of Waltham, Mass. has developed a robot-controlled lance that travels in the center lane of the inner U-tubes to remove the sludge in pressurized-water reactors (PWRs).

transition zone from expanded to unexpanded material. Older units in which the expansion was done by mechanical rolling instead of hydroexpanding may be more subject to such attack.

Failure patterns in vertical, channel-down recirculating U-tube steam generators. Some nuclear power stations have recirculating (U-tube) steam generators. This design uses many fewer tubes than once-through types. Capacity is achieved by providing more units. They are subject to the same kinds of failures as the once-through type. However, another possible source of tube failures is in the U-bends because of inadequate vibration protection or SCC in the smallest-bend region.

The tolerances applied to bending tubes for nuclear steam generators are smaller than those for most other applications. The radius tolerance is ±0.8 mm (±0.03 in). Compare this with requirements of Paragraph 13.5 of Specification SA-688/SA-688M for austenitic feedwater tubes in ASME Code, Section II, which states, "The leg spacing measured between the points of tangency of the bend to the legs shall not vary from the value (2*R*-specified tube outside diameter) by more than 1⁄16 in (1.5 mm) where *R* is the center-line bend radius," and Paragraph 13.6, which states, "The bent portion of the U-tube shall be substantially uniform in curvature, and not to exceed ±1⁄16 in (1.5 mm) of the nominal center-line radius."

Manufacture to the tighter specification for nuclear steam generators employs a technique in which the dies used to make the smaller bends contain the tube and minimize ovality. Combined with lattice-type supports, the stricter requirements reduce the tendency of these U-tubes to fail.

Failure patterns in closed feedwater heaters[3]

The general failure patterns described above apply to closed feedwater heaters as well as process exchangers. However, some additional patterns are unique to feedwater heaters. Chapter 13 (on which Ref. 3 is based) discusses what to look for in failed feedwater heaters during postmortem examinations. Here we deal with discerning patterns in operating ones, prior to plugging, with the goal of preventing further deterioration.

Tube-side flow-induced failures in feedwater heaters. Full-access channel construction is more costly than manway access. Consequently, despite the maintenance advantages of having full access to the pass partitions and tubesheet, more manway-access units are installed than full-access ones. In either, the tubes are usually joined to the tubesheets before the channel is welded to the tubesheet. Most of the time, the channel-to-tubesheet weld and channel barrel must be heat treated. The heat treatment may affect the peripheral tube-to-tubesheet joints.

The ability to inspect and repair outlying joints after the channel barrel is attached is limited by their closeness to the barrel and distance from the manway. Therefore, they are the most likely to be troublesome. Unless the channel barrel is extended to allow radial feedwater inlet connections, the nozzles in hemiheads are positioned at an angle of approximately 30° to the tubesheet face. This directs incoming feedwater at the tubesheet. Despite various strategies to limit the effects of feedwater impingement on the tubesheet, protruding tube ends, and tube-to-tubesheet welds, it may erode these regions. Wormholing occurs if there are cracks or through-weld porosity, or if the crowns of the seal welds are washed away. Since the 6.3- to 9.5-mm (¼- to ⅜-in)-thick weld overlay deposit is harder than the parent metal and the start of postweld tube expanding is about 9.5 to 12.7 mm (⅜ to ½ in) behind the fusion-welded tubesheet facing, axial and radial wormholing take place behind the fusion- or plasma-welded tubesheet facings.

Under some circumstances, the feedwater may flow at twice the design-point rate. If such flow continues for substantial periods, it may wear the tube ends paper-thin. The authors have examined and videorecorded high-pressure, Monel-tubed heaters with such tube ends. To protect against erosion at the inlet ends of the tubes and tube-to-tubesheet joints, inlet-pass tubes of carbon-steel-tubed high-pressure and supercritical pressure heaters are usually fitted with ferrules or wasting-plate shields.

Pulsating feedwater flows and water hammer may damage tube-to-tubesheet connections. Pulsating flows cyclically stress the joints and fatigue the welds. Water hammer may bend even thick tubesheets of high-pressure units elastically or plastically. In addition to various kinds of damage to the pass partitions and channel, pulsating flows and water hammer may crack the tube-to-tubesheet welds.

Tube failures in feedwater heater desuperheaters. In feedwater desuperheaters, peripheral tubes usually fail first. The probable cause of failures just under the steam inlet is a broken or broken-off impingement plate. If this condition is discovered and attended to early, other tubes may not be damaged by flopping-around pieces of the broken plate. If not, tubes deeper in the bundle will also be damaged.

Tube failures in desuperheaters may also be found on either side of the inlet in the vicinity of the impingement plate, at locations where the tubes penetrate cross-flow baffles, midway between baffles or at all these locations. Failures at and midway between baffles in units operated near the design-point flow are probably the result of a baffle system not suitable for holding tube vibration amplitudes to tolerable levels.

If damage has not progressed too far, you can rectify these situations as described in Chap. 11. However, if the heater will continue to be operated substantially overloaded, and it is a condition for which it was not designed, no surgery or alterations will save it from a premature demise.

Tube failures in feedwater heater condensing zones. The following paragraphs discuss some patterns of tube failure in feedwater heater condensing zones.

Tube damage just beyond the exit of the desuperheater. A pattern of tube damage just beyond the desuperheater exit indicates that the heater probably operated at a wet-wall condition.

When a power plant is maintained on line at a small fraction (less than 20 to 25 percent) of the full-load-design basis conditions, the heater design must incorporate the low-load conditions in the desuperheater design to prevent the desuperheater from having so much surface as to cause premature steam condensation on the tubes near the desuperheater exit to the condensing zone.

In wet-wall condition, the surface available for cooling in the desuperheater is larger than that required for the flow. The metal wall temperature may be low enough to eliminate the superheat safety margin the manufacturer provided in response to the purchase specification. Not only is all the superheat removed from the extraction steam but also some latent heat. When condensate droplets form in the presence of the relatively high velocity of the steam exiting the desuperheater, they are entrained and hurled against the tubes and first support plate in the condensing zone with high velocity. As a result, wet steam exits the desuperheater and erodes the peripheral tubes.

Even when the droplet-laden steam does not immediately abrade the tubes to failure, it wreaks havoc with the first and sometimes second and third condensing-zone tube supports. It thins their edges and enlarges the tube holes to the degree that the ligaments between them may be completely washed away. The enlarged holes have the effect of increasing the tube span, leading to vibration failure from interspan collisions at midspan and fatigue.

Wet-wall damage rarely occurs in properly designed heaters operated at load conditions close to or greater than the design point. Heater manufacturers, apprised in advance of projected low-load operation, will adjust the design surface allocated to desuperheating to maintain sufficient residual superheat in the steam exiting the desuperheater to prevent wet wall from occurring under these conditions. The residual superheat needed to prevent wet wall will range from about 15 to 40°C (~25 to 75°F) depending on the steam extraction pressure and flow rate at the specified low load condition.

Tube erosion at inlets of drains from higher-pressure feedwater heaters. Some condensate flashes as it travels from a higher- to a lower-pressure heater. If the drains inlet is located behind the U-tubes, and the tubes are not sufficiently shielded, the U-bends are eroded away progressively from outer to inner tubes. If a deficiently shielded drains inlet is located along a straight leg, erosion will proceed from outer straight legs to inner ones. The tube erosion may be accompanied by wearing away of the heater shell.

A good fix is to install a separate flash chamber between the two heaters. When flashing drains enter behind the U-tubes, the situation may be corrected by relocating the drains inlet to a shell extension. Cut off the shell head, install a shield and stainless-steel-lined or clad cylinder behind the U-bends, and weld the cutoff head to the cylinder.

Condensing-zone tube failures near saturated steam inlets. In units operated at or near design-point flow, failures near the impingement plate are likely to occur if the impingement plate design is inadequate. It may not be wide enough, or may be poorly placed. If it is a perforated-plate design, it may be inadequate for the flow.

Condensing-zone tube failures near the tubesheet. Tube failure near the tubesheet in the condensing zone usually indicates inadequate venting. It may be accompanied by corrosion of the back side of the tubesheet. To prevent progressive deterioration, follow the manufacturer's recommendation for the use of start-up and operating vents.

Another cause of tube failure of tubes subject to SCC in this region is the residual tensile stress in the transition from expanded to unexpanded sections of tubing emerging from the tubesheet. The level of stress is increased when the tubes vibrate even within tolerable vibration amplitudes. The SCC failure may be combined with fatigue resulting from tube vibration.

Subcooler failure patterns in feedwater heaters. Patterns of damage to subcooler tubes are affected by whether the design is full-pass, partial-length, or partial-pass, full-length. Horizontal heaters may be either. However, vertical head-up subcoolers are partial-pass full-length, and head-down ones are full-pass partial-length designs.

A recorded DCA substantially higher than designed may indicate (1) that level was maintained so low that saturated steam entered the subcooler and damaged the tubes and baffles, (2) that saturated steam leaked through the end plate—especially of older heaters with end plates thinner than approximately about 76 mm (~3 in)—and damaged the tubes, and (3) perforations in a tube or tube in the inlet leg within the subcooler and in the outlet leg.

Horizontal full-pass partial-length subcoolers may exhibit a pattern in which failures begin in upper rows and work their way down. The axial position of the failures indicates the cause. Failures located near the end plate were probably caused by uncondensed saturated steam entering the subcooler through the annular spaces between the holes in the end plate and the tubes and eroding the holes in the end plate and baffles.

The latter phenomenon usually happens to units with subcooler end plates thinner than about 76 mm (~3 in). The tubes are not sealed to the end plates of full-pass partial-length subcoolers. Therefore, a condensate seal is required between the condenser and the subcooler to prevent saturated-steam blowby. In *blowby,* steam above the liquid level in the condenser section enters the annular space between the tubes and holes in the end plate. The amount of steam entering the annulus between the tube hole I.D. and the tube O.D. will depend on the pressure differential between the steam side and condensate side of the end plate, the flow area available, and the length of cold tube surface in the end plate available to condense the steam to form a capillary seal. If the end plate is thick enough, all the entering steam is condensed by contact with the cold tube wall and the condensate-cooled end plate. But if the end plate is too thin, the entering steam is only partially condensed and two-phase vapor and liquid will blow through the annular spaces.

An early clue to two-phase flow into the subcooler is increasing DCA with progressively more saturated steam entering the zone. Because of the insufficient tube-wall surface in the end plate, wet steam rushing through the annulus erodes the hole. When the steam makes contact with the condensate in the subcooler, it collapses. The resulting water hammer causes the tubes to fail from fatigue. When steam is able to penetrate several tube holes in the end plate, the resulting water hammer may be severe enough to crack the first baffle. The erosion gradually worsens, with incoming wet steam eventually eroding the subcooler baffles. Improper end-plate drilling with looser-than-standard tolerances contributes to this problem.

You may be able to extend the life of such heaters by reducing the spaces between the tubes and holes. You can do this by hydroexpanding the tubes at a pressure adjusted to just cause the tubes to make contact with the hole, i.e., not make an hydraulically tight seal.

Most current specifications for new and replacement feedwater heaters require subcooler end plates to be at least 3 in (~76 mm) thick and drilled to the special close fit listed in TEMA Standards Table RCB-7.41. For ⅝-, ¾-, ⅞-, and 1-in (~16-, 19-, 22-, and 25-mm)-O.D. tubes, this permits a maximum diametral clearance between the tubes and holes of 0.01 in (~0.25 mm). Applying these criteria ensures there will be a condensate seal in the annular space between the holes and tubes. The seal will prevent saturated steam from flowing into the subcooler, collapsing as it condenses, abrading the tube surfaces and inducing vibration.

Failures in the vicinity of the tubesheet probably result from inadequate removal of noncondensibles at start-up. Air and other noncondensibles trapped at the top of the subcooling zone near the tubesheet, if not vented completely during start-up, will corrode the welds where the flat plate is welded to the semicylindrical enclosure and to the back face of the tubesheet. Failures of either of these welds will cause the DCA to rise precipitously.

When the shroud-to-top-plate welds fail, steam enters the subcooler. This causes the tubes to fail in the top corners from fatigue caused by water hammer resulting from steam collapsing when it contacts subcooled condensate. If the problem is identified early, this kind of failure is usually repairable by rewelding the shroud-to-top-plate welds. You can prevent this from recurring by opening the drain cooler start-up vents at each start-up.

When the weld that joins the top plate of the subcooler to the tubesheet fails, the row of tubes just below the top plate also fails as a result of fatigue. These failures generally start near the middle of the row and work their way outward. None of the schemes for correcting this problem with which the authors are familiar have worked.

In horizontal heaters, if lower-row tubes are failing in the regions of the subcooler inlet and the first baffle, examine records to see if the DCA has regularly been recorded higher than designed. If the level control is set so that the liquid level is consistently below its optimum setting, saturated steam will enter the subcooler inlet during load fluctuations and create an abrasive steam-condensate flow that will chew up the tubes and thin the ligaments in the baffles, bringing on vibration damage.

In nuclear power stations condensate flow in feedwater heaters is enormous. Operators concerned with water induction into the turbines at high liquid levels may maintain levels too low to ensure that all tubes in the subcooler are submerged. This will cause higher-than-design DCA, signaling that there are steam leaks through the end plate. Erosion and enlargement of the holes where unsubmerged tubes penetrate the end plate and tube damage in the rows nearest the flat divider plate will result.

You may forestall and possibly rectify these problems by setting liquid level by optimizing the liquid level by testing. Do this by measuring DCA at various level increments and plotting DCA against liquid level (Fig. 10.1). Choose as a set point a level sufficiently above the height at which the slope of the curve approaches zero (curve approaches the specified DCA asymptotically) to accommodate level changes due to load variations. If damage has not progressed too far,

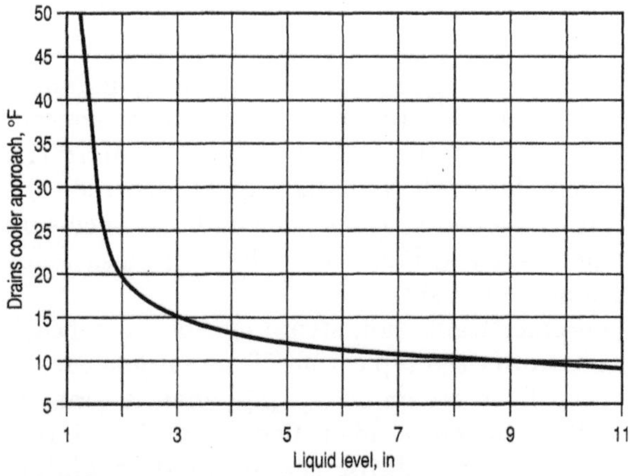

Figure 10.1 Guide to locating normal liquid level in horizontal heaters.

cure the problem by adjusting the level controller to maintain a level high enough to ensure that the snorkel is immersed in condensate under any loads or load changes.

If you find a pattern of tube damage throughout the subcooler accompanied by high DCA, ask the operators if they have heard popping sounds when standing near the subcooler end. This can indicate failures in the welds joining the semicylindrical wrapper to the flat divider plate. If the operators are unsure about what they heard, after plugging and restarting the heaters, put your ear to a screwdriver pressed through the insulation until it makes contact with the surface of the shell in the vicinity of the subcooler and listen for popping sounds.

Tube damage from stray superheated or saturated-steam flow into the subcooler. If a feedwater heater's inlet tube leg is perforated in the subcooler and its outlet leg perforated in the desuperheater or condenser section, superheated or saturated steam will flow into the subcooler *even when both ends of the tube are plugged at the tubesheet.* The operator will perceive this as an increase in the DCA that cannot be cured by raising the condensate level. The repair for this problem is to thimble-plug or sleeve the tube beyond the perforation.

Full-length partial-pass subcooler failure patterns are similar to those of the full-pass partial-length type. View the drawings or the tubesheet channel face of partial-pass ones to find out which cold legs are in the subcooler and which are in the condenser to simplify examining plugging patterns. Failures in welds of the flat divider plate to semicylindrical enclosure produce patterns similar to those in the full-pass partial-length type.

Effects of operating mode

Properly designed exchangers, made from materials resistant to corrosion, operating continuously at steady-state design-point conditions require the least maintenance and have the longest life. Ordinarily, moderate deviations from the design point do not seriously affect exchanger life or introduce requirements for extra maintenance. Wide deviations above or below design points can cause more-rapid-than-expected deterioration, especially if frequent or continuous.

Continuous flow greater than design point. Greater-than-design stream flows cause exchangers to behave as if underdesigned. Elevated flows increase velocities, producing pressure losses higher than predicted for the design case. Pressure loss increases approximately as the 1.8th power of flow velocity. Inlet turbulence effects, such as erosion and erosion-corrosion on the channels, tubesheets, tube ends, pass partitions, and turnaround compartments of multipass units, also increase with velocity.

On both sides of the tubes, higher velocity reduces the tendency to foul and increases the film coefficient of heat-transfer. Because of the higher heat transfer rates, the tube-wall average temperature varies from that calculated for the design.

In fixed-tubesheet exchangers without shell expansion joints, the tube metal temperature change can cause excessive stress in the tubes, shell, tubesheets, and tube-to-tubesheet joints. In exchangers constructed with expansion joints, the flexible element may undergo greater deflections than planned because of the different tube-wall metal temperature. The greater movement can shorten the expansion joint life.

Hotter-than-designed tube metal temperature may degrade or char sensitive fluids, which may lead to tube corrosion. Lower-than-designed tube metal temperature may cause fluids close to their freezing point to freeze on the tubes. Tubes may buckle or bow as a result.

Austenitic stainless-steel and titanium tubes can tolerate linear velocities considerably higher than the TEMA and HEI standards recommend without damaging the protective oxide coating on which they depend for corrosion resistance. However, carbon-steel ones and most nonferrous tubes lose corrosion resistance when the flow velocity is much above the limits listed in the standards.

The acceptable exit velocity of fluid between impingement plates, shell, and tube bundle permitted by the TEMA standards is an average based on dividing the volumetric flow per unit of time by the cross-sectional escape area. Linear velocities in some regions of the escaping stream may far exceed the average even at design point conditions.

Greater-than-design-point flow may damage tubes that design-point flows would not harm. The higher flow also increases the likelihood of shell-flow-induced vibration and erosion of the shell and cage.

Continuous flow less than design point. Continuous flows less than designed make an exchanger act as if it is oversurfaced. Fouling rate varies inversely with flow velocity, making foulants more likely to coat the surfaces than at design flow. As tubes foul, the exchanger appears to approach the design condition. However, the thickness of the fouling layer deposited on the tube surface may be so great that cleaning is impossible or costly. Despite the excess surface, the exchanger will probably have to be cleaned more frequently than anticipated. Tube metal temperature will also deviate from the design metal temperature with results similar to those described above. In process condensers that condense on the shell side, a low flow rate in the shell with design rate in the tubes causes low tube-wall temperatures and reduces the shell-side pressure. The low-pressure gas and vapor flow may induce tubes to vibrate that would not do so at design conditions. Low shell-side pressure may also impede condensate drainage in condensers that depend on gravity flow from the condensate drain.

Deviant stream inlet temperatures. The effects of higher or lower stream inlet temperatures are like those described for undersurfaced and oversurfaced units.

Deviant operating pressures. Safety considerations determine the set pressures of relief devices. Operating pressure must be below the relief pressure setting to prevent leakage through or blowout of the device. Modest pressure elevations within the design limits rarely affect exchanger life. It may be necessary to adjust level-control devices to protect tubes that must remain submerged in condensate when the operating pressure deviates from design.

Cyclic operation. Differences in thermal expansion during start-up and shutdown subject exchangers in cyclical operation to cyclical application of pressure and loading. The direction of thermal loading may change from start-up to shutdown. Exchangers used alternatively to heat and cool (e.g., switch condensers) not only bear such loads but almost always endure reversals in direction of loading. U-tube designs with sufficient separation between bends are the least likely to fail under these conditions. This is because the individual tubes can expand and contract without affecting each other.

Straight-tube designs, on the other hand, may fail in a variety of ways when they are operated cyclically. Expansion joints of single-pass, fixed-tubesheet exchangers must be able to accommodate the anticipated number of cycles and reversals of direction of deflection, but also the on/off application of pressure. Transient differences between the temperature of the tubes in inlet and subsequent passes of cyclically run multipass units are likely to be greater than when operation is steady state. Therefore, these units undergo greater interpass thermal loadings.

These occurrences, combined with the on/off or switch operation, induce repetitive tubesheet flexures, numerous loadings and unloadings of tube-to-tubesheet joints, and bundle bending and unbending. These events lead to patterns of broken tube-to-tubesheet joints in the inlet and outlet passes, SCC failures where expanded tubes emerge from the rear faces of tubesheets in the regions of inlet and outlet passes, and erosion damage where tubes penetrate baffles and supports.

Changes in operating mode before failures

Many exchangers that operated satisfactorily for long periods at moderate rates of deterioration degenerate rapidly after a change in operating mode. When a power plant designed for base-load operation is changed to peaking load service, feedwater heaters not designed for the standby low-load operation may suffer wet-wall damage. Feedwater heaters tubed with carbon steel or 70-30 copper-nickel tubes which will last for 20 years in base-load operating conditions will corrode rapidly in peak-load operation unless the shell side is blanketed at each shutdown with steam or nitrogen to prevent air from coming into contact with the hot, wet tubes. Wet air on the tubes causes formation of a corrosion layer. The layer builds up with each heatup-cooldown cycle until it spalls off. The corrosion begins again and the cycle repeats until the tube fails from through-wall pitting. The changeover to peaking service of a base-loaded station with carbon-steel-tubed feedwater heaters may lead to rapid tube corrosion despite inerting when the heaters come off the line. Acid formed by noncondensibles and wet steam in the regions of the tubesheets rapidly corrodes the tubes, shortening heater life.

More subtle is the effect of large numbers of unanticipated startups and shutdowns. This might be caused by converting an exchanger designed for continuous production to batch operation or starting up and shutting down a plant with overcapacity relative to the market for its products. In these circumstances, fixed-tubesheet units are apt to suffer fatigue cracks in the shell or barrel in the vicinity of the tubesheets or in the welds of the tubesheets to the channel or shell. These failures may spread into the tubesheets. This happens when exchangers designed for a modest number of self-limiting plastic deformations acceptable for a specified number of cycles are cycled more times than specified.

Locating Tube Failure Positions

Maintenance workers are almost always under pressure to restore shutdown units to service as quickly as possible. Therefore, usual practice is to restore production as quickly as possible without locating tube failure positions or taking the time required to get to the root of the problem. Following this procedure may lead to an exchanger's early demise and replacement. Furthermore, it provides no insight to the origin of the problem and corrective measures that might prevent similar difficulties in the replacement. Here are some simple techniques for determining tube failure axial positions.

Preparation for probing tubes

Some preparation is required before tube probing. First, isolate *both sides* of the exchanger from the process or utility streams. The preferred way is to insert line blinds in the lines to and from the unit. However, closing leaktight valves may be acceptable if the fluids on both sides are at temperatures below 50°C (122°F) and not irritating, noxious, or toxic. Where contact with process fluids is hazardous, use double-block valves with the space between them fitted with a vent and drain. *Do not* enter a feedwater heater channel unless the feedwater lines are either double-block-valved as described and precautions have been taken to prevent extraction steam from leaking through tube holes into the channel or the unit is off line.

Nearly always, you must clean the tube interiors before attempting to locate leak positions. If the tube-side stream is not hazardous, irritating, or noxious, it may be feasible to collect corrosion products by inserting an air nozzle into the individual tubes, blowing through and collecting the material blown out in a strainer placed at the other end. Chemical analysis of the collected material may provide clues to the failure mechanism.

As a bare minimum, flush out the process fluids and water-wash the tube side. Blow-dry the tubes with clean, filtered dry air before examining the tube interiors.

Commercial devices for locating tube leak positions

Testers are available for testing individual tubes hydraulically or with air pressure to determine which ones are leaking. But at present there are no commercial devices designed specifically to ascertain where tubes are leaking. You may be able to locate penetrations that are large enough with a borescope. Although the IRIS system described in Chap. 8 can pinpoint even very small penetrations, the cost of hiring a contractor to set it up and probe the tubes makes it an uneconomical option.

The Expando Seal Technologies (EST Inc.) G-150 tube testing gun set shown in Fig. 10.2 is useful for testing tubes for leaks. It consists of an air-injection gun and a tube-plugging gun. The EST literature states that tubes can be tested using air pressure from 40 to 125 psi (~276 to 862 kPa) and that it has interchangeable seal and washer sets that allow testing tubes from 0.250 to 1.2 in (~6.35 to 30.5 mm) I.D. It depends on loss of pressure between the two sealed ends, shown on a gage on the plugging gun to indicate a leak. The EST Inc. G-450 air-activated tube tester (Fig. 10.3) is designed for testing individual

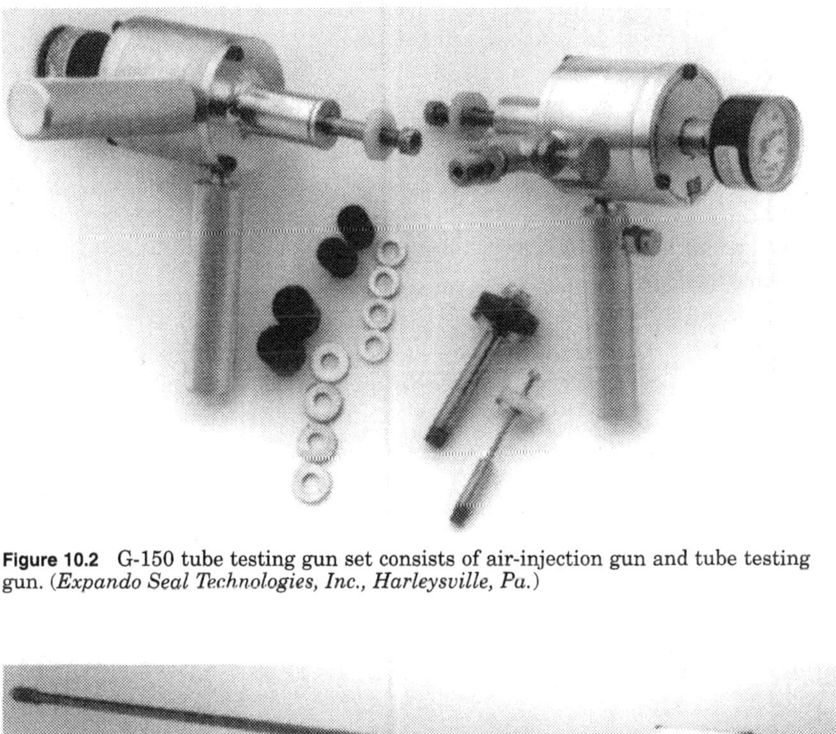

Figure 10.2 G-150 tube testing gun set consists of air-injection gun and tube testing gun. (*Expando Seal Technologies, Inc., Harleysville, Pa.*)

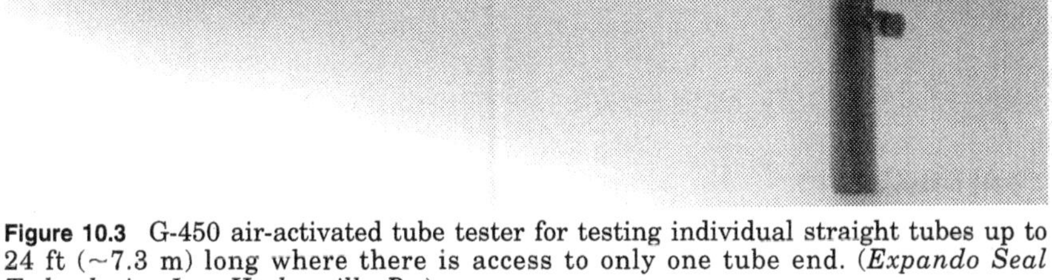

Figure 10.3 G-450 air-activated tube tester for testing individual straight tubes up to 24 ft (~7.3 m) long where there is access to only one tube end. (*Expando Seal Technologies, Inc., Harleysville, Pa.*)

straight tubes in this size range where there is access through only one tube end. It consists of a pull-rod and compression-tube and air-control unit, fitted with a pressure gage. Activating the tester forces the pull rod to compress the remote- and near-end elastomeric seals and fills the tube with air. As with the G-150 unit, leaks are indicated by pressure dropoff shown on a gage. To position the remote-end seal, compression-tube extensions are available in 2- to 6-ft (~0.62- to 1.85-m) lengths.

Pressurizing the shell and traversing the tube with a plug

One method in long-time use to pinpoint the location of a tube leak is to pressurize the shell with air and push a tight-fitting rubber sealing plug into the tube. When a seal is effected, you traverse the length of the tube until the plug passes a perforation. Air leaks out, indicating that the break position has been found. This system's disadvantage is that it locates only one leak position.

Finding leaks with the MGT axial tube leak position finder

The way the MGT device for locating the axial position of a tube leak shown in Fig. 10.4 works is self-evident. Connect an armored air hose or section of pipe connected to the tapped end. Install an air-pressure gage in the air line. Lubricate the O-rings and traverse the tube with the position finder for its full straight length. If you use pipe sections, attach additional lengths as you run the position finder along the tube. When the finder encounters a perforation, the air pressure drops. Mark the position of the leak and continue to traverse until the whole tube has been scanned. If the tube is a U-tube, and no leak is encountered in either leg, you can assume that the perforation is in the U-bend section. Tube interiors must be free of excessive scale or other foulants that would impede the device from traversing the tube.

Borescopy

Borescopy, discussed in Chap. 5, is an excellent way to examine the general appearance of tube interiors. Using flexible probes, available in various lengths, you can find and record on videotape for off-site review and study both axial and radial leak positions (Fig. 10.5).

The configuration of the shell side determines if it can be examined. Pull-through floating-head exchangers (TEMA type T) and most closed feedwater heater shell sides have sufficient space for probing through a nearby nozzle or a hole opened in the shell in the target region. Probes are difficult to control, and the procedure is time-con-

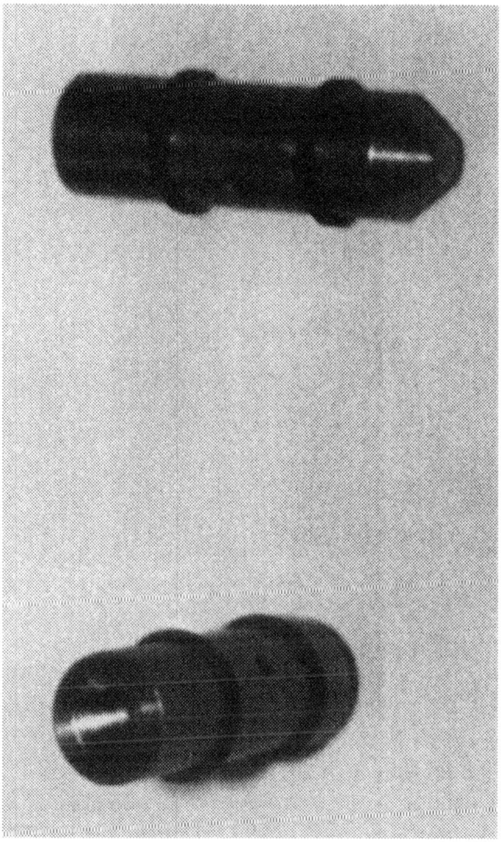

Figure 10.4 MGT Inc. tube leak axial position finder. (*MGT Inc., Boulder, Colo.*)

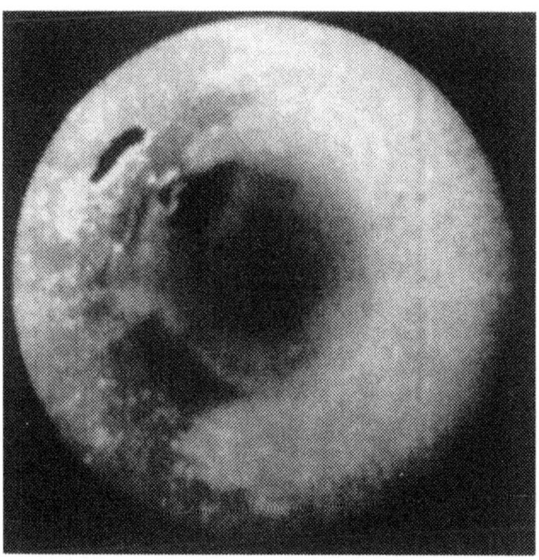

Figure 10.5 Borescope internal tube examination showing perforation. (*Source unknown.*)

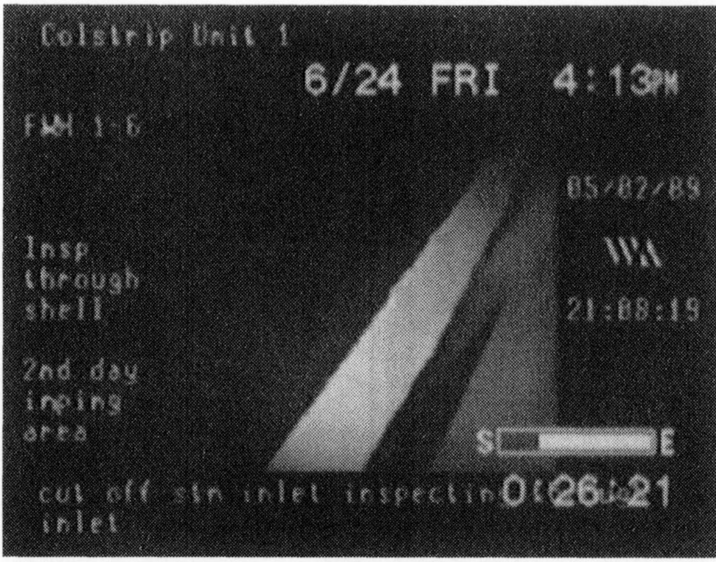

Figure 10.6 Frame from videorecording of borescope examination of feedwater heater impingement area. (*Montana Power Corp., Colstrip, Mont.*)

suming. Nevertheless, you can learn much by probing. An example is discovering broken impingement plates (Fig. 10.6) and vibration damage to the tubes and support-plate holes. Unless the shell interior is too fouled to permit access, no special cleaning is required.

Note that after using a borescope to look inside tube surfaces and surrounding shell regions, it is usually desirable and often necessary to extract a representative number of failed tubes for chemical analysis and metallurgical examination.

Eddy-current examination

Skilled examiners cannot only detect and characterize discontinuities in the tube walls but also determine the degree of wall thinning. They must have a thorough understanding of the signatures of the eddy-current scan. Specimen tubes of material, diameter, and original wall thickness of the ones to be scanned should be on hand. The technician may prepare failure specimens for comparison purposes. Eddy-current examination is usually best performed by an organization specialized in the work. The journal of the American Society for Nondestructive Testing regularly publishes lists of such organizations. It is not always necessary, but for reliable results, hydroblast tubes that have been in service before eddy-current examination.

Disadvantages of eddy-current examining an isolated tube failure are the time required and the cost of bringing in a contractor to perform the scan when in-house equipment and personnel are not available. Therefore, it is seldom performed unless a substantial number of failures have occurred over a relatively short period. Usually eddy-current examinations are scheduled during the plant shutdown, but by that time, it may be too late to save the tubular.

Ultrasonic examination

The Internal Rotary Inspection Scan (IRIS) system, developed for refinery use and licensed by Shell Development Corporation, can give a complete picture of the tube-wall thickness, pits, and perforations of a failed tube. Its present-day application is no longer confined to oil-refinery applications. The system is excellent for examining straight tubes in any service including feedwater heaters and steam surface condensers. Because IRIS requires highly skilled examiners, employ only licensed contractors who can demonstrate their experience.

As stated in Chap. 5, IRIS is time-consuming. Furthermore, you cannot make reliable measurements unless the tube is clean. Therefore, preliminary hydroblasting is usually required. The time required and cost of preparing for and bringing in a contractor make it generally impractical to scan for isolated failures. IRIS is almost always done only during scheduled outages.

Finding Tube-to-Tubesheet Joint Leaks

Probably the most effective way to find tube-to-tubesheet joint leaks is to perform either a halogen or helium sniffer test. These procedures are somewhat time-consuming, and it is almost as effective to bubble-test. Although fluid penetrant examination is a good way to identify where welded joints are leaking, it is rarely done before repairs. Sniffer and bubble tests require pressurizing the shell. This may be inconvenient or costly. Therefore, consider using a commercially available vacuum joint testing gun.

The EST G-650 vacuum tube joint testing gun (Fig. 10.7) is designed for this purpose. The manufacturer's literature states that it uses standard plant air from 40 to 125 psi (~276 to 862.5 kPa). It has interchangeable manifolds and seal washer sets the permit testing tube sizes from ⅜ to 2½ in (~9.5 to 63.5 mm). To test the joint, you insert the gun into the tube end. Press the air cylinder air-control-valve button, and air under pressure fills the cylinder. This expands a seal inside the tube and pulls a second seal firmly against the ligaments surrounding the tube end. You then press the vacuum air-control-valve button. Doing so diverts air through a venturi to create a vacuum around the tube joint area. Reduction of vacuum shown on a

Figure 10.7 EST 650 vacuum-tube joint-testing gun applies vacuum to tube joint area to test tube joint. (*EST Technologies, Harleysville, Pa.*)

gage indicates a joint leak. The tubesheet must be free of weld metal and other accretions where the seal to the ligament metal is made.

Plugging Heat Exchanger Tubes

The most frequent maintenance on tubulars is plugging tubes and repairing leaking tube-to-tubesheet joints. Typically, a tube has been damaged or a tube-to-tubesheet joint failed and it is impractical or too time-consuming to replace the tube or repair the joint. When attempts to repair a tube joint are unsuccessful, it is also usual practice to plug the tube/hole assembly to close the leak.

When to plug

Tubes deteriorate in many ways for a multiplicity of reasons at various rates. Consequently, seldom do all the tubes fail at the same time. Following is a partial list of factors that determine how individual tubes degenerate:

- Differences in tube physical properties and chemical analysis from heat to heat and tube to tube within each heat
- Residual tensile stress in tubes subject to SCC
- How tubes are protected from corrosion and handling damage before installation
- Quality of U-tube bending and control of subsequent bend heat treatment
- Care with which the tubes were loaded into the cage and installed in the tubesheets
- Baffle and support hole drilling
- Position of the tubes in the tube nest
- Adequacy of impingement protection
- Tube support arrangements
- Variations in flow rates and chemical composition of the tube- and shell-side streams

When the connection of the tubes to the tubesheets start to fail, it is likely that there will be many similar failures. Some factors that determine how tube joints fail are

- Kind of joint
- Hole preparation and drilling tolerances
- Tube-to-tubesheet configurations and joining methods
- Tube expanding and joint welding procedures and practices
- Tube and tubesheet metal combinations
- Pressure and temperature loading cycles
- Corrosivity of the tube-side stream
- Tube entrance velocity

Planned life of feedwater heaters may be 40 years. This contrasts with exchangers in the process industries where return on investment may be as short as 3 years and generally harsh operating conditions are not conducive to long lifetimes.

The factors that determine when to replace a feedwater heater are loss of availability (determined by the frequency of unplanned shutdowns for plugging), increase in cost of pumping feedwater through the plugged heater, and heat-rate elevation of the string of heaters due to loss of tube surface.

In the process industries, tubes and tube joints generally degenerate more rapidly and uniformly than in power generation. Therefore, plugging process tubulars is often a stopgap measure pending retubing or replacement.

Because unplanned outages are very costly, to avoid unplanned shutdowns, routine power-station practice is to plug corroded and eroded tubes before deterioration perforates the wall. For this purpose, historical practice in power stations has been to hydrostatically pressurize the feedwater heater tubes to a large fraction of their yield strength in the hopes of blowing out those that have lost more than acceptable thickness. This assumes that you kept a tube heat map and know the mill-tested yield stress of the tubes.

Now it is more frequent practice to examine inside the tubes by eddy current or other suitable means, estimating wall thickness and plugging tubes that have lost 50 percent of their thickness to wear and corrosion before they fail. Caveats are (1) the accuracy of eddy-current and similar wall-thickness measurements is subject to question since it is dependent on the equipment and skill of the operator, (2) results are skewed by tube supports and baffles, and (3) some methods cannot be used with all tube materials.

Good practice when procuring new heaters is to establish a basis for future comparisons by nondestructively testing the tubes to estimate their thickness just after they are installed or before they leave the factory. During scheduled outages, examine the tubes by the same method used to establish the basis and compare the results with the base. You can then trend the rate of wall thickness loss and perhaps decide to plug before a tube fails.

Before removing feedwater heater tubes from service, consider examining with IRIS. Except for exchangers in which through-tube leakage is costly or hazardous, regular tube thickness measuring is not common in the process industries.

Insurance plugging

High-pressure fluid in the tubes, that leaks through a hole the size of a pinpoint, can impinge on an adjacent tube and erode its outside surface. If there is insufficient time to determine which tube the jet struck, it is rational to plug not only the leaker but also all the surrounding tubes as insurance against an unplanned shutdown.

Insurance plugging may remove as many as six sound tubes from service in addition to the one that leaked. To avoid unnecessarily reducing capacity, consider using removable plugs so that you can examine the decommissioned tubes during a future planned outage and restore undamaged ones to service.

Be aware that removable plugs are intended for *temporary* service and may leak if they remain in place too long. This is because they depend on elastomeric rings for leaktightness. Although the plug manufacturers use the most resistant elastomeric materials they can obtain, exposure to high-pressure, sometimes very hot, fluid causes these materials to disintegrate.

Considerations for insurance plugging feedwater heaters. When a tube in a feedwater heater springs a leak, some of the leaking feedwater may flash into steam. The two-phase flow may erode surrounding tubes which could fail between scheduled outages.

Without determining the perforation's radial position, you cannot know which of the six tubes surrounding the leaker could be damaged. Therefore, you would plug all six as insurance against subsequent failures. Time permitting, eddy-current or ultrasonically scan the surrounding tubes and plug only the ones that show evidence of damage. If you do not have time, consider using removable plugs in the surrounding tubes so that you can examine them at the next scheduled shutdown.

Types of plugs

For connection strength and tightness, plugs depend on friction between either the plug and the tube or the plug and the hole or on welds that seal the leaks. The following paragraphs discuss some common types of plugs.

Friction-fit plugs. Resistance to blowout and fit tightness of friction-fit plugs depends on the contact surface area(s), coefficients of friction between the parts, and forces normal to the contact surfaces developed by driving or expanding the plugs into position. Leak rates vary with differential pressure between the shell and tube sides, viscosity of the higher-pressure fluid, fit tightness, and leak path lengths between plugs and tubes, tubes and holes, and parts of plugs to each other. It is difficult, if not impossible, to seal axial leak paths where the plugs are fitted to the tubes and where the tubes are fitted to the holes. Some explosive plug installation contractors claim they can seal leaks when such paths exist. However, the authors have not observed such success.

The blowout load on friction-fit plugs is the product of the pressure and cross-sectional area on which the blowout pressure acts.

Therefore, blowout load increases as the square of the tube inside diameter.

One-piece taper plugs. One-piece taper plugs (Fig. 10.8) are tapered to an included angle of about 5½° or less. They are commercially available in fibrous materials and most common metals. When a supply is not on hand, most maintenance shops can produce them. Fibrous ones absorb water, which causes them to swell. They are used for low-pressure water service. Metal taper plugs are used for most services.

Taper plugs are driven into the ends of tubes joined to tubesheets by expanding to create the required interference fits between the plug and inner surface of the tube, and increase residual fit pressure between the outer surface of the tube and the tube hole.

When a one-piece taper plug is inserted, it contacts the tube along a circumferential ring. Driving the plug deforms the tube and widens the line of contact into a conical surface. A component vector of the driving force, perpendicular to the conical surface, is distributed over the contact area to create the required interfacial pressure between the plug and the tube. Although the pressure declines radially from that at the plug-tube interface, at the tube exterior it is not zero. The pressure at the outside of the tube adds to the residual interfacial pressure between the expanded-in tube and the surrounding metal.

Taper plugs are not effective for plugging joint leaks when tubes are welded into tubesheets because they depend for tightness on

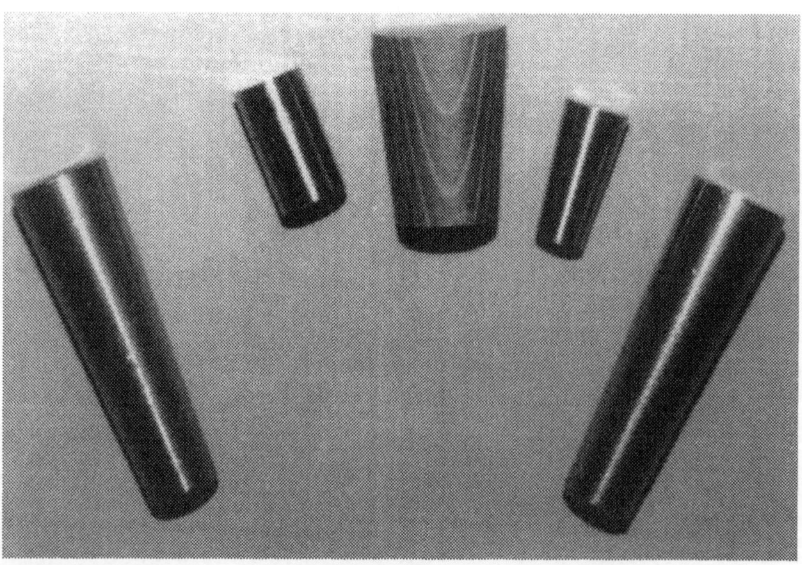

Figure 10.8 Various types of taper plugs. (*MGT Inc., Boulder, Colo.*)

deforming the tube beyond its original expanded condition. Whether the joint is welded only or welded and expanded, the vector of the driving force perpendicular to the tapered surface must first compress the weld to effect a seal. Cracked and porous welds may not be sufficiently compressed to close off leaks. The plug cannot add to residual interfacial pressure between tube and holes because it does not reach the tube expansion zone.

To install one-piece taper plugs, descale the tube end before plugging. Use only enough force to contain the leak. Excessive driving may deform the ligaments between the tube holes and start leaks in adjacent tube-to-tubesheet joints. It is also possible to crack the tubesheet ligaments by overdriving the plugs. If the plug is driven too hard, the pressure on the metal surrounding the tube may exceed the plastic limit ($= 2/\sqrt{3}\sigma_p$, where σ_p = plate yield stress), causing the ligaments to begin to extrude. This is manifested by distortion of the tubesheet surface and the starting of leaks in previously tight adjacent tube-to-tubesheet joints.

These are some of the advantages of one-piece taper plugs:

1. They are readily available commercially.
2. They are inexpensive relative to other types.
3. The plant maintenance facility can usually make them.
4. One-piece taper plugs can be used to provide a backup for welding shut the tube end when friction plugging does not suffice.
5. Tube-side pressure tends to tighten the plugs.

These are some of the disadvantages of one-piece taper plugs:

1. Improper installation can damage ligaments and spring nearby tubes loose. This can be prevented by machining close-fitting tube-end supports somewhat longer than the taper-plug length and inserting them into all the tubes that surround the tube being plugged.
2. A crevice is created behind the plug-tube contact surface.
3. Installation of a one-piece taper plug may further sensitize the tube and the hole to stress-corrosion cracking.
4. Pressure in the shell tends to blow the plugs out.
5. They are ineffective when tubes are welded to the tubesheets.

Two-piece plain plugs. In two-piece plugs (Fig. 10.9), the tapered inner element (pin) is driven into a machined taper on the outer element (ring), expanding it into the tube. To make it possible to insert the assembly into the tube end, initial clearance between ring and tube is

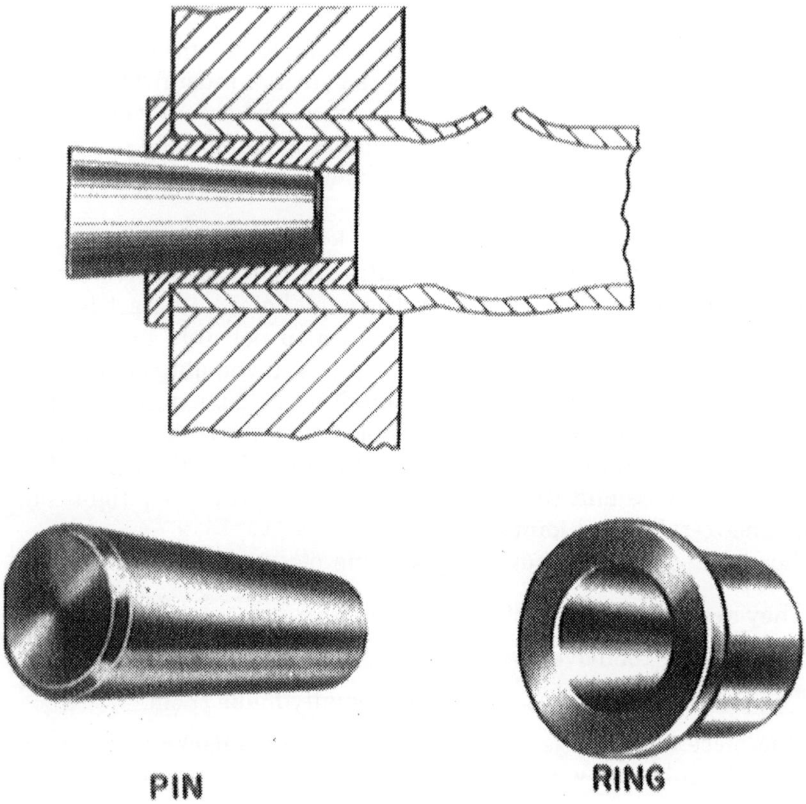

Figure 10.9 Two-piece taper plug showing pin and ring. (*Elliott Company, Dayton, Ohio.*)

required. The theory that applies to expanding the plug into the tube and the secondary expanding of the tube into the tube hole is the same as that which applies when uniform pressure is applied in the tube during expanding. Since the plug remains in place, there is no pressure release. Consequently, the interfacial pressures are the pressures developed by driving.

Tightness requires developing residual interfacial pressure between pin and ring and between ring and tube. But, because the contact surface areas are much greater than those developed by one-piece taper plugs, not as much interfacial pressure is required. The tube end must be free of scale, or tightness will not be attained.

As with one-piece taper plugs, pressure in the channel tends to tighten the seal and pressure in the shell tends to blow the plugs out. A disadvantage of two-piece taper plugs is that the plug may leak at the matched tapers if there is insufficient interfacial pressure.

Serrating the outer surface of the two-piece plug confers these advantages:

1. The serrations may penetrate scale in areas of the tube end that cannot be adequately cleaned.
2. Driving the plug creates high contact forces under the crowns of the serrations. This causes them to dig into the tube and forces the tube to dig into the tube hole to increase tightness.

Explosive plugs. Explosive plugs (Fig. 10.10) may be plain or serrated. They are used with thick tubesheets to seal as much of the tubesheet depth as is desired.[4] These are some of the advantages of explosive plugs:

1. The shock of the detonation usually attenuates so rapidly that adjacent ligaments are not damaged and adjacent tubes are not loosened. However, under some conditions the ligaments may have to be supported by inserting ligament supports into the tubes surrounding the tube undergoing plugging (Fig. 10.11).
2. Explosive plugs can be inserted and detonated where tube ends are inaccessible to plugging by other means.
3. Serrated explosive plugs have been tested at blowout pressures of 10,000 psi (~69,000 kPa) without leakage.

The disadvantages of explosive plugs are

1. Plugging must be done by a qualified technician employed by a plugging or maintenance contractor.

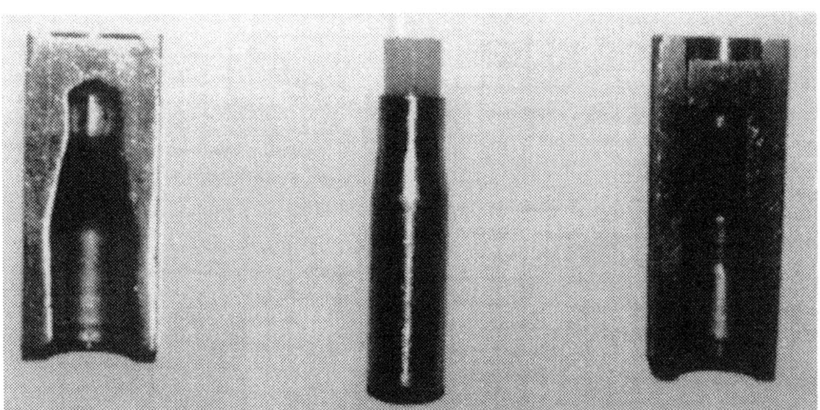

Figure 10.10 Thermal Engineering International, Inc. explosive plug and specimen plugging. (*Photograph by MGT Inc., Boulder, Colo.*)

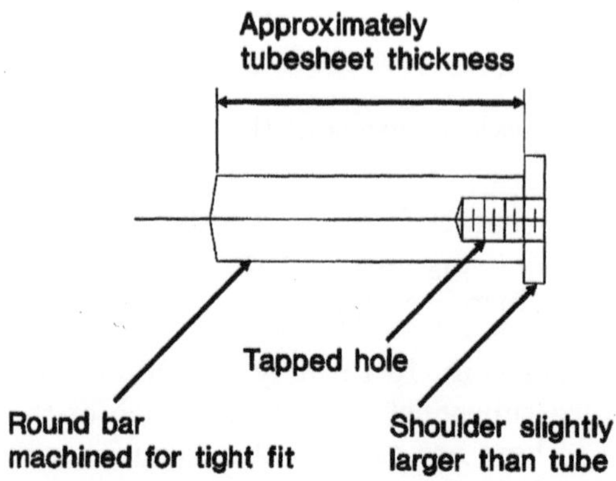

Figure 10.11 Ligament supports to prevent disturbing surrounding tube-to-tubesheet joints.

2. They cannot be used in environments in which the detonation of the explosive plug will cause secondary explosions or fires.
3. The transport of explosive devices is subject to regulations issued by federal, state, and local jurisdictions. Obtaining the necessary permits may be time-consuming.

Breakaway plug, go/no-go gage, and Jimmy brush. The original "Pop-a-Plug,"* or breakaway plug, was patented in 1981 and the patent assigned to Expando Seal Tools, since renamed Expansion Seal Technologies (EST) (Fig. 10.12). Early commercial experience with the plug, the way it works, and other relevant matters were discussed at the Second Symposium on Shell and Tube Heat Exchangers.[5] Since then, its use has become widespread in the power-generation, petroleum, and petrochemical industries.

Expansion Seal Technologies has extensively investigated ways to improve their manufacturing techniques, simplify Pop-a-Plug installation, and increase the scope of its use. Three additional patents have been granted, and Expando now manufactures and markets a descendant called the "Pop-a-Plug II" or P-II. It seals tightly when the pin expands the serrated ring by approximately 0.25 mm (~0.01 in). This provides rings of pressure under the serrations that seal the assembly against plug blowout. Expando has tested some plug sizes against blowout pressures of 14,000 psi (about 96,500 kPa) and performed helium tracer probe leak tests using an instrument sensitivity of 1×10^{-10} std cm^3/s. The tests demonstrated leak rates less than the maximum acceptable leak-rate evaluation of 1×10^{-5}/ std cm^3 permit-

*U.S. Patents 4,393,564, 4,425,943 and 4,436,117. Pop-a-Plug is a product of Expansion Seal Technologies, a unit of EST Group, Inc. (formerly called *Expando*) of Harleysville, Pa.

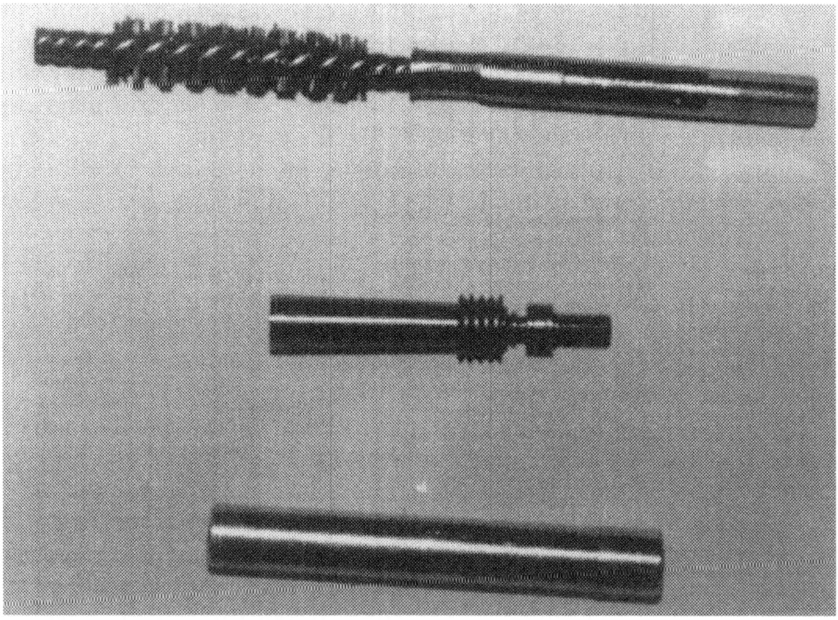

Figure 10.12 EST Technologies Pop-a-Plug II with go/no-go gage and Jimmy brush. (*Photograph by MGT Inc., Boulder, Colo.*)

ted in Paragraph V-1081 of the ASME Code, Section V, "Nondestructive Examination."

Installing Pop-a-Plug II is simple and quick when the face of the tubesheet is readily accessible. Expando provides a kit consisting of a Jimmy brush for cleaning before plugging, a go/no-go gage, and plugs. Their installation video shows other equipment needs: a flashlight, drill or air motor, tapered reamer, close-quarter tool for use where access is limited, pull-rod assembly matched to the proper-sized plug, positioner assembly that slides over the pull rod and positions the plug, hydraulic hose assembly, and small hydraulic ram connected to an air over hydraulic pump.

They list six basic steps: (1) removing weld droop with the drill and tapered reamer (when the tubes are welded to the tubesheets), (2) visual inspection for cracks and pits (no seal can be obtained if the tube end is cracked or if the tube or tubesheet is axially scored), (3) initial measurement of the diameter of the expanded-in tube region, (4) cleaning with the Jimmy brush [which removes scale and rounds the tubes—operating it for 30 s removes 0.127 to 0.178 mm (0.005 to 0.007 in)], (5) a second measurement with the go/no-go gage, and (6) installation of the plug. Although not discussed in their video, soft and sticky deposits should be removed by solvent or detergent washing before commencing the operation.

As stated, the foregoing installation is for tubesheets at the front end. But the Pop-a-Plug can also be installed in tubes in remote tubesheets, such as the outlet end of a single-pass or return end of a multipass straight-tube exchanger. You must make careful measurements to determine where to position the far-end plug.

Figure 10.12 is a photograph of a plug, Jimmy brush for cleaning before plugging, and go/no-go gage used to measure the tube I.D. before plugging.

Breakaway plugs consist of two-piece serrated taper plugs in which the small end of the driver is attached to a rod somewhat longer than the tube (or tubesheet if a thick tubesheet is being plugged at its rear face). The rod is notched or scored at its attachment to the driver so that, under sufficient tension, it will break before the rod yields.

The rod passes through a sleeve that bears against a thrust collar set on the outer face of the near tubesheet. The rod's end is connected to a hydraulic cylinder that bears against the thrust collar.

The appropriate plug is selected, the tube cleaned, the depth of installation measured, and the plug positioned. Hydraulically retracting the rod pulls the tapered pin into the outer matching serrated ring, expanding the ring into the tube. When the tension becomes great enough, the rod breaks away from the tapered pin, leaving the two-piece pin-ring assembly in place.

In addition to the tight control over ring expansion and avoidance of hammering plugs into place, Pop-a-Plugs have these advantages:

1. You can plug thick tubesheets with two plugs, one near the front face and one near the rear to limit further corrosion in the encapsulated tube end and hole.
2. You can plug inaccessible far-end tubesheets from the front end.
3. The plugs have successfully provided leaktight seals at high differential pressures between the shell and tube sides of heat exchangers.

The principal disadvantage of the breakaway plug is that if tightness is not achieved after breakaway, it is difficult to replug. Here is a typical sequence of installing Pop-a-Plug II:

1. Clean the tubes thoroughly in front of and at the position where the plug is to be installed. If the tube is welded into the tubesheet, remove weld droop with a finishing reamer.
2. Measure tube I.D. with a micrometer to set go/no-go plug gage range.
3. Plug gage to determine plug diameter to use.
4. Measure actual tube length.

5. Set standoff ring to fix plug position.
6. Set plug position and compression tube.
7. Retract hydraulic ram until rod breaks away.

Removable plugs for temporary high-pressure service. There are several proprietary or patented removable plugs on the market. They are used when it is necessary to plug a tube and return the exchanger to service, but it is desired to investigate the tube interior at a subsequent outage or plant turnaround. Some of these depend on O-ring seals for tightness. Users are cautioned to investigate the life to be expected from the O-ring seals before installing such plugs.

PowerFect removable plug. Figure 10.13 is a photograph of the PowerFect Mechanical Seal Plug.* The sketch in Fig. 10.14*a* shows the plug ready for insertion; Fig. 10.14*b* shows it installed. PowerFect states

> This plug incorporates gripper "jaw" segments to anchor the plug in position, and a properly selected high-temperature elastomer O-ring to provide the seal. The plug is installed by tightening the "jaws" to contact the internal surface of the tube to be plugged thus enabling the O-ring seal to be radially expanded to isolate leaky tube failures.

The PowerFect removable plug consists of a reverse two-piece, serrated plug in which the outer piece is extended and closed off with a section thick enough to drill and tap. The inner tapered piece terminates at its small end in a section threaded into the front end of the outer piece so as to protrude far enough for a nut to fit on the end. Tightening the nut pulls the tapered part forward, thereby providing

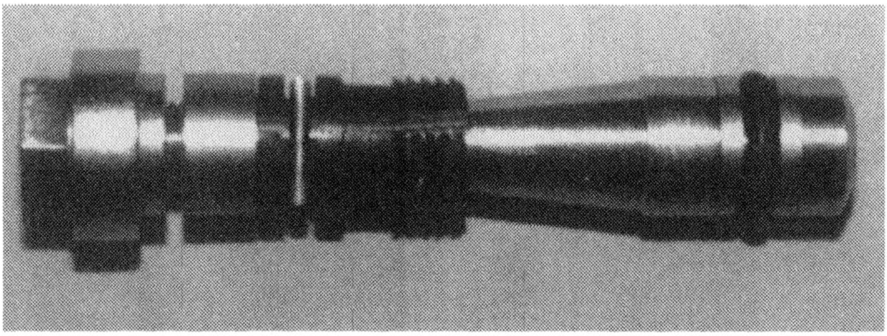

Figure 10.13 PowerFect mechanical seal plug. (*PowerFect, East Hanover, N.J., MGT Inc. photograph.*)

*U.S. Patent 4,474,216.

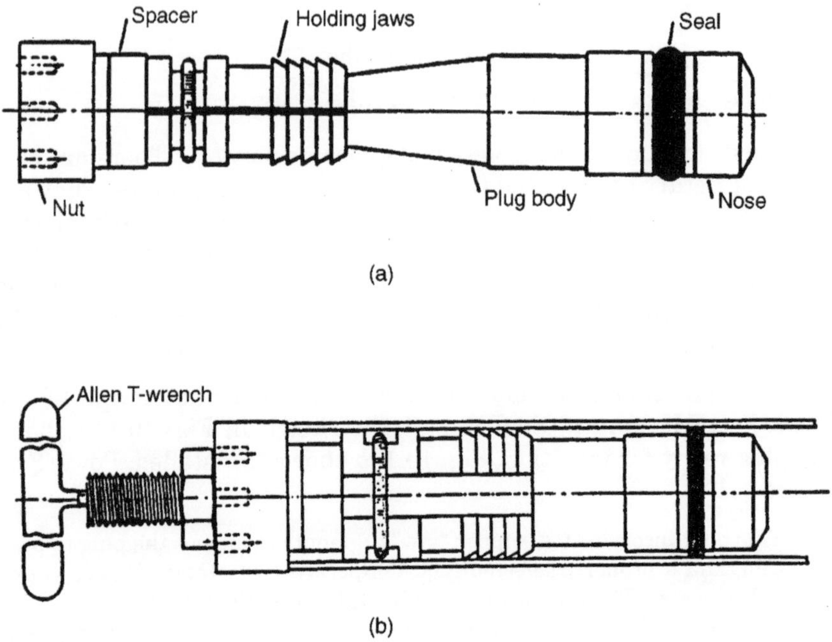

Figure 10.14 PowerFect mechanical seal plug (*a*) ready for insertion and (*b*) installed. (*PowerFect, East Hanover, N.J.*)

expanding force to both pieces. The plug is removed by backing off on the nut and tapping the threaded end to release the expansion.

Conco high-confidence plug. Figure 10.15 shows the Conco high-confidence plug. It has an elastomeric cylinder, an O-ring, sandwiched between two backup rings with tapered driving rings, a washer and nut. Tightening the nut compresses the elastomeric cylinder which

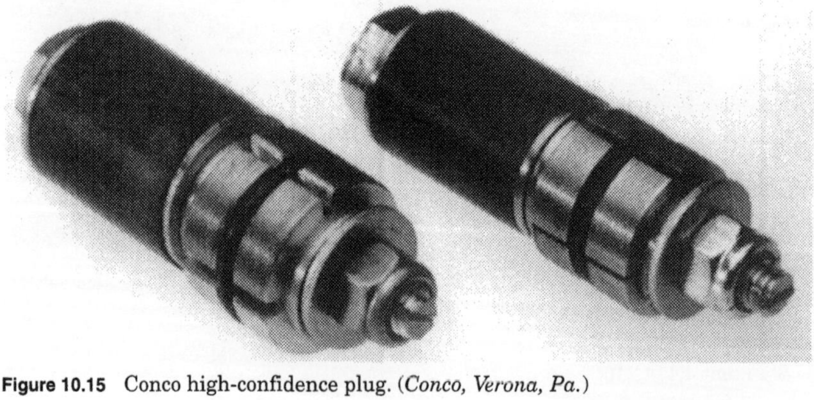

Figure 10.15 Conco high-confidence plug. (*Conco, Verona, Pa.*)

presses against the tube wall. Its length is designed to provide an adequate leak path and its compressibility helps seal against axial scores. The O-ring provides a seal against the high-pressure differential between the feedwater and steam sides.

Torq 'N Seal high-pressure plug. Torq 'N Seal high-pressure plugs* (Fig. 10.16) produced by JNT Technical Services of Little Ferry, N.J., have been widely used to plug high-pressure feedwater heaters. JNT manufactures these plugs in size increments of ten-thousandths of an inch from 0.420 in to 0.950 in diameter in carbon steel and brass. They also offer custom-made plugs in other sizes and alloys such as stainless steels and titanium on a special-order basis. They provide accessories including an adjustable ⅜-in drive ratcheting torque wrench of 0 to 960 in-lb torque capacity with gradations of 10 in-lb increments and three sizes of wrench-to-plug adapters that they call *Torq bits*.

Figure 10.17 shows the installation technique. Conditions to avoid when you install this plug are attempting to seal across the transition from an expanded to unexpanded length of tube, and trying to seal across pits and gouges. Before installing one, be sure to remove weld

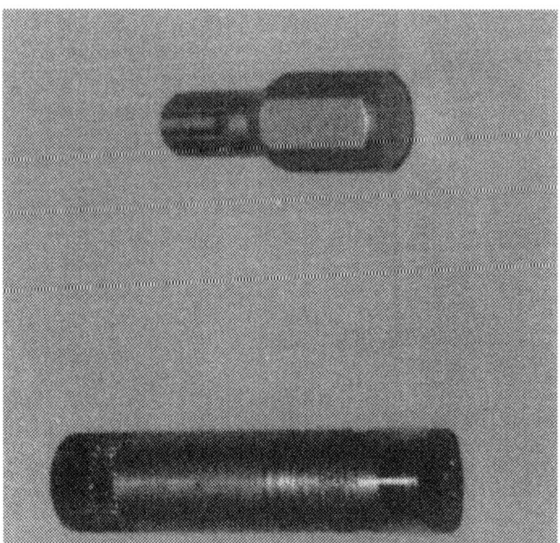

Figure 10.16 Torq'N Seal high-pressure plug with drive. (*JNT Technical Services Inc., Little Ferry, N.J., photograph by MGT Inc., Boulder, Colo.*)

---

*U.S. Patent 5,289,851.

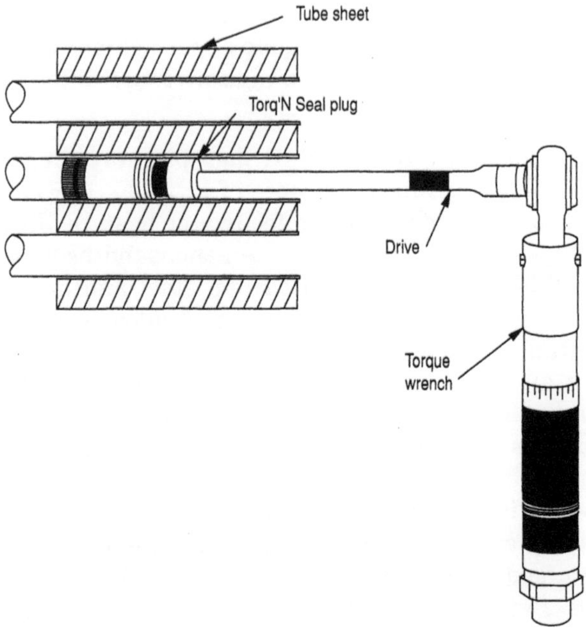

Figure 10.17 Installation of Torq'N Seal high-pressure plug. (*JNT Technical Services, Inc., Little Ferry, N.J.*)

droop, clean the tube where the plug will be installed, resize the tube with a finishing reamer if it is out of round, and measure the tube I.D. Select a standard plug of diameter no smaller than 0.381 mm (0.015 in) less than the tube I.D. To achieve successful sealing requires scrupulously following the manufacturer's recommendations for plug size, drive size, torque levels, and use of compound for holding the torque drive in place.

Torq 'N Seal condenser plug. The Torq 'N Seal condenser plug (Fig. 10.18) consists of a thick-walled, flanged elastomeric thimble into

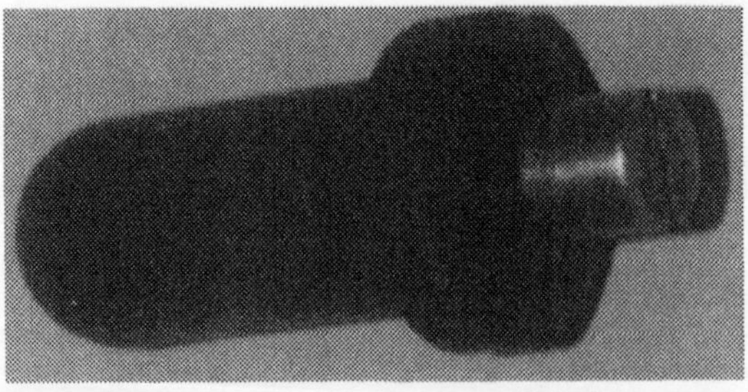

Figure 10.18 Torq'N Seal condenser plug. (*JNT Technologies, Little Ferry, N.J., photograph by MGT Inc., Boulder, Colo.*)

which is fitted an adjusting screw mechanism with screwdriver slot, capped by a screwdriver-slotted cover. The left-hand threaded cover fits snugly into the opening in the flanged end of the elastomeric thimble and protects the threaded adjusting screw. To insert the plug, remove the cover by turning it clockwise. Turn the adjusting screw counterclockwise, thereby stretching the thimble and reducing its diameter. Insert the plug in the tube and turn it clockwise. The thimble returns to its original length and diameter, with its exterior bearing firmly against the tube interior. Replace the protective cap by turning it counterclockwise.

Elastomeric front and rear seals. Elastomeric front and rear seals (Fig. 10.19), separated by a compression tube and compressed hydraulically, may be used to seal tubes in relatively low-pressure service. The whole assembly remains in place.

Temporary homemade rubber plugs. Figure 10.20 illustrates a homemade plug that can be inserted in the end of a failed tube to seal it off during bubble testing. The one illustrated is held tight under 100 psi (~690 kPa) shell-side air pressure.

Thimble plugs. *Thimble plugs* are closed-end thin-walled cylinders with flanged-over lips on their open ends. The O.D. of the lip is no greater than the tube O.D. plus approximately 50 to 75 percent of a ligament thickness. The O.D. of the cylinder is a maximum of 1 mm

Figure 10.19 Conco EX-4 expanding tube plug with elastomeric front and rear seals. (*Conco Systems, Verona, Pa.*)

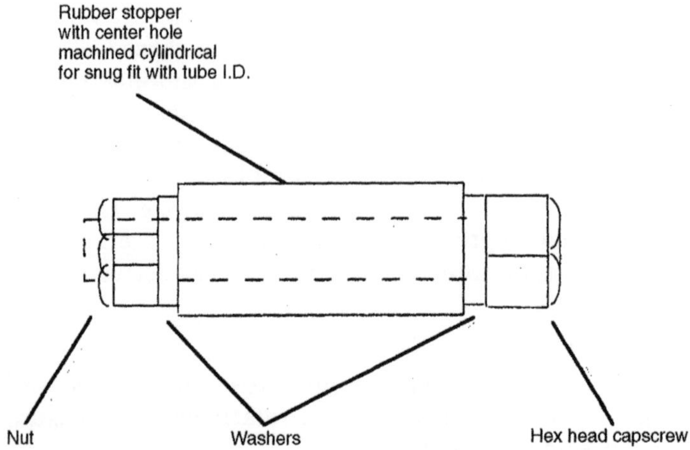

Figure 10.20 Temporary homemade rubber plug.

(approximately 0.040 in) smaller than the tube I.D. The depth of the thimble is great enough to allow insertion of a hydraulic expanding mandrel or a tube rolling tool used to expand the thimble into place. Thimble plugs may also be manufactured with external serrations or threads.

MGT thimble plug. The MGT thimble plug* (Figs. 10.21 and 10.22) consists of a machined thimble expanded into place by an expansion-control pin. In one version, the exterior of the thimble is serrated. The

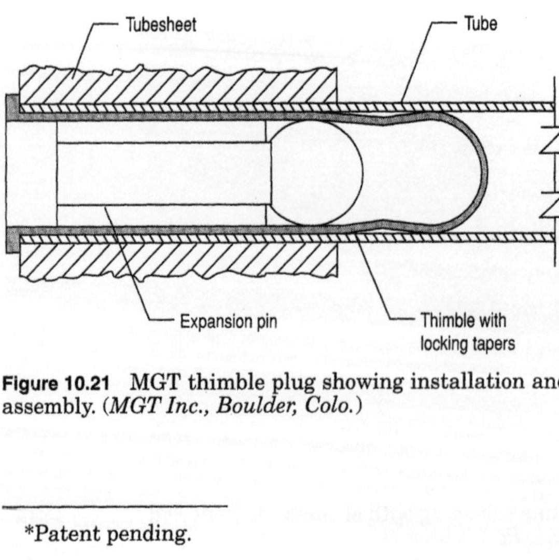

Figure 10.21 MGT thimble plug showing installation and assembly. (*MGT Inc., Boulder, Colo.*)

*Patent pending.

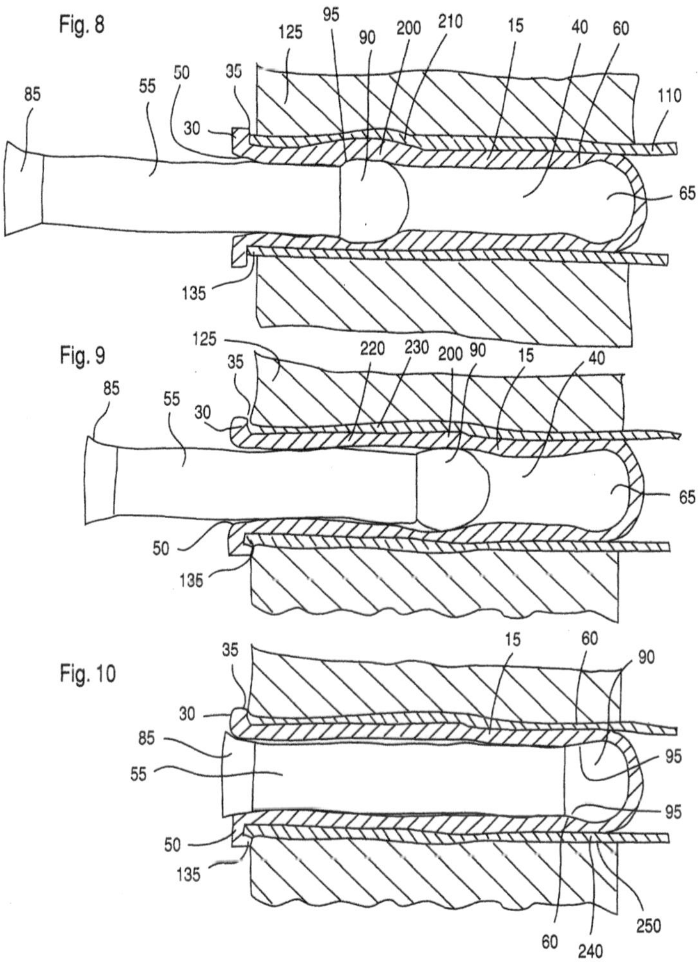

Figure 10.22 These sketches were used for the patent application for the plug sketched in Fig. 10.21.

front end of the thimble is flanged to bear against the tube end and machined to receive the tapered drive end of the expansion control pin. At the closed end, the thimble inside diameter transitions through a taper to a chamber that receives the head of the expansion control pin.

The expansion-control pin consists of a machined rod with an enlarged head at its front end and a taper at its drive end. The drive-end taper matches the taper on the front end of the thimble. The transition of the rod to the enlarged head at the pin's front end matches the taper of the transition at the thimble's front end.

The expansion pin head is just enough larger than the thimble inside diameter that driving the pin into the thimble expands the thimble tightly into the tube. The pressure of the thimble on the inside of the tube contributes to increasing any residual interfacial pressure between the tube and the hole. When the pin reaches the end of its travel, the matching tapers of the head and thimble lock the pin into place. The length of the pin and positions of the tapers in the thimble are dimensioned so as to ensure that the matching tapers of pin and thimble at the drive end bear tightly against each other.

The design of the MGT thimble plug permits calculating the theoretical interfacial pressure that remains between the thimble and the tube after the pin is driven into place. As the head end locks into place, it creates a tensile force on the pin that pulls its tapered drive end into tight, sealing contact with the thimble's open end. The MGT plug is as simple to install as a taper pin or two-piece taper pin and ring. Driving the expansion pin into place does not deform the surrounding ligaments enough to loosen the joints of adjacent tubes to the tubesheet.

Far-end or remote-position sealing with steel wool and lead. Inner tubesheets and tube regions behind the front tubesheet may be sealed off by inserting a long rod into the tube as far as it can go. The rod may butt against the rear cover or against the bend of a U-tube. A sleeve, of a length selected to position the plug, is slid down the rod as far as it can go.

Steel wool and lead are wrapped into a roughly cylindrical plug that fits snugly over the rod. The plug O.D. is somewhat smaller than the tube I.D. A second sleeve is slid over the rod to compress the plug. This kind of stopgap plugging has fallen out of favor because of the health hazard associated with lead.

Welding plugs. Plug high-pressure equipment such as feedwater heaters with welded-in plugs. The plant maintenance force can machine these out of round bar stock compatible with the tube, tubesheet, and weld rod metal and do the installation. One variety consists of a snug-fitting, round bar approximately 50 mm (~2 in) long. The procedure is to clean the tube interior where the plug is to be installed with a Jimmy brush or finishing reamer. Clear the tube end of contaminants and oxides with an end mill or modified twist drill. Press the bar into the tubes and set it about 6 mm (~¼ in) below the front face of the tubesheet. Then seal it in place by fillet welding. You may fill the recess with weld to provide a smooth surface across the tube end and bar-plug face.

A preferred alternative is to cut out the tube for the lesser of about half the depth of the tubesheet or 50 mm (~2 in), clean the hole and

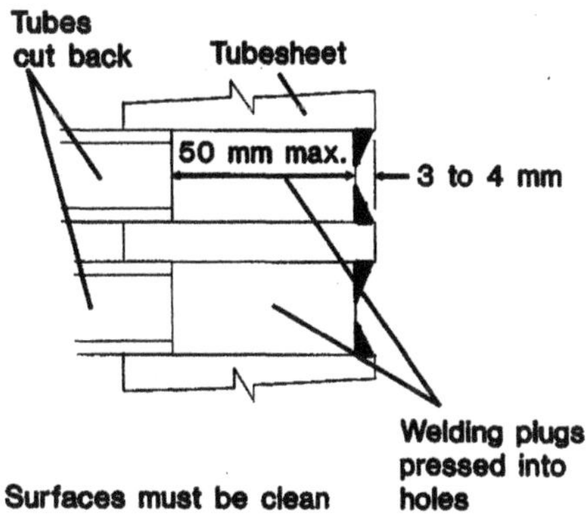

Figure 10.23 Plug for welding to tubesheet set below tubesheet face.

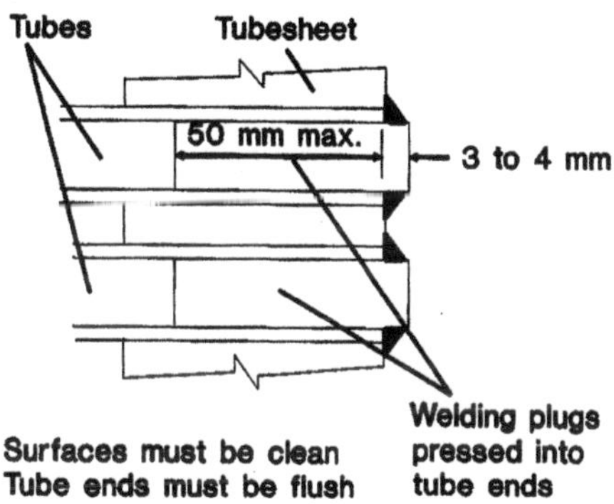

Figure 10.24 Plug for welding to tubesheet with plug protruding beyond tubesheet face.

tube end, and press-fit a weldable round bar plug into the hole until it bears against the end of the tube, with its other end approximately 3.2 mm (~⅛ in) below or above the tubesheet face. Then weld the plug to the tubesheet. When the high-pressure stream is in the tubes, this alternative prevents wire-drawing damage to the tubesheet. Figures 10.23 and 10.24 show configurations with the plug set below and above the tubesheet face.

General effects of tube plugging

Nomenclature (in consistent units) that applies to the following discussions is as follows:

Q = heat flow

U = overall coefficient of heat transfer

W = flow of hot fluid

w = flow of cold fluid

C = specific heat of hot fluid
c = specific heat of cold fluid
T_1, T_2 = hot-fluid inlet and outlet temperatures, respectively
t_1, t_2 = cold-fluid inlet and outlet temperatures, respectively

Plugging an exchanger reduces its heat-transfer surface and cross-sectional area for tube-side flow, thus reducing the unit's capacity. The surest, and perhaps quickest, way to examine how plugging affects performance is to rerate the exchanger using the reduced number of tubes available for tube-side flow. Heat-transfer and pressure-drop computer programs are widely available for performing the required calculations. You have to understand thoroughly how a program works and its options for changing parameters. You must be able to adjust a program to allow inputting the number of unplugged tubes for tube-side flow. At the same time, you have to arrange it to accept input for shell-side flow for the full number of tubes. With these changes, you can fairly well estimate the tube-side, shell-side, and overall heat-transfer coefficients. Using the individual resistances, you can estimate the metal temperatures of plugged and unplugged tubes to enable you to examine structural effects. To reiterate, calculate the shell-side and tube-side film coefficients using the full tube count. Then calculate the new performance, inputting the new tube count to calculate the new tube-side film coefficient. Calculate the new overall coefficient using the new tube-side film coefficient and the shell-side coefficient you determined with a full tube count. This will permit you to determine the performance with the reduced tube count.

Working with the Electric Power Research Institute (EPRI) of Palo Alto, Calif., Karta Technology Inc. of San Antonio, Tex. has developed a Heat Exchanger Workstation that can be used to examine the effects of plugging feedwater heaters and other tubulars used in power generation.

General effects of plugging on duty. Plugging does not materially affect the shell-side film coefficient of heat transfer. Loss of shell-side heat-transfer capacity, based on its film coefficient, is linear with the reduction in surface in single-zone shell arrangements. In multizone feedwater heaters and process exchangers that have tubes flooded to achieve subcooling, the effect on capacity depends on the location of the plugged tubes in the tube field.

The tube-side film coefficient of heat transfer in turbulent-flow forced convection will increase in accordance with the following equation:

$$h_{t\text{plugged}} = h_{t\text{new}}\left(\frac{N_{\text{new}}}{N_{\text{new}} - \mathbf{N}_{\text{plugged}}}\right)^{0.8} \tag{10.1}$$

where $h_{t\text{new}}$ is the film coefficient in tubes with all tubes unplugged and $h_{t\text{plugged}}$ is the film coefficient in tubes after plugging.

The resistance to heat transfer from the hot to cold fluid comprises the sum of the shell-side film resistance R_s $(= 1/h_s)$, outside fouling resistance R_o, metal resistance R_m, inside fouling resistance R_i, and inside film resistance R_t $(= 1/h_t)$. Resistances R_i and R_t are referred to the outside tube surface as R_{io} and R_{to} by multiplying R_i and R_t by tube O.D./I.D. ratio. In all but exceptional cases, the reduced after-plugging tube-side film resistance $(= 1/h_{t\text{plugged}})$ hardly increases the overall heat-transfer rate. For example, in a water-cooled steel-tubed condenser with 1.25-mm (0.049-in)-wall tubes, with water in the tubes, typical resistances to heat transfer are

	h•m²•°C/kcal	h•ft²•°F/Btu
Water film resistance	0.004638	0.00095
Water-side fouling	0.009765	0.00200
Metal resistance	0.0007812	0.00016
Steam-side fouling	0.0014647	0.00030
Steam resistance	0.0032712	0.00067
Total resistance*	0.0199201	0.00408

*For the purposes of this estimate, the inside film coefficient and inside fouling resistance have not been referred to the outside, i.e., not multiplied by the O.D./I.D. ratio.

This is equivalent to an overall coefficient of about 50.20 kcal/h•m²•°C [~245 Btu/(h•ft²•°F)]. If 10 percent of the tubes on the water side were plugged, because of increased flow velocity, the tube-side resistance would decrease to approximately 0.00426 h•m²•°C/kcal (~0.00087 h•ft²•°F/Btu). The overall resistance would decrease to approximately 0.019545 h•m²•°C/kcal (~0.00400 h•ft²•°F/Btu) to give an overall coefficient of 51.16 kcal/(h•m²•°C) [~250 Btu/(h•ft²•°F)] or an increase of about 2 percent. The tube-side film coefficient will vary only slightly when material in the tubes is being vaporized or condensed.

Since improvement of the overall heat-transfer coefficient is modest, removing tubes from service in single-shell-side-zone heat exchangers reduces Q proportionally and reduces the temperature change (ΔT and Δt) on the shell and tube sides, respectively, and increases the mean temperature difference. The effect on tubulars fitted with dams, enclosures, or shrouds that isolate groups of tubes for gas cooling, condensing, and subcooling depends on which tubes are removed from service.

Effect of plugging on pressure drop. Shell-side pressure loss remains pretty much constant.

For forced-convection, turbulent tube-side flow the pressure drop through the tubes in any pass will increase as shown in the following:*

$$\Delta P_{plugged} = \Delta P_{new}\left(\frac{N_{new}}{N_{new} - N_{plugged}}\right)^{1.8-2} \tag{10.2}$$

where ΔP_{new} = pressure drop through the tubes in the pass with all tubes unplugged
$\Delta P_{plugged}$ = pressure drop through tubes in the pass after plugging
N_{new} = number of tubes in the pass of a new unit
$N_{plugged}$ = number of tubes plugged in the pass

Note that, as shown in Eq. (10.2), the exponent may vary from 1.8 to 2.

To estimate the pressure drop after plugging, subtract from the measured or calculated new pressure drop the nozzle and return pressure losses. These are approximately 1.54 velocity heads per nozzle and 4 velocity heads per pass as shown in the following equations:

$$\Delta P_n = \frac{\rho V^2}{870}\ \text{kPa} = \frac{\rho V^2}{6000}\ \text{psi} \tag{10.3}$$

$$\Delta P_r = \frac{\rho V^2}{335}\ \text{kPa} = \frac{\rho V^2}{2310}\ \text{psi} \tag{10.4}$$

where ΔP_n = pressure loss per nozzle, kPa, psi
ρ = density, kg/m^3, lb/ft^3
V = velocity, m/s, ft/s

This will provide the approximate pressure drop attributable to flow through all the tubes of the unplugged exchanger, since it neglects tube entrance and exit losses.

Assume that the pressure drop per pass is the total tube pressure drop divided by the number of passes. For each pass with plugged tubes, increment the pressure loss using Eq. (10.2). Add the resulting pressure drops in the individual passes and the nozzle and return pressure losses.

The pressure loss through the tube side of a severely plugged exchanger or feedwater heater may exceed the available pressure drop or pump capacity. If a unit must remain in the tube-side stream, even at reduced thermal performance, you can install a bypass fitted with a flow-control device between the channel inlet and outlet con-

*For most exchangers, the exponent in Eq. (10.2) is 1.8.

nections. If the unit is of multipass tube-side design and installing an external bypass is unacceptable, you can bypass tube-side flow internally by drilling, burning, or arcing holes in the pass-partition plates.

A flow calculation is required to estimate the size of hole required to bypass enough fluid to achieve the required reduction in pressure loss. Partition plates are generally at least 6.35 mm (¼ in) thick, and burning and arcing reasonably round holes is difficult. Therefore, the estimate of an appropriate orifice constant to use in the calculation is at best only a guess. You can surmount this problem by making a large hole in the first pass-partition plate and fitting into it a plate drilled and machined to a sharp-edged orifice sized for the desired flow. A reasonable assumption for the orifice constant is 0.6.

Effects of plugging on corrosion rates. In single-pass exchangers with hot fluid flowing in the tubes, the improvement in inside film coefficient, which increased velocity confers, causes the average tube metal temperature to rise. If the tube corrosion rate is temperature-dependent, plugging will cause an increase in corrosion rate in the unplugged tubes.

In multipass exchangers with hot fluid in the tubes, the corrosion rate depends on the pass in which the tubes are plugged. If all the plugging is in the inlet (hot) pass, the tube-side film coefficient will increase only in the hot pass. The average metal temperature of the hot-pass tubes will increase, thereby increasing the corrosion rate of hot-pass tubes. This will be compounded by the greater erosion of the inner tube wall resulting from the higher velocity. Because the outlet temperature from the hot pass is higher than it would be if no tubes were plugged, the metal temperatures of the tubes in succeeding passes will also rise, possibly increasing their corrosion rates.

Exchangers in which tube-temperature-variant corrosion is confined to a discrete location may be candidates for partial retubing. In multipass ones, the pass-to-pass average metal temperatures and thermal coefficient of expansion of the replacement tubes relative to those in the other passes must be examined.[6] The restrained difference in axial expansion between the hot-pass tubes and the others must be within the structure's tolerable limits.

You can estimate the metal temperatures of the active tubes in each pass using the method of Ref. 6, which is paraphrased in the following equations:

$$\sum R = R_{to} + R_{io} + R_m + R_o + R_s \tag{10.5}$$

and

$$\sum' = \sum - \mathrm{R_o} - \mathrm{R_{io}} \tag{10.5a}$$

For the fouled condition,

$$T_m = T_b \pm \frac{R_o + R_s}{\sum R}(T_b - t_b) \tag{10.6}$$

$$t_m = t_b \pm \frac{R_{io} + R_{to}}{\sum R}(T_b - t_b) \tag{10.7}$$

For the clean condition,

$$T_m = T_b \pm \frac{R_s}{\sum R'}(T_b - t_b) \tag{10.6a}$$

$$t_m = t_b \pm \frac{R_{io}}{\sum R'}(T_b - t_b) \tag{10.7a}$$

$$T_w = \frac{T_m + t_m}{2} \tag{10.8}$$

where* ΣR = sum of the resistances (= $1/U$)
T_b = bulk temperature of the fluid outside the tubes
t_b = bulk temperature of the fluid inside the tubes in the pass under consideration
T_m = metal temperature on the outside tube wall of the active tubes in the pass under consideration
t_m = metal temperature on the inside tube wall of the active tubes in the pass under consideration
t_w = tube metal temperature of the active tubes in the pass under consideration

Structural effects of plugging. When you restart a unit after plugging, the tube metal average temperature in plugged tubes will come to approximately the shell-side fluid bulk temperature. The metal in the unplugged ones will come to an average temperature somewhere between the shell fluid bulk temperature and the mean tube-side fluid temperature. The latter depends on the film, fouling, and metal resistances.

In straight-tube exchangers, plugged tubes at temperatures substantially different from those of the unplugged ones will generate axial forces in the bundle. The force magnitudes and directions depend on the average plugged and unplugged tube metal tempera-

*In Eqs. (10.5) to (10.8).

tures, thermal coefficients of expansion, and cross-sectional areas. Unless plugs are distributed uniformly throughout the tube field, the load on individual tubes surrounded by ones that have fluid flowing inside will be greater than the load calculated from the overall plugged and unplugged tube cross-sectional areas. These thermally generated loads may crimp plugged tubes and break their joints to the tubesheets.

When most of the plugs are located in a hot inlet pass of a multipass straight-tube bundle, the higher tube metal temperature of the unplugged, hot-pass tubes may have the additional effect of banana-shaping the whole bundle. Within the hot pass, the difference in metal temperature between the in-service and plugged tubes applies tensile loads to the tubes remaining in service and compressive loads to the plugged ones. These loads may crimp the hot-pass plugged tubes and break tube-to-tubesheet joints.

Effect of plugging multipass exchangers. Thermal engineers strive to design multipass exchangers to have the same number of tubes in each pass but rarely succeed. The precision of correlations used to establish tube-side film coefficients is fuzzy, and the error introduced in the logarithmic mean temperature difference (LMTD) is not considered to be excessive in turbulent flow if the number of tubes in each pass is within a tolerance of ±5 percent of the total number of tubes divided by the number of passes. Plugging a multipass exchanger may shift the balance beyond this tolerance. In this circumstance, determining plugging effects requires analysis of the heat-transfer relations on a pass-to-pass basis.

You can approximate overall performance in straight-tube exchangers when the number of passes is a multiple of 2, making these assumptions: (1) there is no change in the shell-side film coefficient h_s and (2) there is no change in fouling resistances R_o and R_i or tube metal resistance R_m. Unknowns are $h_{t\text{plugged}}$, Q, U, T_2, t_2, and LMTD. If a computer output sheet or hand calculation of the original design is available, establish $h_{t\text{plugged}}$ for each pass using the calculated $h_{t\text{new}}$ in Eq. (10.1). If not, estimate it using one of the methods available in the open literature (see, e.g., Ref. 7). Correct this to $h_{t\text{plugged}}$ and calculate the overall coefficient U for each pass. Calculate the available surface in each pass from the number of unplugged tubes. For each pass, calculate UA. Determine the weighted overall coefficient U_w from the following equation:

$$U_w = \frac{U_1A_1 + U_2A_2 + U_3A_3 + U_4A_4 + U_nA_n}{A_1 + A_2 + A_3 + A_4 + A_n} \tag{10.9}$$

The subscript indicates the pass number, and n is an even number.

The following equations describe the heat transferred (duty):

$$Q = UAf(\text{LMTD}) \tag{10.10}$$

$$Q = WC(T_1 - T_2) \tag{10.11}$$

$$Q = wc(t_2 - t_1) \tag{10.12}$$

where f is the LMTD correction factor[8] and other quantities are as defined previously.

Effects of plugging feedwater heaters

Except for ones designed for straight condensing, closed feedwater heaters consist of two or three heat exchangers arranged in series in a single shell. In some plant arrangements where the feedwater flow rate is more than about 1,400,000 to 2,270,000 kg/h (~3,000,000 to 5,000,000 lb/h), multiple strings or trains are used to enable the plant to keep running at or near full load, even when some heater is out of service because of tube leaks. Feedwater flows in series through the tubes of the heaters that receive the lowest-pressure extraction steam to those that receive the highest-pressure steam extracted from the turbines. Flow is countercurrent. Cold feedwater enters the tubes where the coldest shell side stream exits the shell. Heated feedwater exits the tubes where steam enters the shell. Desuperheater sections are gas-to-water exchangers that transfer sensible heat from superheated steam to feedwater,* condenser sections transfer latent heat from pure saturated steam to feedwater, and subcooler sections are water-to-water exchangers that transfer sensible heat from condensate and drains to feedwater.

Over the lifetime of a feedwater heater, tubes fail for the reasons discussed previously. If you can tolerate the plugging frequency, increase in pressure drop, and degradation of thermal performance, you should keep the heater in service and monitor these parameters to plan ahead for heater replacement. Following are some guidelines.

Plugging and feedwater heater pressure drop. Neglecting nozzle and turnaround losses, the increase in pressure drop due to plugging is as described in Eq. (10.2). A prevalent rule of thumb is to replace a heater when 10 percent of the tubes are plugged. However, the authors have seen heaters kept in service with more than 25 percent plugged.

*A modest amount of residual superheat remains in the steam as it exits the desuperheater into the condenser to prevent "wet wall" damage to the tubes. Some noncondensible gas is also present. These are neglected in thermal calculations.

When the rate of failure is tolerable and not increasing unacceptably, but the increase in pressure drop approaches the limit of its tolerance, consider bypassing a portion of the feedwater with an internal bypass in the channel pass partition. The following derives a means of calculating the required bypass flow and orifice size.

Calculating required feedwater bypass flow. This derivation is based on consistent units:

W = total feedwater flow

W_B = bypass flow

N_{new} = previously defined number of new tubes

N_{plugged} = previously defined number of plugged tubes

ΔP_{new} = previously defined pressure drop in new condition with W flow

ΔP_B = pressure drop through N_{new} tubes with $W - W_B$ flow

$\Delta P_{B\text{plugged}}$ = pressure drop through $N_{\text{new}} - N_{\text{plugged}}$ tubes with $W - W_B$ flow

V = linear velocity through N_{new} tubes with W flow

V_B = linear velocity through N_{new} tubes with $W - W_B$ flow

$V_{B\text{plugged}}$ = linear velocity through $N_{\text{new}} - N_{\text{plugged}}$ tubes with $W - W_B$ flow

The linear velocity through the tubes is proportional to the feedwater flow divided by the number of active tubes:

$$V \propto \frac{W}{N_{\text{new}}} \qquad V_B \propto \frac{W - W_B}{N_{\text{new}}} \qquad V_{B\text{plugged}} \propto \frac{W - W_B}{N_{\text{new}} - N_{\text{plugged}}} \tag{10.13}$$

As stated previously, the change in pressure drop from one flow condition to the next varies as the linear velocity to the 1.8 power, and $\Delta P_B/\Delta P_{\text{new}} = (V_B/V)^{1.8}$, $\Delta P_{B\text{plugged}}/\Delta P_B = (V_{B\text{plugged}}/V_B)^{1.8}$, $\Delta P_B/\Delta P_{\text{new}} = [(W - W_B)/W]^{1.8}$, $\Delta P_{B\text{plugged}}/\Delta P_B = [N_{\text{new}}/(N_{\text{new}} - N_{\text{plugged}})]^{1.8}$.

Substituting and rearranging, we obtain

$$W_B = \left(1 - \left\{\frac{\Delta P_{B\text{plugged}}}{\Delta P_{\text{new}}} \cdot \frac{1}{[N_{\text{new}}/(N_{\text{new}} - N_{\text{plugged}})]^{1.8}}\right\}\right)^{0.556} \tag{10.14}$$

The following example, taken from an actual feedwater heater, illustrates how to use Eq. (10.14).

A feedwater heater has 5,239,643 lb/h (~2,376,660 kg/h) of feedwater entering the tubes. At this flow rate the pressure drop is 12.2 psi (~84.18 kPa = 8578.1 kg/m^2) at the average temperature between inlet and outlet [480°F (249°C)]. If 10 percent of the tubes are plugged and the maximum allowable pressure drop is 13 psi (~89.7 kPa = 9140.6 kg/m^2), how much flow must be bypassed?

When 10 percent of the tubes are plugged, the number of total tubes divided by the number of remaining active tubes 1/(1−0.1) = 1.111. Raised to the 1.8 power, this is 1.2086, and the inverse is 0.827. $\Delta P_{Bplugged}/\Delta P_{new}$ is 13/12.2 = 1.066. Multiplied by 0.827 this becomes 0.881, which—raised to the 0.556 power—is 0.932. Subtracted from 1, this is 0.068. Multiplying the feedwater flow of 5,239,643 lb/h (~2,376,650 kg/h) by 0.068 gives a required bypass flow of 356,296 lb/h (161,613 kg/h). You can estimate the size of the sharp-edged orifice that will meter this flow through the pass partition using the following assumptions. The gravitational constant is 9.81 m/s^2 (= 32.29 ft/s^2), the orifice constant K is 0.6, and the water density = 1000 kg/m^3 (= 62.4 lb/ft^3).

The general equation for a sharp-edged orifice is

$$G = KA_o\sqrt{2g\ \Delta p\ \rho} \qquad (10.15)$$

where G = mass flow
A_o = cross-sectional area of orifice
Δp = pressure drop across the orifice
ρ = density of the fluid flowing at the temperature at which it enters the orifice

Substituting W_B for G and $\Delta P_{Bplugged}$ for Δp and extracting the diameter from A_o, Eq. (10.14) in consistent units becomes

$$D = 11.5\sqrt{\frac{W_{Bplugged}}{(\rho\Delta P_{Bplugged})^{0.5}}}\ \text{mm} \qquad (10.16)$$

$$D = 0.0297\sqrt{\frac{W_{Bplugged}}{(\rho\Delta P_{Bplugged})^{0.5}}}\ \text{in} \qquad (10.17)$$

The current common practice is to burn a hole in the pass-partition plate that has the same cross-sectional area as that of the plugged tubes. This will maintain or possibly reduce the pressure drop across the heater. In either case, you can calculate the pressure drop using the hydraulic diameter, in millimeters (inches), to perform a fanning friction factor calculation. However, if you maintain or reduce the pressure drop by bypassing more feedwater than is required, the colder feedwater will cause the next heater in the string to draw more steam from the turbine next or increase the load on the boiler if the internal bypass is located in the top heater.

Plugging effects on feedwater heater thermal performance. How plugging alters thermal performance depends on whether the unit is a low-pressure single-zone external subcooler, a low-pressure two-zone condensing and subcooling heater, a straight condensing heater, a heater with a desuperheating and condensing zone, or one with a desuperheater, condensing zone, and subcooler.

The locations of the plugged tubes and design arrangement have some effect. The impact of plugging on performance of heaters with full-pass partial-length subcoolers is somewhat different from that on those with partial-pass full-length ones.

Plugging subcooler tubes of three-zone heaters raises the DCA since less surface is available to cool the drains. If there is a desuperheater, the outlet legs of plugged tubes of units with full-pass partial-length drains coolers pass through the desuperheater, reducing the surface available for extracting superheat. Some superheated steam will be desuperheated and condensed in the condensing section. The condensing section picks up some of the duty loss because of the subcooling surface reduction.

Plugging a tube in a full-length partial-pass subcooler reduces DCA more than plugging one in a full-pass partial-length subcooler since more surface is removed because of the longer tube length. The outlet leg of a tube plugged in a full-length partial-pass subcooler will pass through the condenser or the desuperheater if one is included in the design.

A number of installations of external boiler feedwater subcoolers and low-pressure condensing and subcooling feedwater heaters in condenser necks are designed with fixed tubesheets. Because condensate pumps often surge on load rejection, designers avoid constructions in which pass partitions are gasket-sealed to the tubesheet or channel cover. Therefore, although these may be conventional TEMA type N channel designs in other respects, the internal pass partitions have covers and are not sealed to the channel covers. Manway-access channels welded to the tubesheets are also used. In these exchangers, the channel barrel is integral with the tubesheet and the manway is located in the elliptical or hemihead. Fixed-tubesheet heaters may have flanged-and-flued or flanged-only shell expansion joints to accommodate differential expansion between the shell and tubes.

In straight condensing units and external fixed-tubesheet single-zone subcoolers, the effects of plugging are as described above. In low-pressure, fixed-tubesheet, condenser-subcooler feedwater heaters, removing surface from the subcooler will cause an increase in the DCA.

Substantially reducing the available heat-transfer surface in a heater reduces its outlet feedwater temperature. This becomes a colder-than-designed inlet temperature to the heater drawing steam from the next-higher extraction stage, reducing its tube-wall temperature.

The colder tubes extract more steam from the turbine. This effect can cascade up the line and increase the load on the boiler.

In the extreme, a heater may be so fully plugged that it must be bypassed because of pumping considerations. The next-higher pressure heater will then draw extraction steam substantially greater than its design flow. For this reason heater specifications should include provision for the effects on upstream heaters of bypassing downstream ones.

To analyze the effects of plugging on the thermal performance of a given heater, start with the manufacturer's feedwater heater specification sheet. This shows the feedwater velocity at 15.55°C (60°F), duty, mean temperature difference, heat-transfer rate and surface for each zone, total tube count, tube diameter, and gage. Sometimes it also has the film coefficients. Assume a tube-side fouling resistance of 0.0352 m^2•°C/kW (0.0002 h•ft^2•°F/Btu) inside the tubes, referred to the outside. For desuperheaters and drains subcoolers, assume shell-side fouling resistance of 0.0528 h•m^2•°C/kW (0.0003 h•ft^2•°F/Btu). Extract the fouling and metal resistances from the transfer rate for each section. If the original tube-side film coefficient was not shown on the specification sheet, calculate it based on the design-point flow. Correcting it to the O.D., calculate the shell-side film coefficients that the manufacturer included in calculations of overall coefficients.

Using Eq. (10.1), calculate the plugged-tube film coefficients. Recalculate the overall coefficients. Assume an outlet temperature from the first zone. Calculate the heat transferred and mean temperature difference that this temperature produces. Iterate this calculation until the heat transferred on the basis of $U_{z1}A_{z1}\,\Delta T_{z1}$ equals the heat transferred on the basis of $wc(T_{o1}-T_{i1})$, where (in consistent units) w is the feedwater flow through the tubes, c is the specific heat, z_1 indicates the first zone, T_{i1} indicates the inlet feedwater temperature, and T_{o1} is the assumed outlet temperature from the first zone. When these are in balance, T_{o1} becomes the inlet temperature to the next zone. Repeat the procedure for the next zone and for the following one if there are three zones. This will enable you to calculate the overall duty, and estimate the DCA and TTD in the plugged condition.

Plug maps

Whenever you plug a tube, be sure to note the location, axial position, date, and suspected cause of the failure on a layout of the tubesheet. If possible, the layout should be to full scale. Figure 10.25 illustrates a typical plug map for a U-tube bundle.

A plot of the cumulative number of plugs against months of operation provides a clear picture of the trend of failures requiring plugging. You can do this by simply using a standard spreadsheet pro-

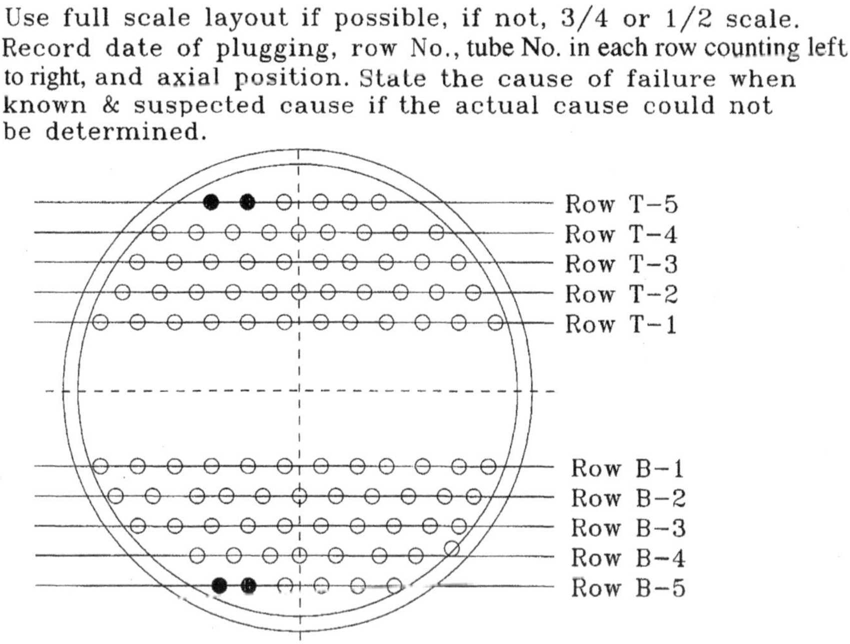

Figure 10.25 Typical plug map for a U-tube bundle.

gram. Plot the number of tubes plugged at each outage or shutdown, whether planned or forced, and the cumulative number of plugs against the months of operation. (Fig. 10.26). This will graphically reveal the history of forced outages for plugging and the rate of increase in plugged leakers.

Typically there is an initial period characterized by a bout of plugging. This leads to investigation to find and correct the causes of the failures. Then, if the problems are corrected or if the failure mechanism is self-limiting, there is usually a long period when failures match anticipated deterioration rates. When the slope of the plot of cumulative number of plugs versus time increases sharply, it is probably time to prepare to retube, rebundle, or replace the unit.

Computerized recording and trending programs such as Karta Technology's Tubestat program are excellent for recording and trending inspection results and plugging.

Plugging procedures and practices

Following are some recommended procedures and practices for properly plugging tubular heat-transfer equipment.

Preparations. For good results, carefully prepare the exchanger. Here are some things to do.

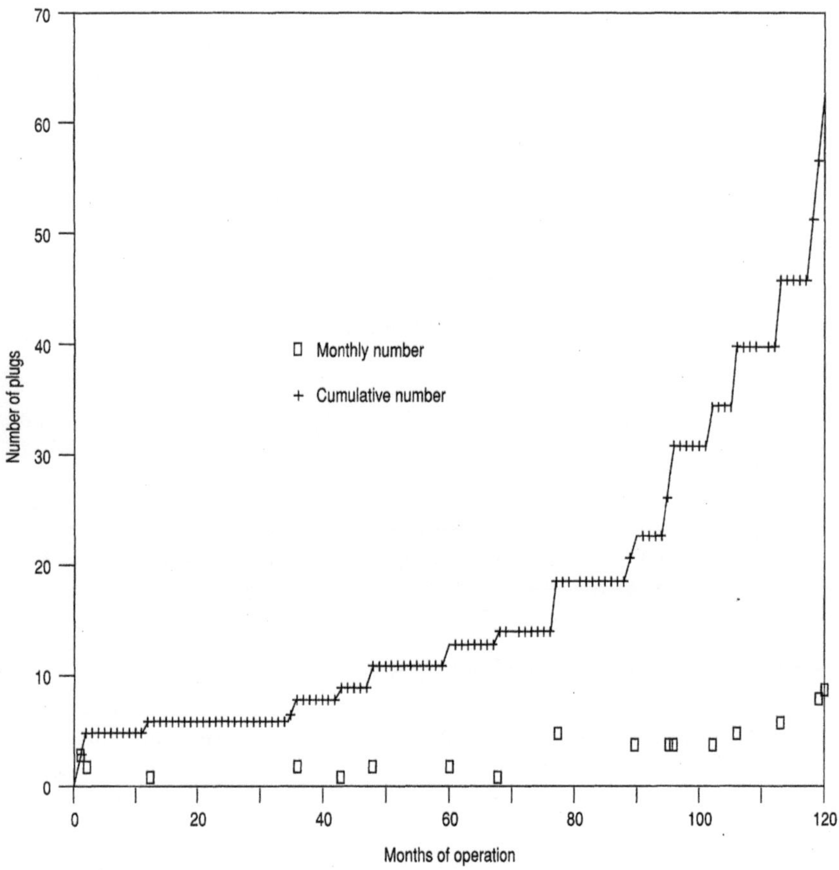

Figure 10.26 Plot of cumulative number of plugs versus months of operation.

Vent tubes of process heat exchangers to prevent pressure buildup. Polymerized fluid and corrosion products sometimes seal perforations in process exchanger tubes after they are plugged. Entrapped shell fluid can generate enormous pressure in the plugged tubes. Workers may be injured when taper plugs loosened during maintenance or inspections are ejected. If the tubes are burned, sawed, or abrasively cut during retubing or scrapping operations, entrapped material may burn or explode. Therefore, it is prudent to perforate or slot tubes to be plugged just behind the tubesheets with a hole or slot large enough to make sure that there will always be free flow of shell fluid into and out of the plugged tube.* Adequate vents and drains in plugged tubes

*U.S. Patent 5,597,276 has been issued to Yokell for a device for venting tubes behind the tubesheet.

of process heat exchanger are necessary for safe operation during the remaining life of the exchanger and safe dismemberment after you consign it to the scrap heap.

For the same reason, before burning or sawing tubes in the bundle, or attempting to extract tubes from an exchanger with a hydraulic extractor, be sure to remove the plugs from *both* ends of the tube. Although this will not prevent explosions and fires in pockets of unpolymerized liquid trapped between lengths of solidified material, it will eliminate the prospect of loosened plugs becoming projectiles because of heat-induced pressure or explosions. Readers are cautioned to remove plugs from retired process exchangers whether they are stored in the "bone yard" or shipped to the scrap yard.

The usual way scrap dealers reduce tubulars to sections small enough to be handled by their customers is to torch-burn with oxy-acetylene, oxy-mapp gas, or oxypropane cutting torches. Torch cutting will ignite any flammable or explosive material trapped in tubes from which plugs have not been removed prior to burning.

Scrap-yard workers must be appropriately supervised and warned not to allow cutting gas to flow freely into the bundle for more than a few seconds before sparking the igniter. These gases are denser than air and will accumulate in the bundle from bottom to top as they flow unignited out of the torch. If there is substantial gas flow before oxygen is turned on, sparking the igniter will cause a flashback or mini-explosion that can hurt a worker, even one wearing protective clothing and goggles.

Do *not* vent the tubes of feedwater heaters. As explained previously, if steam leaks into the outlet leg of a heater bundle and out of the inlet leg into the subcooler, it will raise the DCA. Therefore, we recommend sealing the leaks in at least one of the legs by sleeving or thimble plugging to prevent stray steam from entering the subcooler.

Cleaning. All your plugging procedures should require thorough cleaning of the place that will receive the plug. With plugs that depend on friction for holding power and tightness, the plug-to-tube seal and the tube-to-hole seal depend on interfacial pressure between plug and tube and between tube and hole and the coefficient of friction of the parts in contact. Scales and hard particles of foreign matter may reduce the effective interfacial pressure by interspersing incompressible brittle substances between the parts. These materials may also have more lubricity than the metals. Contact surfaces soiled by greases or fluids will also slide more readily over each other than will clean ones.

Choose compatible materials for welding plugs. The requirements for fusion welding plugs soundly to the tube and tubesheet are the same as for fusion welding tubes to tubesheets. These are weld compatibility of the plug, tube and tubesheet material, adequate root area, and cleanliness. In addition, the parts being sealed by welding must have approximately similar thermal coefficients of expansion. Prepare by verifying material compatibility, thermal coefficients of expansion, and design of the welds. Establish a cleaning procedure that includes brushing and volatile solvent washing or steam cleaning the surfaces followed by drying with filtered air.

Have a tested qualified welding procedure on hand and welders available whose qualifications in the procedure are current.

Considerations for fusion welding plugs. Taper plugs driven into tubes leave essentially no root gap between the plug and tube contact surfaces. Welding-generated gases escape through the weld puddle, causing porosity. Surface dirt or coatings can cause nonmetallic inclusions, porosity brought on by vaporization of fluids that adhered to the tube end, and incomplete fusion.

Crack-, porosity-, and inclusion-free tungsten arc welding (GTAW or TIG welding) requires near surgical cleanliness. But the metal arc process is generally far more tolerant of impurities and foreign matter. Therefore, after cleaning as well as is practical, fusion-weld plugs to tubes and tubesheets by the metal arc process using the smallest available coated rod size. (For the same reason, we recommend the metal arc process for repair welding of parts that cannot be adequately cleaned such as leaking tube-to-tubesheet welds.)

Stabilizing tubes. In your procedure, include a requirement for stabilizing damaged tubes to prevent ones that become severed from whipping against and damaging surrounding tubes. Whenever a tube corrodes through or fails because of mechanical damage or vibration-caused abrasion, it is apt to break completely at the failure location. For this reason we recommend stabilization as a standard practice in any plugging procedure.

The authors have successfully used the plug and stabilizer shown in Fig. 10.27 to do a quick combination plug and stabilization of $\frac{5}{8}$-in-O.D. $\times$ 0.049-in (approximately 16 $\times$ 1.25-mm) tubes. An alternative, when the break is similarly located, is to install in-tube, welded-in, round-bar plugs that extend well past the inlet nozzle. Devices for stabilizing tubes damaged at locations remote from the tubesheet consist of jointed sections and sheathed cables.

The authors have often been asked about stabilizing tubes by stuffing them tightly with wire cable. We do not consider this to be the best long-term solution. Cables are made by wrapping twisted strands of wire around an inner fiber core. They are intended to carry tensile loads. Therefore, unless they are fastened at their ends in a

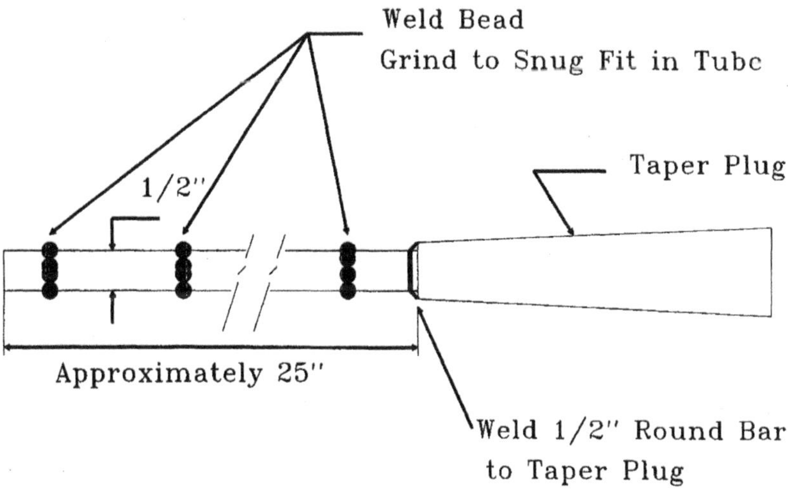

Figure 10.27 Plug and broken-tube stabilizer.

way that will maintain tension, they will permit broken tubes to move laterally to some degree, abrading the outer wires. The thin wires in the cable expose considerable surface to the shell stream. Depending on the shell stream's corrosivity, the cable may deteriorate to the extent that it no longer supports the broken ends.

However, cable stabilization is considerably less expensive than stabilizing with jointed rods and can prevent broken tubes from whipping about and destroying adjacent unbroken ones. It has been successfully used for a considerable period for stabilizing tubes in nuclear steam generators.

Supporting ligaments. Plugging tubes with taper plugs is generally done in a great rush to get the unit back on stream as soon as possible. Maintenance people are apt to drive the plugs too vigorously on the assumption that more force will make for a tighter plug. However, the ligaments that surround the tube are very easily moved, and it is a common experience to find that the plugged tube is tight but expanded tube-to-tubesheet joints in surrounding tubes have begun to leak. You can prevent this by using the ligament supports shown in Fig. 10.11.

Contractors install explosive plugs. They carefully select charge size, usually test specimens before production plugging, and have considerable experience with the technique. The explosive force attenuates exponentially with distance from the center of the explosion, so explosive plugging should cause no damage to the ligaments. Nevertheless, the authors have witnessed ligament cracking during explosive plugging. Therefore, make sure that the contractor's procedure includes supporting the ligaments as illustrated in Fig. 10.11, especially when tube yield strength is high relative to that of the tubesheet.

Plugging units in lethal service. When you prepare purchase specifications for exchangers in lethal service as defined in ASME Code, Section VIII, require the exchanger manufacturer to submit a recommended plugging system. It should include a description and drawings of the plugs to be used and the procedure for installing them. Permanent plugs for this service, regardless of whether for high pressure, should always be either (1) welded ones, with plugs, tubes, tubesheets, and weld deposit fully compatible with the tubesheet and tube material or (2) explosively installed if there is access and it is safe to set off explosions in the area.

Make sure that the plugging, nondestructive examining, and post-plugging testing procedures are acceptable to the AI responsible for on-site inspection. We recommend visually examining the welds with high-intensity light and magnification in the range of 30 power followed by fluid penetrant examination.

The NBIC discussed in Chap. 6 does not directly address welded plugging. The National Board has issued an interpretation stating that welding of a plug to seal tubes, because of failures, reduced wall due to corrosion, or for operating reasons in a boiler or pressure is considered a repair. Procedures for plugging radioactive tube ends are beyond the scope of this work.

Testing after plugging. It is advisable to bubble-test at the tubesheet face after plugging to make sure of the tube-to-tubesheet tightness and of the plug to the tube. If you have plugged by cutting back the tube and welding a plug to the hole, the purpose will be to test the integrity of the weld of the plug to the tubesheet. The pressure applied to the shell side need not be much more than 350 kPa (~50 psig). Sometimes a higher pressure may be desirable. However, under no circumstances take shell-side air test pressure higher than the 80 percent of the MAWP stamped on the nameplate.

Ferrules and Sleeves

Ferrules are installed in tube ends to protect against entrance erosion caused by entrance turbulence. They may also be installed to carry hot tube-side fluid past the position of the rear face of the tubesheet when the tube-side inlet temperature is too high to permit direct contact with tube and tubesheet metal not bathed by shell-side fluid.

Figures 10.28 and 10.29 illustrate two types of ferrules. These contrivances are snug-fitting and are flanged over to a diameter slightly

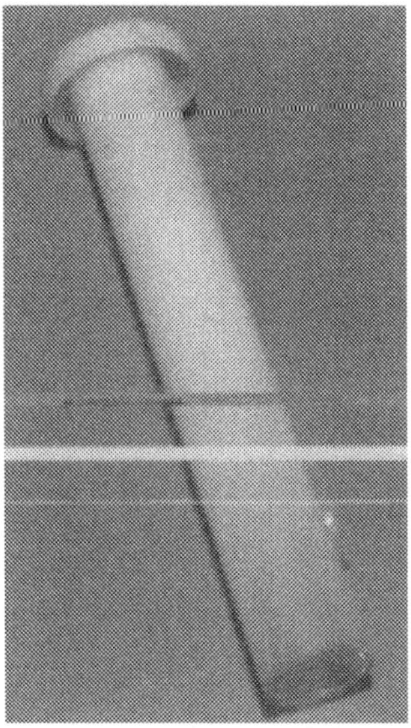

Figure 10.28 Plastic ferrule. (*Photograph by MGT Inc., Boulder, Colo.*)

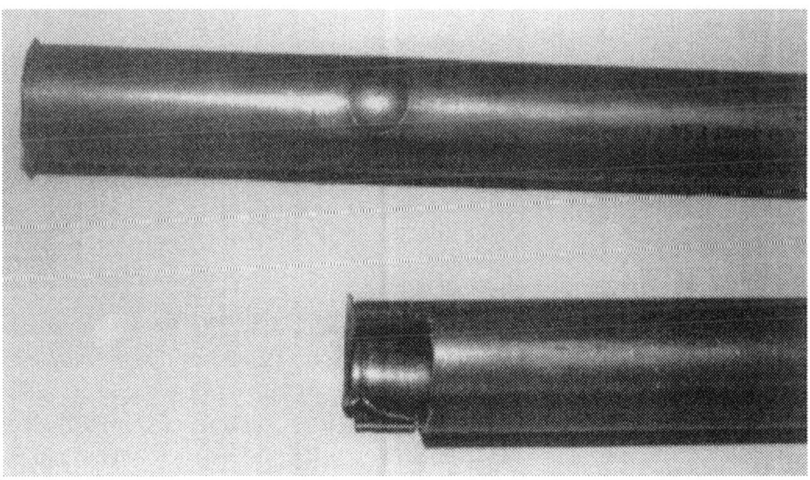

Figure 10.29 Section through installed metal ferrule and appearance of expanded-in metal in a tube hole. (*Photograph by MGT Inc., Boulder, Colo.*)

greater than the tube O.D. When crevice corrosion is not likely to occur, they are simply pressed into place, which is common for the plastic variety (Fig. 10.28). However, if crevice corrosion is likely, metal ferrules are used that can be expanded to make an interference fit with the tube (Fig. 10.29). A technique sometimes practiced with metal ferrules to avoid crevice corrosion is to gas-tungsten-arc-weld the inner end of the ferrule to the tube before expanding.

Sleeves are thin, close-fitting cylinders that can be slid into an exchanger tube with a localized perforated or thinned tube to bridge the damaged region.

A consideration when installing ferrules and sleeves is that the fluid regime does not stabilize after entrance into the tube for a distance of 10 to 20 tube diameters, and that further restricting the flow can cause additional downstream turbulence. Sleeves for bridging damaged regions of tubes are therefore are about 25 mm (~0.01 in) thick in order to cause minimal downstream turbulence. Ferrules, which are usually thicker, should have their I.D.s tapered out to a thickness at the inner end of about 25 mm (~0.01 in).

Most sleeving contractors expand the sleeves into place. However, the sleeve may also be gas-tungsten-arc-welded to the tube I.D. In the *controlled strain system* hydraulic pressure is applied inside the sleeve, in much the same fashion as in hydroexpanding tubes into tubesheets. However, instead of controlling the expanding pressure, the volume of water fed into the mandrel is controlled. This limits the deflection of the sleeve or ferrule, thereby preventing overexpanding and consequent tube bulging.

Heat and Erosion Shields

The authors have had successful experience with the design, manufacture, and installation of shields to protect against high tube-side inlet fluid temperatures and erosive inlet fluids. These are described in the following paragraphs.

Sleeve shields

To protect tubes against high-temperature inlet fluids and erosion, weld sleeves into a thin plate drilled on the tubesheet template (see Fig. 10.30). Slide all the sleeves into the tubes simultaneously. To accomplish this, it may be necessary to slide "pickup rods" consisting of lengths of round bar into the sleeves. Expand the sleeve ends to intimate contact with the tube, or gas-tungsten-arc-weld and expand.

Perforated-plate shields

When channel inlet nozzles are so placed that entering fluid erodes the tube ends and tubesheet (most often found in closed feedwater heaters), consider installing a perforated-plate shield (Fig. 10.31). Here, the strategy for protecting against further damage is to fabricate and install a drilled plate segment into the inlet compartment. The segment is drilled with holes on the same drilling template as the tubesheet, with holes drilled to the I.D. of the tubes. The back side of the holes in the plate seg-

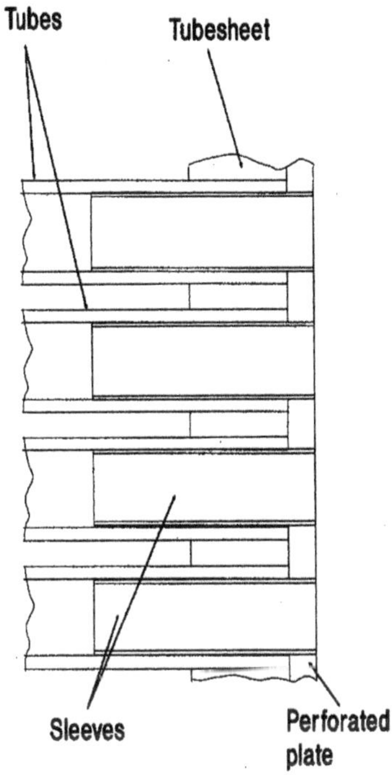

Figure 10.30 Sleeve shield.

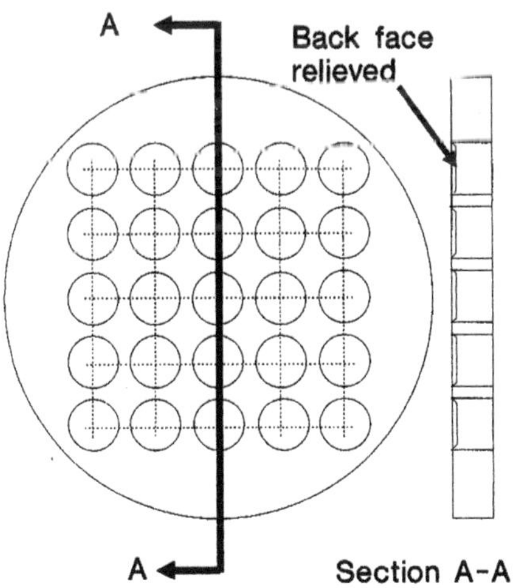

Figure 10.31 Perforated-plate shield.

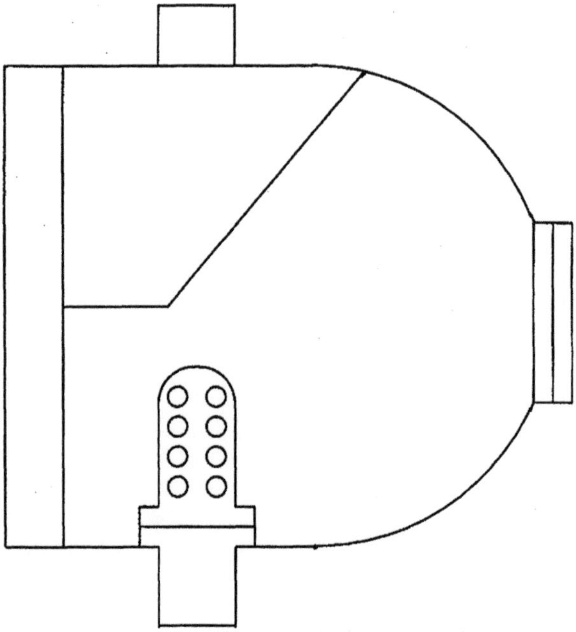

Figure 10.32 Feedwater distributor to reduce erosion on tube ends and joints.

ment is relieved to clear any tube or weld protrusion. The plate is then fastened to the tubesheet and serves as a replaceable wasting plate.

Figure 10.32 shows an alternative means of protecting the inlet-pass face of the tubesheet against erosion by installing a perforated-pipe diffuser to spread the inlet flow.

References

1. S. Yokell, "Troubleshooting Shell-and-Tube Heat Exchangers," *Chem. Eng.,* July 25, 1983.
2. A. Lohmeir and S. D. Reynolds, Jr., "Carbon-Steel, Feedwater-Heater Tube-to-Tube-Sheet Joints," ASME paper 65-WA/PWE-10, ASME, New York, Aug. 1965.
3. S. Yokell and C. F. Andreone, "What to Look for in Failed Feedwater Heaters," *Power* 74–76, March 1995.
4. Editors, "Explosive Welding Does Bang-up Job in Heater Repair," *Welding J.,* Feb. 1980.
5. J. E. Martin and G. Jorgensen, "Testing and Plugging Heat Exchanger Tubes to Withstand 5,000 P.S.I. by Removing One Head Only," Shell and Tube Heat Exchangers, *Proc. 2d Symp. Shell and Tube Heat Exchangers,* Sept. 14–16, 1981, Houston, Tex.
6. G. K. Pase and S. Yokell, "Interpass Temperature Effects in Multipass Straight Tube Exchangers," *Chem. Eng. Progress,* July 1983, pp. 51–53.
7. D. Q. Kern, *Process Heat Transfer,* McGraw-Hill, New York, 1950, p. 99.
8. *Standards of the Tubular Exchanger Manufacturers Association,* 7th, TEMA, Tarreytown, N.Y., 1988, pp. 111–112.

Chapter

11

Shell-Side Repairs and Alterations

Introduction

The first part of this chapter discusses the causes of shell-side heat exchanger failures. We follow with a discussion of procedures and practices for altering and repairing heat exchangers.

Shell-Side Failures and Their Causes

Shell-side failures occur for many reasons. The causes of failures in the shell and tube sides are interrelated. Often shell-side failures will manifest themselves in tube failures that stem from impingement-plate failures or excessive tube vibration. Early identification of the failure mode will usually allow repairs to the exchanger that correct the condition. Early identification is possible only if complete records are kept of all maintenance activities.

Relationship between thermal and mechanical causes of failure

Mechanical and thermal problems are usually related. Differences in metal temperatures in the shell and tubes may cause distortions and leaks in tube-to-tubesheet joints, flanged closures between tubesheets and shells, and between tubesheets and bonnets or channels. The following thermal problems lead to mechanical difficulties:

1. Oversurface
2. Undersurface
3. Maldistribution

Certain types of shell-side failure are discussed in the following paragraphs.

General corrosion

General corrosion often takes place in aggressive shell-side environments. The attack is in the form of pits and grooves on the shell and bundle surfaces. These are some of its causes:

1. The materials of construction selected are not satisfactory to resist the corrosive process fluids, causing general corrosion on the *shell-side* surfaces.
2. An unanticipated change was made to process fluids more corrosive than the material of construction can resist. Locations of attack are the same as those described in item 1.
3. The operating flows are greater than the exchanger was designed for. This leads to shell erosion in the vicinity of the tube supports and/or baffles and inlet-nozzle regions and erosion-corrosion of the baffle and tube surfaces, and may cause harmful tube vibration.
4. The operating temperature of the shell-side fluid is higher than was anticipated, increasing the rate of attack.

Shell corrosion at baffle and support locations may be more severe than elsewhere. Tubes and baffles corrode more at the sites where the tubes penetrate than at other places. This may be due to a combination of crevice corrosion, and fretting and pounding caused by tube vibration. Frequently, it is possible to observe rings of erosion-corrosion in the shell at the locations of the baffles or tube supports. Such damage may be caused by the high velocity of fluid in the shell as it leaks between baffle and shell, or it may be a form of crevice corrosion.

A typical example of general corrosion failure on the shell side is exfoliation corrosion of 70-30 Cu-Ni (copper-nickel) tubes, which are sensitive to some water treatment and dissolved noncondensible gases. Another is pitting of carbon-steel tubes which are aggressively attacked when air is introduced to the heat exchanger while the tubes are warm and wet (usually during a shutdown). Concentrated (98%) sulfuric acid and oleum, which are generally harmless to carbon steel, become very corrosive if they are diluted. Many other acid solutions become very aggressive to some materials when the water content rises to certain levels.

Nozzle-to-shell weld cracking

Cracks may occur in heat-affected zones adjacent to welds when large [>150°C (~300°F)] temperature cycles occur as the heat exchanger goes into and out of service. The temperature changes cause thermal stress ratcheting until cracks develop or until the welds and parent metal fatigue. These cracks also develop at the juncture of the shell to the tubesheet when the design does not provide for adequate flexibility.

Shell-flow-induced vibration failures

Shell-flow-induced tube vibration causes the following kinds of damage:

- Leaks due to through-wall abrasion at baffle or support penetrations. (Such leaks may also be a corrosion phenomenon or result from a combination of corrosion-erosion and vibration.)
- A pattern of leaks between baffles or supports or tubesheets and first support or baffle. This pattern indicates that vibration caused interspan collisions that damaged the tubes between baffles or supports, wearing them in a characteristic pattern.
- Leaks at U-bends.
- Cracks in tubes behind tubesheets.
- Failures in welds of welded-only tube-to-tubesheet joints.
- Fretting of tubes where they penetrate the baffles or tube supports.
- Wear and pounding of the baffles where the tubes pass through the baffle or support holes.
- Stress-corrosion cracking (SCC) induced by tensile stress levels enhanced by tube vibration.
- Fatigue failures in the region of the tubesheets.
- Impingement protection failures.
- Failures of zone-enclosure plates in feedwater heaters.

Impingement-plate failures and failures in shell inlet region

Impingement plates may break away when attachment welds are inadequate or of poor quality. The impingement plate may be too small to protect the tubes in the shell inlet-nozzle area. The impingement plate may be too thin or made of the wrong materials to withstand the high impact force of the incoming fluids.

TEMA Standards Paragraph RCB-4.6 and its associated subparagraphs provide limitations intended to prevent or minimize erosion of tube bundle components at fluid entrance and exit areas. The HEI Standards have similar criteria for flow areas under shell nozzles associated with setting impingement-plate dimensions. Paragraph 3.7 in the HEI Standards for Power Plant Heat Exchangers requires impingement plates to be sized assuming a minimum angle of diffusion of 15° from the point at which the nozzle penetrates the shell. Paragraph 2.9 in the HEI Standards for Closed Feedwater Heaters requires a minimum angle of 45°. Paragraph RGP-RCB-4.62 of TEMA Standards, Section 10, "Recommended Good Practice," has methods for determining approximate shell and bundle areas for some common configurations. The latter paragraph cautions the user that results are somewhat approximate because of nonuniform location of tubes at the bundle periphery, the presence of untubed lanes through the bundle, and the presence of tie rods, spacers, and/or bypass leak devices.

The standards point out that other configurations should be given special considerations, listing, as examples, nozzles located near the bends of U-tube bundles, nozzles attached to the shell in a semi- or full tangential position, rod-type impingement protection, perforated impingement devices, impingement plates that are not flat or are positioned with significant clearance off the bundle, and annular distributor shells.

Despite the manufacturer's best efforts to follow the requirements and good practices of the standards, the provided protection is sometimes not capable of preventing erosion at bundle inlet and outlet regions. The velocity used to determine ρV^2 (where ρ is the fluid density and V is the average linear velocity of the entering fluid, in consistent units) of the entering and exiting stream is the *average* linear velocity determined by dividing the flow volume per second by the cross-sectional area. However, some regions of the stream may flow at considerably higher velocities than the average. Such stream components may erode parts even when the ρV^2 criterion is met. This is of special concern with two-phase streams.

Following are some ways to fix inadequate impingement-plate problems:

1. If the bundle is removable, extract it. Replace the impingement plate on the bundle with a wider one, and perhaps one that wraps around the bundle. Extract and replace the damaged tubes.

2. Plug the failed tubes (*a*) and those in one or two succeeding rows in the vicinity of the inlet, stabilizing the tubes to prevent them from whipping around and damaging other ones if they should become completely severed or (*b*) cut out a circular section of the shell centered on the inlet nozzle. Make the hole large enough to accept the large end of a reducer to the small end of which the inlet nozzle is welded. You can combine this with the fix described in option 1. If you can get to the impingement plate, replace it with one at least 25 mm (~1 in) larger in diameter or rectangular dimensions than the greater of either (i) the large-end diameter or (ii) the projection of the cone of diffusion from the transition of the smaller to larger diameter of the

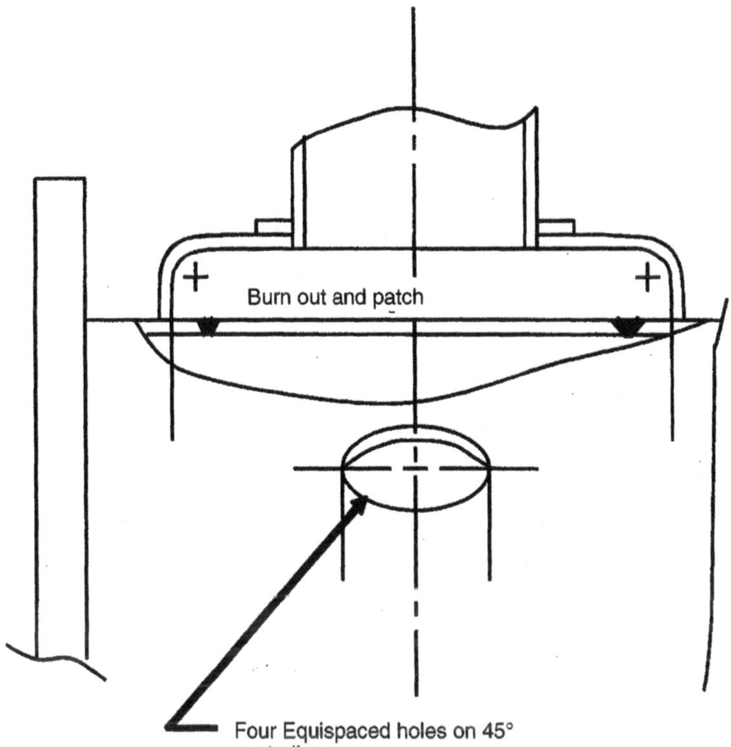

Figure 11.1 Correcting inadequate inlet impingement protection by cutting out shell and impingement plate, patching shell, and installing a distributor belt.

reducer. To do this, there must be enough room between the inlet opening and inlet-end tubesheet or shell flange.

3. Refer to Fig. 11.1. Cut out the inlet nozzle. Patch the hole. Cut four equally spaced holes into the shell equal in size to the inlet nozzle. Place the holes on 45° centerlines so that they straddle the patched hole in the shell. Construct and weld a shell distribution belt to the shell that overlaps the region of the holes cut into the shell. Install the cutout inlet nozzle in the distribution belt. This procedure also permits replacing the damaged tubes in straight tube units of all types and in peripheral tubes of removable U-tube bundles.

Failures in feedwater heaters, desuperheaters, and subcoolers

Desuperheating-zone shrouds may fail in the steam inlet area and at the welds of the longitudinal plate where it is welded to the shroud, shell skirt, or tubesheet. The welds may fail because the manufacturer fillet-welded the partial cylinder to the flat plate and tubesheet instead of using full-penetration welds, or because of poor weld quality. (These welds are not pressure-retaining welds. Consequently, the welders who do the welding may not be as skilled as those qualified under the governing pressure safety code.)

The baffle arrangement may be inadequate to prevent tube vibration—especially at overload conditions—leading to tube failures and baffle damage in the desuperheater. It is noteworthy that the TEMA Standards caution users that their limitation criteria for impingement protection have no correlation with tube vibration and refer the designer to Section 6 for information regarding tube vibration.

If there is more heat-transfer surface in a desuperheating zone than is required to desuperheat the steam under low load operating conditions, some steam will condense in the desuperheater. The wet steam erodes the tubes, baffles near the shroud outlet, and tube supports in the condensing section of the heater (see Fig. 11.2).

Subcooling-zone enclosures can fail at the welds of the longitudinal plate to the shroud, shell skirt, or tubesheet for the same reasons as welds fail in desuperheaters. Typical subcooler failure locations are at the subcooling-zone inlet and through the zone end plates of horizontal heaters with a short full-pass subcooler design. The tubes are not

Figure 11.2 Erosion in tube supports downstream of desuperheater exit, resulting from wet-wall condition in desuperheater.

in hydraulically tight contact with the holes in the end plate. If the end plate is not thick enough to condense any steam passing through the end plate to the subcooler, the end plate and tubes passing through it are eroded by the wet, partially condensed steam that flows through the annular space between the tubes and surrounding holes.

Shell expansion joint failures

Appendix 26 of Section VIII, Division 1 of the ASME Code is a mandatory appendix with rules for design and construction of bellows type (thin-walled) expansion joints.[1] Appendix CC, which is presently nonmandatory, but which will eventually probably become mandatory, has rules for the design and construction of flanged-and-flued and flanged-only expansion joints (thick-walled expansion joints). Both require designing for the anticipated magnitude and directions of deflection and for the number of cycles of operation. If the user fails to specify the cycles, the appendixes require the expansion joint designer to design for a minimum of 100 cycles. Appendix CC and Section 5 of the TEMA Standards, devoted to flexible shell elements of the type covered in ASME Code, Section CC, are somewhat similar and overlapping.[2] The TEMA formulas are based on these assumptions:

- Applied loadings are axial.
- Torsional loads are negligible.
- The flexible elements are sufficiently thick to avoid instability.
- The flexible elements are axisymmetric.
- All dimensions are in inches, and all forces are in pounds. (With some effort, the data in the TEMA Standards may be converted to SI units. A nonmandatory appendix in the ASME Code provides guidelines for converting the units.)

Expansion joints fail for these reasons:

1. Either (*a*) the user failed to specify the cycles of operation and the unit operates for more than the cycles the designer assumed (many failures result from users not telling heat exchanger manufacturers the number of cycles of expansion and contraction anticipated over the exchanger's life) or (*b*) the user-specified number of cycles is less than the cycles encountered during operation.

2. The user underestimated or understated the range of deflection to be accommodated by the expansion joint or did not provide the manufacturer with sufficient information to estimate the range. A major cause of expansion joint failures is not designing for the most

severe operating conditions. If the user does not apprise the manufacturer of how the unit will be started up and shut down, alternative normal operating conditions, anticipated temperature excursions, and idle operation with the tube-side stream flowing and shutoff shell-side stream or empty shell, an expansion joint may be installed that cannot operate safely at the most severe condition. It is important to realize that joint cycle life directly depends on the magnitude of joint deflection.

3. The joint was precompressed or prestretched beyond the design precompression or preextension. When workers installing an expansion joint precompress or preextend it and the design has not provided for such compression or extension, it will fail much sooner than expected. A similar result occurs when the joint manufacturer's design is based on planned precompression or preextension and the workers release it when they install the joint. For these reasons, bellows expansion joints, which are very easily deflected, should be fitted with shipping bars, which are bridges across the flexible element that maintain welding-stub–welding stub length. The shipping bars should not be removed until the unit has been installed, but must be removed before start-up.

4. An unanticipated torsional or angular deflection occurred either in operation or during heat exchanger manufacture. Unless the expansion joint manufacturer is told in advance of any anticipated angular or torsional deflections, the joint will be constructed to accommodate only axial deflections. If there is a possibility of angular deflection in a horizontal exchanger, be sure to provide a means of restraining it. Two general types are used for such restraint. One consists of welding three or four equally spaced plate or bar extensions perpendicular to the shell on either side of the expansion joint. Male machined bars or pipes are welded to one extension and slip-fitted into female sleeves on the opposite ones. This permits axial movement but restrains the tendency to sag. The other system consists of welding three or four equispaced plate or bar extensions along the shell axis as shown in Fig. 11.3. The section that is sandwiched is slotted. A pipe sleeve fits into the slot and through holes in the outer bars. Bolts with flat washers under the head and nut are passed through the sleeve. The three or four alignment guides permit axial expansion but prevent the shell halves from sagging.

5. Excessive pressure was applied, causing squirm in multicorrugation thin-walled joints and permanent deformation in thick-walled ones. When a multicorrugation bellows is subjected to more pressure than it can handle, it squirms. This means that spaces between the crowns of adjacent corrugations are not uniform. Tolerable squirm is limited in the various applicable standards. Both types of expansion joint elements may also bulge when they are subjected to excessive pressure.

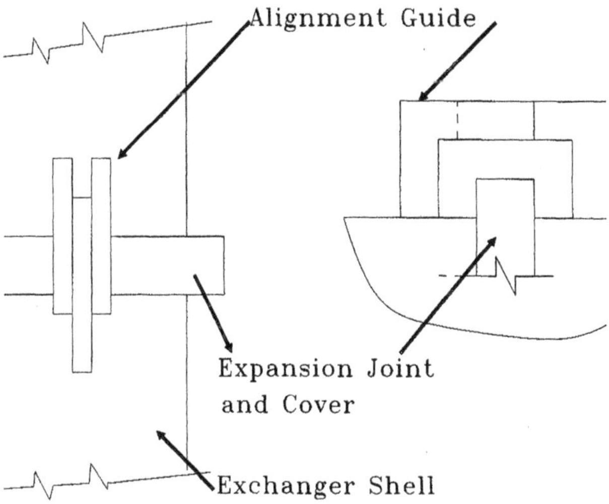

Figure 11.3 Schematic of bar-type expansion joint alignment guide.

6. The joint material was subject to stress-corrosion cracking (SCC) when exposed to the fluid environment and subjected to tensile stress resulting from pressure and deflection within the TEMA and ASME Code design limits. Joints often fail from SCC when fluids high in chloride ion concentration are entrapped. Most bellows are fabricated from austenitic steels which are sensitive to SCC in such environments under tensile stress.

7. Either (*a*) the fabrication quality was deficient (poorly fabricated bellows joints generally fail at welds of flexible elements to welding stubs; improperly formed flanged-and-flued-head joints usually fail in the knuckles); (*b*) the joint was improperly installed, creating unanticipated angularity, twist compression, or expansion; or (*c*) a stress riser was created on a thin-walled joint by a dent or a weld drag.

8. Piping loads on shell- and tube-side nozzles created bending moments causing angularity for which the joint was not designed.

Longitudinal baffle failures

Welded-in longitudinal baffles usually fail at the weld area because the welds are inadequate. In some constructions multilayered flexible seals that curve into a fit with the shell are welded to removable-bundle shell-side pass partitions. When these devices are damaged during installation or corrode prematurely, the resulting bypassing reduces exchanger performance. In a small number of removable-bundle exchangers with multipass shells, the bundle pass partition fits into channels formed by welding pairs of strips to the shell. The seal is effected by rectangular, braided gaskets fitted between the strips and on which the pass-partition edges bear. If the gasket material shrinks or becomes hard, fluid will bypass the partition with consequent performance deterioration.

Lining failures

Nozzle liners or shell liners will fail if the welds joining the dissimilar metals are not made correctly or the difference in thermal coefficients of expansion of the weld metal and parent metals induce excessive stress in the welds. Note that residual weld shrinkage stresses and thermal stress are additive.

Failures caused by process and operating problems

The origin of many failures is often attributable to other causes or events unrelated to the mechanical design and conditions used for the thermal design basis. We discuss some of these causes in the following paragraphs.

Inadequate venting. During operation, unvented inerts can accumulate somewhere in the exchanger where they blanket surfaces, thereby reducing performance. Trapped gases can cause corrosion of highly stressed tubes and in heat-affected zones of welded parts. Accumulations under tubesheets of vertical units can also lead to ligament cracking due to excessive heat flux resulting from unsteady-state heat transfer. Air entering an exchanger after shutting down will aggressively attack and corrode wet carbon steel and 70-30 copper-nickel tubes. Repeated start-ups and shutdowns increase the corrosion layer until it spalls off during a heatup cycle, exposing fresh metal to attack. Eventually pin-hole failures occur.

Condensate flooding. Unremoved condensate floods heat transfer surfaces, thereby reducing the heat-transfer rate. Sediments and corrodants concentrate in the bottom of the vapor or steam space and on the tube supports, leading to mechanical damage.

Insufficient blowdown in vertical steam generators. In vertical steam generators, unless the blowdown rate is high enough to prevent such buildup, the accumulations can cause tube denting at the support plates and corrosion at the tubesheets. Flow velocity across the tubesheet should be at least 1 m/s (~3 ft/s).

Low liquid levels. In horizontal exchangers, low liquid levels will result in uneven temperature between upper and lower tubes. This can cause tube buckling and tube-to-tubesheet joint leaks in straight-tube horizontal and U-tube units in which the plane of the tubes is horizontal. In U-tubes arranged in the vertical plane that have one leg in the flooded portion of the shell and the other in the vapor space, differences in the thermal growth of the hot and cold legs can cause the tubes to twist the tube supports and fret in the tube holes as they cycle from hot to cold.

In feedwater heaters, low liquid levels are generally a problem only if they have subcooling zones. When the liquid level gets too low, a mixture of steam and condensate drains enters the subcooling zone. The steam immediately collapses when it contacts the cold tubes, causing a water hammer. Repeated hammering will cause the tubes to fail.

Hydraulic hammer. Hydraulic hammer that results from steam-trap or condensate-accumulator problems can knock tubes loose from the tubesheet and contribute to tube vibration problems. Figure 11.4 shows how severely hydraulic hammer can deform a thick tubesheet in a high-pressure feedwater heater.

Instrument control problems. Instrument and control problems may lead to vibration failures or failures that result from excessive cycling. Failures in pressure-reduction and pressure-relief valves can cause severe distortions in the pressure envelope.

Figure 11.4 High-pressure feedwater heater tubesheet deformed by hydraulic hammer.

Piping problems. Plugging caused by foreign matter in pipelines may create pressure- and temperature-related failures. If the piping is improperly connected to the unit (e.g., inlet piping connected to the outlet connection), tube metal temperatures may vary significantly from those of design, thereby causing mechanical failures. If a shell inlet coolant stream is inadvertently piped to the top of an exchanger shell and coolant flows at a rate insufficient to fill the shell, the unit is likely to fail because of temperature differences between cold and hot tubes and shell and tube metal. Always arrange piping of liquids to and from heat exchangers to ensure that all the heat-transfer surface is always full on the tube side and flooded on the shell side. Ordinarily there is enough backpressure in the exit piping to maintain flooding, but sometimes you may have to install a loop seal in the piping.

Start-up and shutdown practices. Start-up and shutdown practices are frequent causes of failure. It is always necessary to consider the effects of the temperature changes to the metal during these times. This is especially important where a shell is welded to the tubesheet. Be sure to introduce both streams in a manner that provides for a relatively even temperature at these welded junctures.

Thick tubesheets must not be cooled so rapidly that excessive thermal stress develops between the surfaces exposed to coolant and interior metal. The authors have seen tube-to-tubesheet joints and ligaments in supercritical pressure feedwater heaters destroyed by cooling the channel too rapidly.

Deviations from design point. Excessive deviations from design-point conditions can cause failures when the designer is not aware that these conditions can occur. An example is a fixed-tubesheet exchanger in condensing or reboiling service in a distillation process. If the start-up procedure requires steaming out the fractionation column, alert the exchanger manufacturer to be aware of the anticipated coincident metal temperatures. This may affect the expansion joint calculations.

Feedwater heaters operate with uncontrolled steam extraction. Consequently, bypassing a heater in a string of heaters can cause more than twice the design-basis steam flow to enter the next heater in the string. This is generally not a serious problem in heaters designed for condensing only, or condensing and subcooling, but if the heater has a desuperheating zone, this condition can lead to tube failure because of vibration.

Of equal concern is the temperature gradient across the tubesheet face. If the temperature rise across the heater is very large [~55 to 85°C (~100 to 150°F)], thermal stress will build up in the tubesheet at the outlet pass and cracks will develop at the juncture of the channel to the tubesheet and the pass partition to the tubesheet. Similarly, when a string of parallel feedwater heaters is bypassed, only those with desuperheating zones are likely to suffer from tube vibration.

Inadequate level control. When level control is inadequate in a horizontal steam generator, upper rows of tubes may not be flooded. The metal temperatures of the "dry" tubes may be much higher than those of the flooded ones. This difference in temperature can cause mechanical failures.

Poorly placed level-control positions can lead to mechanical damage. In kettle-type reboilers and evaporators, the liquid level in the shell should always be at a level where all the tubes are submerged in the liquid pool. The liquid surface must be enough below the vapor outlet nozzle to provide adequate disengaging space to avoid the vapors transporting the liquids. A standpipe is usually required because the boiling action in the shell is too violent to give a steady indication.

Inadequate level control in boiler feedwater heaters is a principal cause of operating problems and mechanical failures. Level-control connections in feedwater heaters have to be carefully reviewed to verify that they will not be adversely affected by any high-velocity flows which would give a false signal to the level controller. One should always be aware that the surface of the fluid between the condensate inlet and outlet is not flat but is wavy, and that the level-control position and means for isolating the instrument from turbulence must be selected with this in mind.

Mechanical causes of failure

Following is a discussion of some mechanical causes of failures.

Excessive piping loads on nozzles. Nozzle loads need to be properly defined in advance of building an exchanger. Piping can place severe loads on the exchanger nozzles. This is especially a problem with steam lines feeding a heat exchanger. The exchanger manufacturer can easily provide nozzle designs that will satisfy most nozzle load requirements. However, unless the purchaser specifies the nozzle loads, the manufacturer's nozzle design will simply satisfy the pressure vessel code's reinforcement requirements for the specified design pressure.

Although it is true that nozzles designed only to meet pressure requirements are usually capable of bearing piping-imposed loads, purchasers must either evaluate the nozzle load capacity or negotiate the manufacturer's agreement to additional piping load requirements.

Impingement attack. Impingement attack can be the result of

- Inlet nozzles too small
- Inadequate impingement protection
- Insufficient escape area for incoming fluids

The first step in seeking a remedy is to check the ρV^2 values at inlets and outlets and correct nozzle sizes as necessary. Second is to examine the pattern of failure. If the pattern shows damage from stray high-velocity inlet stream components, replace the impingement plate with a wider one that impedes these streams from directly striking the bundle.

Cavitation at outlets. Cavitation at shell outlets causes pitting damage to tubes in the region of shell outlet and tube failures. It is caused by discharging the shell fluid down an unrestricted vertical pipe. You can fix the problem by installing a vortex breaker in the outlet, providing a trap in the outlet piping, and installing a standpipe in the outlet piping.

Inadequate bundle support at the free end. Inadequate support at the free end of a U-tube or floating-tubesheet bundle can cause these failures:

- Bundle distortion at the free end
- Tube failures at the free-end support
- Tube-to-tubesheet joint leaks in floating heads
- Baffle damage at rearmost tube support
- Scored packed floating head barrels
- Tube failures due to abrasion in the unsupported U-bend span

Some of the things to do to fix the problem:

- Stabilize and plug failed tubes.
- Replace broken tubes.
- Stake the tubes to provide adequate support.
- Install sectional grid baffles for support.
- Replace the bundle with one that is adequately supported.

Vibration-damage failures. Some of the causes of damaging vibration are as follows:

- Excessive shell stream inlet velocity
- Operating at lower-than-design vapor pressure
- Excessive span length between adjacent tube supports
- Inadequately supporting tubes at the U-bend end
- Changing tube material and/or gage without investigating the effects of the different natural frequency on the tubes' tendency to vibrate
- Imposition of external vibration-forcing energy

Some remedies:

- Install deresonating baffles.
- Lace tubes with wires.
- Stake the tubes with flat bars or commercially available stakes (Fig. 11.5).
- Screw in helical stabilizers (Fig. 11.6).
- Change the fluid flow character or the flow pattern.
- Retube with a tube material and thickness that has a suitable natural frequency.
- Eliminate the external vibration-forcing energy or isolate the unit from it.
- Replace the bundle with one designed to be vibration-free under the existing service conditions.

Packed-floating-head failures. When packed floating heads fail, fluids may leak to the atmosphere, tubes may be distorted, and the bundle may be bent. Some causes of packed floating-head failures:

- Scoring of floating barrel during bundle installation
- Excessive tightening of gland follower
- Cocking of gland follower
- Hardened, dried-out packing
- Interpass temperature differences

These are some fixes:

- Remachine scored surfaces to 63 rms or smoother.
- Replace packing with appropriately lubricated packing. Tighten gland only enough to achieve a controlled leak of one to 5 drops per minute.

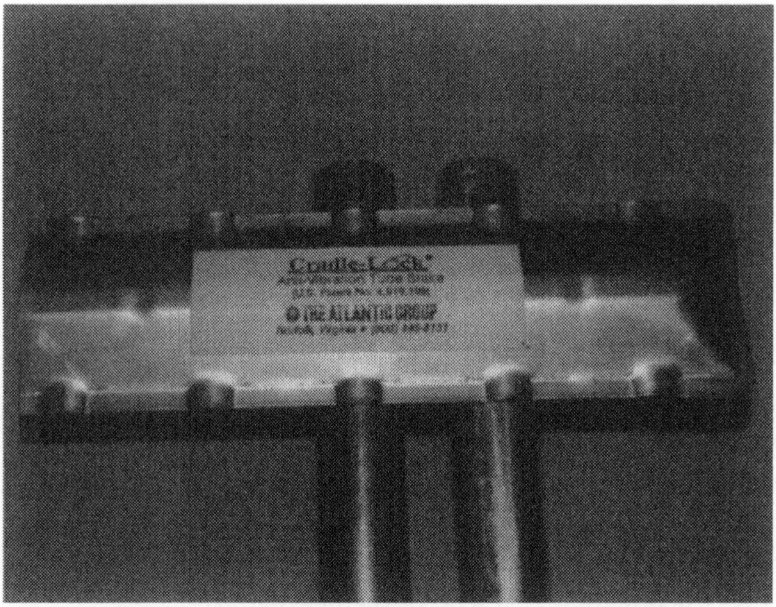

(a)

(b)

Figure 11.5 (*a*) Atlantic Group–patented cradle lock antivibration stake. (*Atlantic Group, Norfolk, Va., photograph by MGT Inc., Boulder, Colo.*) (*b*) Cradle lock antivibration stakes installed. (*Atlantic Group, Norfolk, Va.*).

Figure 11.6 Patented helical antivibration stake. (*Dr. Frank Eisinger, Clinton, N.J., MGT Inc. photograph.*)

- Wrap barrel in a protective sheath when installing the bundle.
- Install and tighten gland followers so as to avoid cocking.

Procedures and Practices for Repairing and Altering Heat Exchangers

In this section we suggest some specific procedures and practices to use in making repairs and alterations.

Repairing cracked nozzle-to-shell welds

Nozzle-to-shell welds may crack because of excessive loads, cyclical loads, or thermal stresses. Welds that join nozzles to the shell in the vicinity of fixed-tubesheet–shell welds may crack because of cyclical plastic deformation of the shell. Before making any repairs, find out why the weld cracked. The repair procedure assumes that the parent metal is not so badly damaged that it is "broken-hearted." This is factory-floor talk for metal fatigue. If the nozzle failed because of excessive piping loads, for which the unit was not designed, reinforce the nozzle. This may be considered an alteration under the previously discussed inspection codes.

There are two essential elements to repairing cracked nozzle-to-shell welds: (1) chipping or grinding out the cracked region to the bottom of the crack for the whole crack length and (2) making sure that the region to be rewelded is free of all foreign matter. If the nozzle is large, with a substantial weld, find the ends of the crack and drill into the weld to its root to prevent chasing the crack further. To find the ends of the crack, use magnetic particle or fluid penetrant examination. Reexamine the damaged region after grinding, chipping, and filing to make sure that the flaws are removed to the bottom of the crack and to its ends.

Unless the original welding procedure was for gas tungsten arc welding (GTAW), it cannot be used for the root pass of the repair weld without acceptance of the AI. Some repair organizations prefer GTAW because it provides a dense base for any subsequent metal arc welding. However, absolute cleanliness is required. Furthermore, if the root pass cannot be examined, and an internal backing ring is required, it is not permissible to substitute GTAW without a backing ring. In any event, examine the welds interpass with fluid penetrant or magnetic particle.

If the pressure vessel code rules for the original construction required preheating or postweld heating on the basis of shell thickness and material for identical new construction, the inspection code rules will require it for the repair or alteration. The specifics will have to be discussed with the repair organization and the AI.

If the nozzle is small, the best procedure is to remove all of the weld, clean the hole edge and nozzle neck, and proceed as when installing a new nozzle.

Reinforcing nozzles

The main reason for reinforcing a nozzle on an existing exchanger is to make it capable of withstanding a load that had not been foreseen when the exchanger was designed. This situation comes about principally from piping designers using exchangers as anchors. However, unanticipated loads are sometimes imposed when exchangers are stacked.

Small nozzles are sometimes reinforced to protect them from damage during exchanger maintenance. For this purpose, three or four thin gussets welded to the nozzle neck and shell are generally satisfactory (Fig. 11.7). Be sure to bevel the base of the gusset to clear the nozzle-to-shell weld. Leave a wide enough space so that any liquid that runs down the nozzle neck is not trapped.

The kind of reinforcement to install on larger nozzles depends on the magnitude and direction of the forces to be resisted. The exchanger manufacturer will have taken advantage of excess metal in the shell, nozzle, and attachment welds to replace the metal removed in the hole. If the excess metal was not sufficient to replace the entire area, external reinforcing pads will have been added within the geometric limits that the pressure-vessel code rules for nozzle reinforce-

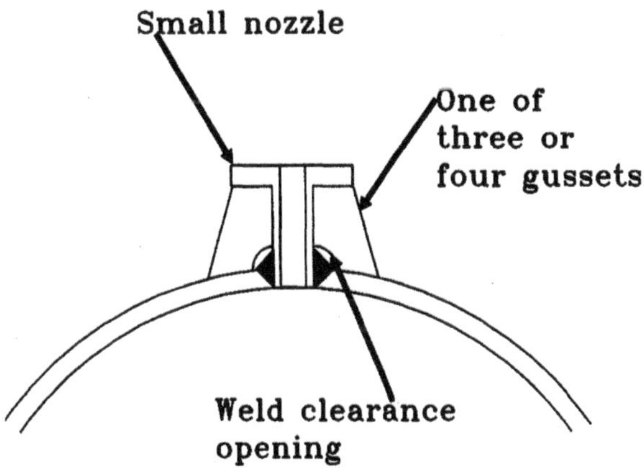

Figure 11.7 Small-nozzle reinforcement.

ment permit. Therefore, it is seldom feasible to add additional reinforcing pads.

As a result, other means have to be used to strengthen the connection against additional non-pressure-related loads. If the load is a bending or axial one, adding gussets as for small nozzles will usually be satisfactory. Severe torsional ones are harder to deal with. One way is to fabricate a split cone, and weld it in halves to the nozzle neck and shell. Because of inspection requirements, the cone must have openings large enough to see the nozzle attachment welds and the welds of any reinforcing pads.

The materials and welding must meet the inspection code requirements, and the procedure must be satisfactory to the AI.

Installing new nozzles

Process and operating problems may require installing additional nozzles or relocating existing ones. For example, when shell-side pressure drop is excessive, one way to reduce it may be to convert the shell configuration to divided flow. This requires careful consideration of the thermal design. It generally also requires removing the existing shell inlet and outlet, patching the shell, and installing a larger centrally located inlet and two outlets.

This major surgery must be done in accordance with the applicable inspection code. Nozzle removal, hole patching and producing the new openings to receive the new nozzles must be done in a way that will not harm the tubes. Probably the most successful technique is to arcair-gouge. This may present some problems with metals into which it is undesirable to have carbon dissolve. For such metals, grind away the arced edges the greater of one metal thickness or 12.5 mm (~½ in).

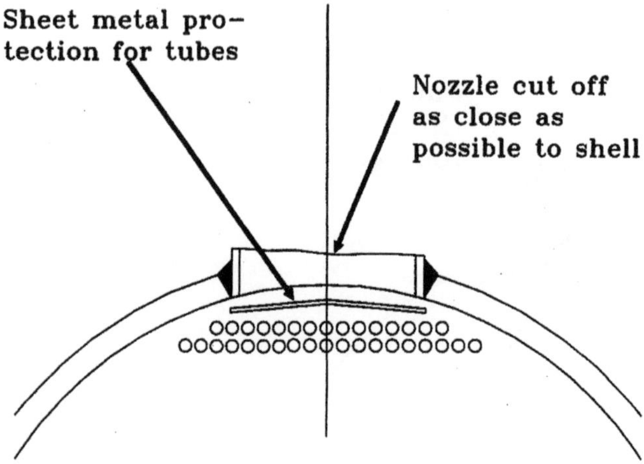

Figure 11.8 Protecting tubes when removing a shell nozzle.

In effecting the removal, cut off the nozzle perpendicular to its axial centerline as close to the shell as possible (Fig. 11.8). This will permit sliding a protective light-gage sheetmetal shield into a position that will protect the tubes when removing the nozzle neck stub.

Patching shells

It was stated above that to convert a shell to divided flow, the original inlet and outlet would be removed and the shell patched. In addition, local corrosion is sometimes so severe as to require cutting out a shell section and installing a patch. On occasion it is necessary to blank off an opening, but undesirable to use a blind-flange cover. In either case, a shell patch must be installed.

Figure 11.9 illustrates acceptable and unacceptable ways to make the patch. Openings to be patched are preferably circular, obround, or oval. Avoid rectangular ones because stress concentration at the corners contributes to tearing when the shell is under pressure. If you must use a rectangular shape, radius the corners with the greater of 20 mm (~¾ in) or 3 times the shell plate thickness. Grind all arced or burned edges of the opening smooth and prepare the edges of the patch and the opening for single-vee welding. The weld that joins the patch must be a full penetration one. Either an internal backing strip that remains in place or a consumable one is acceptable.

Lap-welded patches are not acceptable except for very small openings. Even there, they are undesirable and possibly hazardous because the shell stream can reside in the crevice between the lapped patch and the shell.

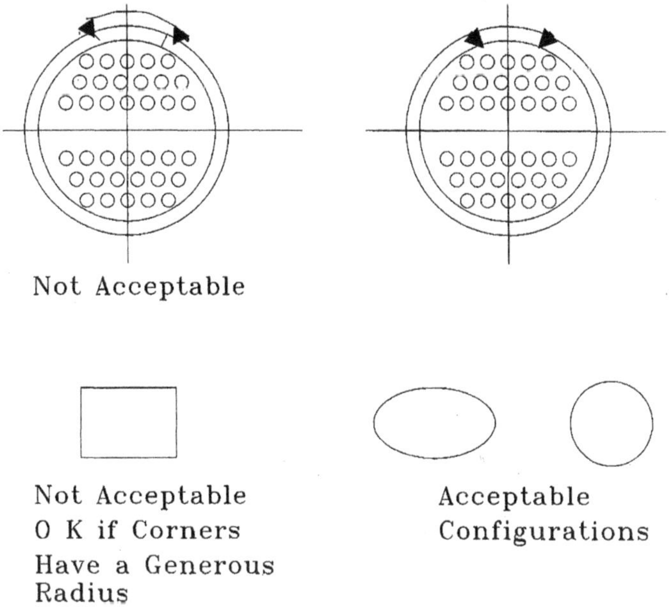

Figure 11.9 Unacceptable and acceptable shell patches.

Dealing with inadequate impingement protection

The following summarizes steps to take to deal with inadequate impingement protection:

1. Increase the inlet nozzle size.
2. Replace the inlets with enlargers.
3. Install distribution domes and larger impingement plates.
4. Install a distribution belt; see Fig. 11.1 for sizing.
5. Replace the tubes in inlet region with rods.

The strategy of replacing tubes in the inlet region with rods is very effective if reduced heat-transfer surface is acceptable. Insert close-fitting round bars into the first two rows of tubes in line-of-sight of the inlet nozzle from the nearest tubesheet face and extending several inches past the next baffle or support after the nozzle. Fix them in place by welding them to the tubes in the region of the tubesheet. In a more elaborate installation, drill out the tubes for about 15 mm ($\sim\frac{5}{8}$ in) at the tubesheet face. Weld a 10-mm ($\sim\frac{3}{8}$-in)-thick disk to the end of the rod. Insert the rod until the disk bottoms out on the cutaway tube end. Then weld the disk to the hole exposed by cutting away the tube.

Dealing with expansion joint failures

Following is a recommended sequence of actions and a procedure for dealing with expansion joint failures.

1. Determine what caused the failure. Investigate for the following:
   a. Stress-corrosion cracking (SCC). SCC occurs mostly in thin-walled joints made from austenitic steels. An environment high in chloride ion content is especially likely to cause an SCC failure.
   b. Fatigue failure. This appears as cracks in bellows roots and crowns and in the knuckles of flanged-only and flanged-and-flued-head joints.
   c. Expansion joint corrosion. Determine whether corrosion is general throughout, or only in the region of undrained liquid or unvented vapor.
   d. Erosion.
   e. Excessive axial deflection. This can cause failures at the crown and contribute to early fatigue-induced failure.
   f. Unplanned angular deflection. The joint appears to have been compressed in one quadrant and stretched in the opposite quadrant.
   g. Exposure to excessive pressure. Thin-walled or low-profile multicorrugation expansion joints squirm under excessive pressure. High-profile thin-walled joints and flanged-only and thicker-walled flanged-and-flued-head joints exhibit bulged sides under excessive pressure.
2. Inspect the whole exchanger for damage. When an expansion joint fails, other parts may be affected by the expansion joint failure, or may have been damaged by the same mechanism that damaged the expansion joint. Examine the following:
   a. The tube-to-tubesheet joints.
   b. The tubes.
   c. The tubesheets.
3. Plan a replacement less likely to fail. Consider these alternatives:
   a. Making the replacement joint from a material less sensitive to corrosion or to stress-corrosion cracking.
   b. Installing a higher cycle-life joint.
   c. Providing an internal sleeve to protect against erosion.
   d. Providing a joint with greater angular deflection capability.
   e. Providing a joint designed to withstand higher pressure. Be aware that expansion joints are less flexible when designed for high pressure than for low pressure. The effect of the greater spring constant on the tubesheet design thickness must be examined.
4. Remove any impediments to free expansion of the shell.

5. When failure has resulted from angular deformation not considered in the original design, provide shell alignment guides to prevent sagging. It is possible to design expansion joints that accept loading due to angular deflection. However, bundle life will probably be shortened if the shell is permitted to sag.
6. Replace single corrugation U-shaped, thin-walled bellows expansion joint with omega-shaped one with vent and drain.
7. Decide on the type of repair.

Repairing and replacing thick-walled expansion joints

A failed flanged-only or flanged-and-flued-head element can be replaced in sections. Under proper conditions, the work can be done with the exchanger in its site. However, a better job will result if the exchanger is taken down and brought to the repair shop. Here is a suggested way to proceed:

Prepare a written procedure for presentation for approval to the Authorized Inspector and Authorized Inspection Agency responsible for on-site inspection of the exchanger. Incorporate the following steps:

1. Recalculate the joint in accordance with the TEMA Standards paragraphs and ASME Code, Appendix CC for flanged-only and flanged-and-flued-head expansion joints. Perform any redesign necessary to accommodate the actual operating deflections (if you know them) and the projected cycles of operation. Since you are dealing with a unit that has been in service, use the operating logs to find out how many cycles the joint endured until it failed. This will provide a sound basis for your projection. Discuss any redesign with the manufacturer. If the manufacturer is no longer in business and the joint is to be an exact replacement, measure the joint, taking special care to determine the knuckle radius at the transition of the straight flange to the flat portion and the transition of the flue to the flat portion. Make a sketch of the joint whether or not it is to be an exact replacement. In the United States and Canada, order it from a pressure vessel or heat exchanger manufacturer who is a holder of a Certificate of Authorization to stamp pressure vessels with the U symbol. Require the manufacturer to furnish a U-2 partial data report.

2. Procure the replacement joint from the manufacturer beveled for fitting and welding to the shell with a single-butt or J groove depending on the joint thickness. Use a J groove for thicknesses greater than 13 mm (~½ in). Advise the manufacturer that you will be fitting consumable backing rings to the shell for welding from the outside only. Instruct the manufacturer to make the following preparations: (*a*) saw or abrasive-cut the joint in half along a major diameter; (*b*) prepare the two halves for rejoining with single bevels or J grooves depending on the thickness; and (*c*) fit along one-half of the sawed-apart joint a backing strip that will remain in place.

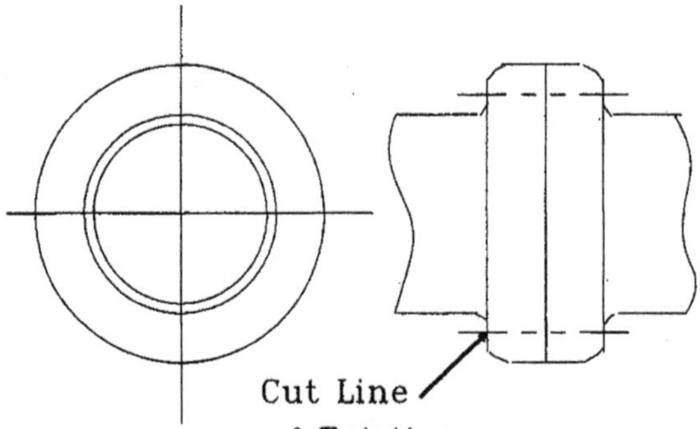

Figure 11.10 Replacing a flanged-only or flanged-and-flued expansion joint.

3. When the replacement joint is on hand, and after you have cleaned the exchanger to a clean and safe state for burning and welding, weld a stabilizing bridge to the shell that fits across the existing joint from beyond the joint-to-shell weld on either side. Arc or burn the extended part a short distance beyond its attachment to the shell (see Fig. 11.10).

4. Then (*a*) slide a protective light-gage cylinder into the open shell and arc or burn off the remaining parts of the old joint at the weld line. Grind or arc the shell edges for welding; (*b*) place consumable welding rings onto the shell ends prepared for welding and tack into place; and (*c*) slide the two halves into place, fitting the edges over the backing bar. Weld with a suitable welding procedure. Nondestructively examine and hydrostatically test as agreed on with the AI. Examining the seams by ultrasonic testing is generally acceptable.

Replacing thin-walled expansion joints

The following sketches illustrate two common options and a third, less common, for replacing thin-walled expansion joints:

1. Cut the tubes at each tubesheet as shown in Fig. 11.11. Shorten the shell to accommodate the loss of tube length. Replace the joint with a more suitable one. Relocate connections as necessary. Trim the tube ends and rebuild to the shorter length. The constraints on this option are (*a*) it requires considerable labor; (*b*) capacity is reduced; (*c*) piping must be relocated; (*d*) tubesheet-to-baffle space is shortened or spacers must be trimmed and baffle pitch adjusted—in the latter case, shell-side pressure drop must be recalculated; and (*e*) the process is suitable only when the tubes are uncorroded, undamaged, and expensive.

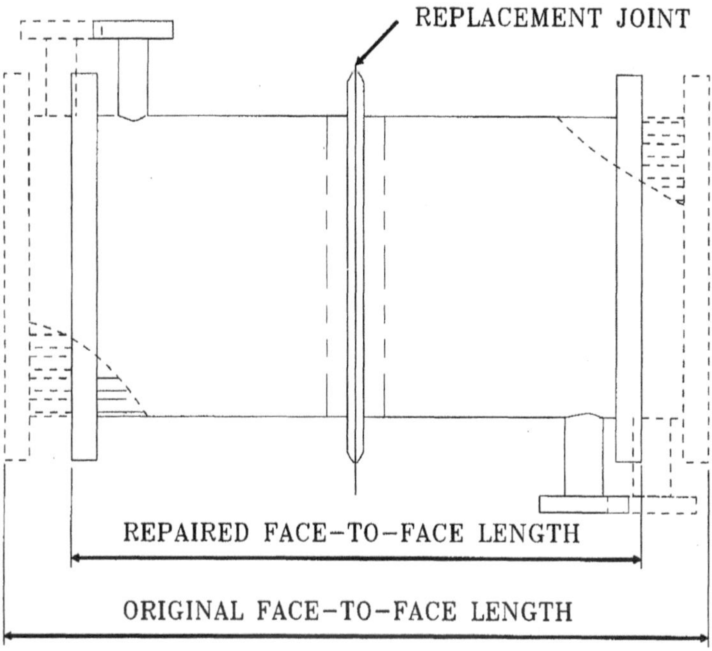

Figure 11.11 Replacing a failed bellows expansion joint by cutting off the tubesheet and shortening the unit.

2. Remove the tubes and failed expansion joint and rebuild with new tubes—the economics of retubing and rebuilding must be compared with those of procuring a complete replacement tubesheet-to-tubesheet. Then install a larger-diameter expansion joint as shown in Fig. 11.12. The I.D. of the joint must be large enough to pass over the O.D. of the tubesheets. The shell nozzles must be cut off to allow the larger expansion joint I.D. to clear, and reweld afterward. Rings must be welded to the shell as shown to meet the expansion joint I.D. The capability of the tubesheet to withstand the pressure, using the diameter and spring constant of the replacements bellows, must be recalculated. If the tubesheet is not adequate, the MAWP must be reduced.

3. A third option, applicable to high-profile, thin-walled joints, is to procure a clamshell kit from the manufacturer and have a repair organization in possession of an R symbol or the expansion joint manufacturer install the replacement joint. The authors have witnessed remarkably fine field welding on such installations, with restoration of the exchanger to its original capability. In some exchangers repaired this way the original stainless-steel type 304 bellows was replaced with Inconel, which apparently cured the SCC problem.

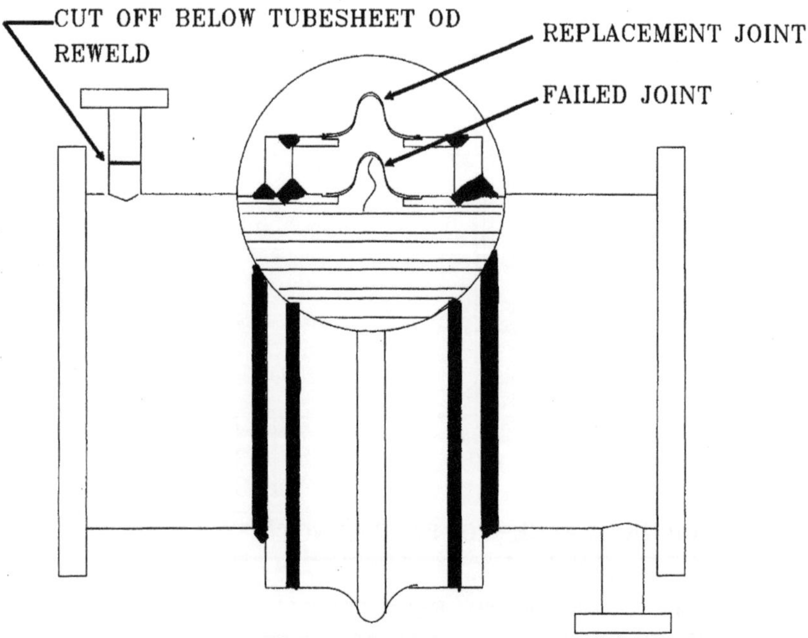

Figure 11.12 Installing a larger-diameter bellows expansion joint over one that failed.

A clamshell repair kit consists essentially of a joint made in two halves that, when welded together, produce an integral joint with two longitudinal weld seams. If you procure such a repair kit and your maintenance department or a repair organization does the work, the manufacturer will warrant only that the material supplied was in accordance with your purchase specification. However, if the manufacturer does the work in the field or in the manufacturer's factory, it is possible to obtain a full guarantee of the materials and quality of work for the replacement.

Repairing feedwater desuperheaters

It may be more economical to repair a feedwater heater desuperheater than to rebundle or replace the unit. Circumstances that favor this course are (1) tube failures occur mostly in the desuperheater at a discrete axial location, such as under the steam inlet; (2) the wall thickness of the unplugged tubes is essentially undiminished; (3) there are sufficient unplugged tubes to permit the heater to operate substantially within acceptable tube-side pressure loss and terminal temperature limitations; (4) economic evaluation indicates a substantial cost savings if the existing unit can be repaired in place; (5) there is space available to perform the work on site; and (6) outage time is available in which to do the work.

An example of such repairs is the replacement of a shattered impingement plate in a nickel-copper-tubed high-pressure heater with which one of the authors was involved. A heater in a string had suffered a tube leak in the desuperheater. Borescopy revealed the leak to be approximately at the position of the steam inlet. This led to the conclusion that there was an impingement protection failure.

Borescopy was then performed on the shell side and disclosed that the impingement protection had broken off and fragmented into pieces. The author sketched a repair and wrote out a suggested repair procedure that he estimated could be done within the remaining outage time. The suggested repair was to cut out the section of shell containing the steam inlet. Remove the remnants of the damaged parts. Manufacture and install replacements. Reinstall the removed shell section, heat-treat the welds, perform NDT on the welds, and hydrotest. For want of a better descriptive term, we call such cutouts "pumpkin sections" because of their resemblance to the part cut out of a pumpkin when preparing a Jack-o'-Lantern.

Fortuitously, a qualified repair organization was working on site. After a brief conference with the on-site repair organization supervisor and other concerned people and a telephone conference with the heater manufacturer's repair organization, the station manager decided to proceed. He took all necessary steps to make sure that best efforts would be made to restore the unit to service at completion of the scheduled outage.

The repair required having the manufacturer's detail fabrication drawings of the zone construction and impingement plate attachment. Because the station did not have the manufacturing details of the zone and impingement-plate attachment, it was decided to have the manufacturer fly a supervisor to the site with detail drawings. The work, which conformed with the requirements of the *National Board Inspection Code* (NBIC), was successfully completed during the remaining outage time.

To facilitate the borescope inspection that preceded the decision to make the repair, a section had been cut out of the steam inlet line and laid to one side for reinstallation. Referring to the fabrication drawings, replacement parts were produced locally. These included a complete replacement impingement plate and a section of shroud of size sufficient to replace any of the existing shroud that might not be reusable. At the same time the field people of the qualified repair organization, were cutting an approximately obround section out of the shell that encompassed the region of the steam inlet and impingement protection.

Here is how the work on the heater was done. The insulation was removed and the exposed shell marked to show the line along which the pumpkin section was to be cut out. The shell section was parted by oxyacetylene burning. The work was done from both sides of the parting line so as to facilitate grinding a weld bevel on the removed and stationary parts for subsequent rewelding.

The broken-off pieces of impingement plate were fished out of the enclosure. Inspection revealed no damage to the desuperheater shroud. The fragment of the impingement plate was ground off. The replacement impingement plate was previously welded on the I.D. The shroud was stitch-welded on the O.D. The edges of the opening in the shroud and replacement section were ground to suitable bevels for welding and the section welded into place using the metal arc process.

The attachment welds of the curved part of the zone enclosure to the flat plate and the welds joining the close-off plate to the shroud and shell were examined and found to be undamaged.

A backing strip was fitted to the underside of the intact part of the shell and the pumpkin section was fitted into place. The shell was maintained as round as possible by means of two $\frac{3}{4}$-in (~19-mm) strong backs fabricated from plate, flame-cut to match the shell O.D. and welded to the shell where the pumpkin section was to be cut out. Fitting clips and hydraulic jacks were used successfully to restore required roundness.

The pumpkin section was welded into place after the AI and utility accepted the qualified welding procedures that the repair organization proposed. CooperHeat* blankets were placed and the section of shell that had been worked on was heat-treated as required by the NBIC. Because of lack of access, the pumpkin-section-to-shell weld was not radiographed. Instead, the AI accepted ultrasound weld examination as permitted by the NBIC. The shell side was hydrostatically tested to the maximum allowable working pressure (MAWP) stamped on the nameplate. By this repair, the useful life of this heater was extended more than 5 years.

Postweld heat treatment

Both the NBIC and the API Inspection Codes require postweld heat treatment when the original pressure vessel code rules would require it for the material type and thickness. When a welded shell is under construction, it is a relatively simple matter to furnace-heat-treat.

---

*Cooper Heat's corporate offices are at 1021 Centennial Ave., Piscataway, NJ 08854. They have many regional offices in North America.

However, when doing repairs or alterations to a heat exchanger shell on site, the situation becomes more complex. Fortunately, it is usually necessary to heat-treat not the whole shell, but just the region of the welds and any formed parts.

Consider the following example. The feedwater heater from which the pumpkin section was cut out and rewelded to the shell was heat-treated as required by Section VIII, Division 1 of the ASME Code edition that prevailed at the time of its manufacture. To perform the heat treatment, the contractor wrapped thermal electric blankets around the shell to bring the temperature to the required [~600°C (~1100°F)] stress-relieving temperature. The contractor had to satisfy the AI that a sufficient number of properly placed thermocouples was used. The time at temperature and rate of cooling have to be demonstrated to conform with ASME Code requirements.

References

1. *ASME Boiler and Pressure Vessel Code,* Section VIII, Division 1, American Society of Mechanical Engineers, New York. New editions are published every 3 years (1995, 1998, etc.). Addenda, Code Cases, and Interpretations are published in the intervening years.
2. *Standards of the Tubular Exchanger Manufacturers Association,* 7th ed., 1988, TEMA, Tarreytown, N.Y.

Chapter

12

Retubing and Rebundling

Introduction

We open this chapter with a discussion of general considerations that apply to retubing and rebundling. We cover specific retubing techniques and procedures for extracting and reinstalling bundles and disassembling exchangers, inspecting their condition, and assessing the reusability of the various parts. We follow by surveying in-place retubing and shop retubing methods. We close the chapter with a discussion of rebundling. Since many procedures, practices, and techniques are the same regardless of whether a bundle is to be retubed or replaced, readers concerned with rebundling should refer to the discussions on retubing.

General Considerations

Consider the following when you contemplate retubing a heat exchanger or bundle or replacing a bundle:

- Safety requirements
- Reasons to retube or rebundle
- Feasibility of retubing
- When to replace individual tubes, partially retube, or fully retube
- The merits and disadvantages of in-place and shop retubing

General safety requirements

Heat-transfer equipment should be retubed or rebundled only if it can be done safely. The first step is to investigate how safely to access the unit and handle the heavy pieces of hardware and equipment used in doing the work. Whether in place, in the plant maintenance facility, or in an off-site shop, there should be a formal safety plan. It should address all aspects of the work and be reviewed with everyone who will work on the exchanger before the job is started. The following paragraphs discuss some things that must be done for safe retubing and rebundling.

Depressurization. Most in-place retubing is done when the line of which the exchanger is a part or the whole plant is shut down and presumably pressure is at atmospheric. Shop retubing is always assumed to be done under atmospheric conditions. Nevertheless, always make sure before opening a heat exchanger that it is completely depressurized. Always check to see that *all* the vents and drains are open.

Draining, flushing, and blowing out. Many heat exchangers handle dangerous fluids, such as acids, hydrocarbons, or radioactive fluids. Before opening the unit, drain it, flush out process stream residues, and blow out the water or solvent used to flush.

Cleaning. The hazardous nature of some fluids requires cleaning all surfaces free of residues with which the work crew can come into contact. Be sure to examine the Materials Safety Data Sheets (MSDS) and make them available to the workforce beforehand.

Flush the unit thoroughly with a liquid that will absorb and scour away as much of the residues of the process streams as possible. This may be water or a solvent specifically suited for the process stream material. Flushing may be by continuous recirculation or by batch washes. If the latter, flush at least 3 times. Examine the effluent water and continue flushing until chemical analysis shows the content of hazardous fluids and washout solvents to be acceptable. Three flushes are usually sufficient to reduce the concentration of process stream residues to acceptable levels.

When the washing is complete, open the drain connections on both sides of the exchanger. Blow out each side with compressed air only if there is no hazard of combustion or explosion; otherwise, use compressed inert gas. Make sure that no one is in the way of the exit compressed-air stream. If the process fluids were noxious, hazardous, or lethal, be sure to arrange for safely conducting the exit air away from the site.

Prudent repair organizations usually require chemical cleaning or hydroblasting and a certificate stating that the unit is hazard-free before accepting a unit for retubing and should also be provided with MSDSs. Dispose of the wash effluents in accordance with applicable jurisdictional requirements. If your site does not have a facility for waste treatment, collect the wash solvents and water and ship them to a facility set up to properly dispose of them.

Removing plugs before retubing. Plugged tubes may fill with hazardous shell-side fluids. The principal reason for plugging tubes is through-wall leaks. Therefore, it is possible that some material flowing in the shell will have leaked into the tubes. Residues of materials that leaked into the tubes may polymerize or partially polymerize and trap unpolymerized monomer. Flammable or explosive trapped substances can be extremely hazardous to workers sawing or burning through the tubes. Thoroughly perforating the tubes behind the tubesheet before plugging reduces the hazard of polymerized-material entrapping monomer. Nevertheless, it remains a possibility.

If shell-side fluid has leaked into the tube, and the leak is subsequently sealed by fouling or polymerized material, the interior of the plugged tube may be filled with material under pressure. If the tubes are plugged with taper plugs, they may blow out with considerable force when disturbed. Therefore, caution workers removing taper plugs not to stand in front of the tubesheet, and provide a backstop to prevent the projectiles from injuring people in their trajectory.

When material trapped in a plugged tube is flammable or explosive, any attempt to cut or burn through the tube can lead to an explosion or fire. In an incident known to the authors, a major oil refinery shipped a bundle for repair to a shop. The shop elected to cut off the tubes behind the tubesheets without first removing the plugs. The heat of the saw ignited material in a plugged tube, singing the worker and blowing the plug out with enough force to injure him badly.

If plugs cannot be removed to permit cleaning inside the tubes, all persons working on the equipment should be forewarned about the possibility of blowout or fire, and the facility should take appropriate precautionary measures.

Additional safety requirements for retubing in place

Some additional safety requirements are specific to retubing in place. The tasks to be performed are the same as for off-site retubing, but the location, condition of the exchanger, and its service history demand additional protection. Some things to cover are discussed in the following paragraphs.

Isolating the heat exchanger before in-place retubing. In-place retubing generally occurs during a plant outage, but if it is done while the plant is on line, the unit must be isolated from the process and utility streams. The most positive means of isolation are line blinds. If it is impractical to install them, isolate the streams with double-block valves. Install vent and drain valves between the double-block valves to relieve the pressure and carry away liquid in the event the upstream valve leaks. Make sure that the bleed valves are open at all times during the work.

Cooling to a safe and healthful working temperature. Do not open a heat exchanger that has not cooled enough so that you can touch it with an ungloved hand without having to pull away.

Controlling emissions. Removing and replacing the tubes and remaking the tube-to-tubesheet joints may produce substantial amounts of loose particles. These, and drill lubricant and welding dust, are pollutants that must be contained so as not to adversely affect nearby equipment, personnel, and the environment. It may be necessary temporarily to enclose the area. Some enclosures required by welding processes for attaching tubes to be done properly may serve the additional purpose of containing emissions.

Disposing of hazardous materials. Be vigilant to dispose of hazardous materials safely. Dump these substances in ways that will prevent fires from starting and workers from exposure. These may be solvents used to clean the tube holes, lubricants used for drilling out plugs, and hazardous fluids remaining from the heat exchanger service. Require workers to wear face masks or respirators suitable for tasks that may release dangerous air pollutants.

Wearing protective clothing. Workers, who must grind or chip to remove tubes, repair ligaments, or perform similar chores, must wear face masks for eye protection. Because such work is noisy, provide ear protection and require that it be used. Because other work is usually going on during an outage, hard-hat protection is necessary to protect workers against accidentally being struck by falling objects. People entering manway-access channels should also wear hard hats to prevent head injuries.

Using air and steam hoses safely. It is frequently necessary to blow away foreign matter with an air hose. It is unsafe to use one that does not have a safety nozzle that limits the force of the escaping air. In the United States these are OSHA-approved. Elsewhere, use similarly approved nozzles. Maintenance people often object to using safety nozzles because the limit on the air blast interferes with the cleaning they are trying to accomplish. However, it is better to take a little more time than for a worker to suffer an injury. Specifically require using approved nozzles and specifically forbid the use of hoses fitted with nipples and valves that can be opened to full-line air pressure.

In most places where exchangers are installed, steam is available at various pressures. It may seem like a good idea to clean tubesheets with a blast of steam from a hose. However, it is hazardous. If you permit it, make sure the hose is made for steam service and restrained from flying about to eliminate the possibility of a passerby being exposed to the live steam.

Providing adequate lighting, ventilation, and cooling. In some exchangers (e.g., hemihead feedwater heaters), access to the tubesheet is by a manway only. Work room and vision in the channels are significant problems. Lighting the work region, whether in a full-access channel or a manway-access one, must be done safely. Even when the bulbs are shielded, 115-V drop lights are not safe because even heavy electric cords can be broken or abraded. Furthermore, the work area is often wet, making for the possibility of electrocution. Instead use low-voltage high-intensity lighting. In manway-access channels provide miner-type battery-operated headlights.

Make sure that OSHA (or other safety code) requirements for testing the atmosphere and safely entering closed spaces are scrupulously followed. Since burning, arcing, grinding, and welding generate heat, dust, and fumes, arrange for a clean, breathable air supply when work is being done inside a channel.

Reasons to retube

These are some reasons to retube:

1. So many tubes have been plugged that performance is unacceptable.
2. You have to plug so often that shutdown, maintenance, and start-up costs cannot be tolerated.
3. You need to improve performance—perhaps to debottleneck a process—and you can do it by replacing bare tubes with fin tubes or high-flux ones.
4. A process change makes the existing tube material unsuitable.

Feasibility of retubing

It is practical to retube a heat exchanger or bundle only if other major components of the exchanger are sufficiently free from wear and corrosion to justify the expectation that they will last at least as long as the replacement tubes. Inducements to retube are the possibilities of using replacement tubes to increase capacity with finned or enhanced surface tubes, or to increase corrosion resistance by substituting a more resistant alloy.

From the viewpoint of economics, it is feasible to retube only when the cost of retubing is sufficiently below the cost of a replacement or new bundle to make it attractive—usually no less than 10 percent. In your evaluation take into account that retubing cost properly belongs in the maintenance-repair budget, but you may have to fit replacement cost into the capital one. The costs must include the costs of freight, removal, reinstallation, and the costs of lost production, and procurement of replacement power or product during the outage for retubing or replacement.

The costs of retubing will be different for in-place and factory retubing and will depend on whether the work is done by the plant maintenance force or subcontracted to a repair organization. In-place costs to include in your evaluation are the costs of staging, safety watches, and similar costs.

Deciding whether to replace individual tubes or to partially or fully retube

Decisions about retubing require analyzing the causes of the failures and the benefits and costs of retubing. Table 12.1 lists some factors to consider.

When to replace individual tubes. These are some things to take into account when deciding whether to replace an individual failed tube:

1. Unless there is an identifiable, isolated cause for the failure of a single tube, others will probably fail soon after. Replace an individual tube only if you can determine that the failure was unique to the tube and did not result from the fluids to which it was exposed, its position in the exchanger, and the conditions for which it was designed.
2. An advantage of extracting an isolated, failed tube is the ability to examine it visually, metallurgically, and chemically. Rather than plugging the hole in the tubesheet(s), it is usually desirable to replace the tube.

TABLE 12.1 Factors to Consider in Making Retubing Decisions: Costs and Benefits

Cost of doing the work in place
Benefits of doing the work in place
Cost of pulling a removable bundle or taking down a fixed-tubesheet exchanger
Cost of preparations for safely performing the work, such as cleaning, blinding piping connections, and erecting scaffolds
Cost of having the maintenance force do the work (assumes the maintenance force is capable of doing retubing)
Benefits of having the maintenance force do the work
Cost of having the work done in a heat exchanger repair shop
Benefit of having the work done by a heat exchanger repair facility

3. If hydraulic tube-pulling equipment is available, and there is no impediment to tube removal, in-place tube replacement takes only about ½ h more than plugging.
4. There is less potential for damaging the tubesheet in replacing a tube than in plugging it.
5. It is very hard, and may be impossible, to extract and replace individual U-tubes not in the outer tube rows.
6. Replacement of individual tubes in an off-site repair facility is so uneconomical that it is not feasible.

When to partially retube. Consider partial retubing for very localized groups of failures. It is only worth doing if you can address what made the tubes fail and provide replacements that minimize the potential for further failures. An example is peripheral tubes in a surface condenser where high-velocity wet steam has eroded the tubes in direct line of flow. In this circumstance, sometimes changing tube material or providing tube protection can eliminate recurrence of this kind of failure.

Tubes that fail prematurely because of inadequate impingement protection, and which are stabilized and plugged, protect ones deeper in the bundle. Most such damage occurs soon after the equipment is placed in service. After an initial series of disheartening failures, the rate declines sharply. Then it slowly increases over the life of the exchanger as the tubes corrode and wear. If you can correct the impingement problem, readily extract the failed tubes and install replacement tubes without disturbing the rest of the structure, then replace these tubes. In any event, investigate why the tubes failed.

Consider partial retubing under the following conditions:

1. All failures are confined to a discrete part of the tube field or a specific heat of tubing.
2. The failed tubes are accessible and can be extracted without damaging the baffles, tubesheets, or tubes that remain in place. U-tube bundles are generally not good candidates for partial retubing because of the difficulty of tube removal.
3. Failure was caused by corrosion in an inlet or outlet pass resulting from high attack rate at the inlet temperature.
4. Failure was caused by a correctable fluid maldistribution.
5. There is a reasonable life expectancy for the rest of the bundle.

When to fully retube. Weigh the following when considering full retubing:

1. The cost of replacing with a new bundle or exchanger compared with the cost of retubing
2. Benefit of having all new parts in a new bundle or exchanger
3. Benefit of the possibility of upgrading capacity and material by retubing
4. Time required for retubing compared with procuring and installing a new bundle or exchanger
5. Cost of downtime for bundle or exchanger replacement
6. Cost of downtime for retubing

Most full retubing and rebundling require being able to pull removable bundles or having enough work room to pull or push out the tubes. The success of the retubing depends on how well you can reestablish a tight, strong tube joint. Therefore, retubing in a repair facility at the plant or off site may be more effective than work done in place.

Deciding whether to contract to a repair organization or use the plant maintenance force

When considering retubing, it is necessary to decide whether the plant maintenance force should do the work or to contract it out. If the work requires cutting into the pressure envelope and rewelding, or removing old tube-to-tubesheet welds and welding in the new tubes, the inspection code and jurisdictional requirements may dictate hiring a Repair Organization; with some exceptions if there is a change to the tubing gage, material, or type, the NBIC requires the work to be done by an Alteration Organization. Unless the plant has a user's R symbol, these requirements will dictate the choice.* However, when retubing does not involve welding and the replacement tube specification is identical with that of the original tube most jurisdictions will permit the plant workforce to do the work.

If tube replacement is part of a scheduled outage, it is generally more cost-effective to contract out the work. The several nationally and internationally available Repair and/or Alteration Organizations and many local R-symbol holders have work crews with much retubing and refurbishing experience. However, they may send a knowledgeable supervisor to the site to hire and supervise locally available workers. This is acceptable if the local hires are competent, are apprised of any hazards, and agree to abide by the plant's security

*See Chap. 6 for the NBIC definitions of Repair and Alteration Organizations.

rules and safety requirements. Either way, contractors can usually do the retubing without affecting the outage schedule.

On the other hand, plant maintenance crews are responsible for all kinds of maintenance. Caring for heat exchangers is only part of what they do. During a scheduled outage, their availability may be stretched thin. Furthermore, it will usually take more time for the plant crew to do the work because supervisors and workers with experience will be rusty. Those without it will be learning on the job.

Whenever possible, to evaluate the costs, take not-to-exceed-price bids from contractors and compare them with the maintenance force's estimated cost. Assuming that the purchasing department will buy the tubing, exclude its cost from consideration. Note that most contractors will be reluctant to offer not-to-exceed-price bids because of the possibility of unforeseen problems. However, it is generally possible to negotiate on the basis of the work going smoothly, with a time-and-materials adder for any additional work that must be done. Include a contingency cost in your estimate.

Another factor in determining whether to employ a contractor is the cost of equipment required to do the work. Hydraulic tube extractors are relatively inexpensive, but if they are used only occasionally or just for the one job, you should charge their cost to the work. On the other hand, contractors, who are generally equipped with such equipment, amortize its cost over its useful life. This is less likely to be the case for tube rolling equipment because it is fairly common to have it on hand to reroll expanded tubes when there are tube-to-tubesheet joint leaks. Unless your maintenance department already has hydraulic expanding equipment, consider the cost of renting it and training your personnel or contracting the tube expanding.

Advantages and disadvantages of in-place and shop retubing

In-place retubing is done most often to straight-tube units. However, U-tube exchangers and feedwater heaters may be retubed in place, when circumstances make it advantageous to do so. It may just be impractical to pull out and transport a unit to another work location. Feedwater heaters installed in the condenser neck, for example, are usually either retubed in place or rebundled because of the difficulty in removing the shell. Often, what the time in-place retubing requires is more than is available in a normal maintenance shutdown schedule.

To examine the merits of in-place retubing, it is important to assess favorable conditions and the costs, benefits, and economics of contracting to a repair organization or using the plant workforce and safety considerations. In-place retubing may involve extracting tubes and, for closed feedwater heaters, removing and reinstalling their shells. Therefore, we discuss techniques and procedures specific to in-place retubing below.

Conditions favorable to in-place retubing. The following conditions are favorable to retubing in place:

1. The life of the tubes in a large heat exchanger, surrounded by impediments to removal, is ending and the basic unit has not deteriorated, but needs refurbishing.
2. The unit is so large that it is very costly to remove it from the site.
3. An extended shutdown for major plant equipment overhauls and replacements is planned, the time required for in-place retubing is available in the outage schedule, and it is hazardous or costly to take the unit down.
4. The shell-side fluid is relatively clean so that tubes can be easily extracted without damaging baffles and supports.

If a heat exchanger can be simply retubed in place, and there are no complications, the cost of refurbishing may be as little as 50 to 60 percent that of buying a replacement. There are also additional costs savings for not taking it down and installing the replacement. Sometimes the plant arrangement makes replacement installation costs so prohibitive that only retubing in-place is practical.

Another condition that favors in-place retubing is the capacity to debottleneck a system in which the exchanger is the limiting equipment. Exchanger capacity and performance may often be increased by retubing with extended or enhanced tubing. The only serious downside risk is the possibility that it may be difficult to reestablish the tube joint integrity to the same level as that of the new-equipment tube joints. This will be a problem if the tube joints have been reworked many times.

Risks of retubing in place. The major risk of in-place retubing is encountering unanticipated problems that impede or render impossible doing the job or consume more time than allotted in the outage. You cannot fully determine potential problems until after the work has begun. There may be difficulty right at the start when trying to pull the tubes. The following example from the authors' experience illustrates what can happen:

> An owner requested a field service organization to examine some failed tubes in a vertical 12-ft-diameter × 20-ft (~4.7-m × 6.2-m) tube-length, fixed-tubesheet catalytic maleic anhydride reactor. Molten salt from the shell had leaked into quite a few of the tubes and poisoned the catalyst. The work asked of the field service organization was to pull the tubes for failure analysis and replace them with new ones. Before installing the

replacement tubes, they were to drop a borescope through the tube holes to look at the cross-flow baffles where the tubes had been extracted.

The owner drained the salt and flushed the shell with water. New tubes were brought in. Scaffolding was erected and all equipment for doing the work delivered to the site. This included equipment for trepanning out the tube-to-tubesheet welds and preparing the holes for rewelding, hydraulic tube extractors, drills for thinning the tube ends if the hydraulic extractors proved to be inadequate, and hydraulic jacks to be applied to cross-pieces to be welded to the tubes to permit them to be lifted. Tube-driving tools and sledge hammers were also on hand.

Despite the repair crew's best efforts, it was impossible to budge the tubes. After a conference with the owner, it was decided to burn a window into the shell to see if the cause of the problem could be found. When the shell section was cut out, it was found that the salt, tubes, and baffles in the failed tube region had fused into a large mass. The reactor was taken out of its site and shipped to a repair shop for rebuilding.

More common discoveries after detubing are broken baffles, deteriorated tie-rods and spacers, and damaged or eroded impingement plates. Worst of all is to find that you cannot reuse the tubesheet. Therefore, restrict in-place retubing to exchangers where you do not anticipate serious problems with tubesheets or baffles.

Conditions favorable to shop retubing. Retubing a bundle or exchanger in an exchanger factory or repair facility almost always takes less time and is often of a higher quality than that done in place or on the plant site. This is because shop facilities are suited specifically to exchanger and pressure-vessel construction, repair, and alteration and the shop personnel are skilled and practiced in heat exchanger and bundle construction alterations and repairs. Shops usually have the right equipment to manufacture replacement baffles, supports, impingement plates, and other parts. Almost always they have better handling, bundle and tube removal equipment, and welding and testing facilities than do operating companies. They also employ welders who regularly use qualified welding procedures for most kinds of welding required to retube, repair, or refurbish a bundle or exchanger. Furthermore, repair-shop workers do the tasks required on a daily basis. Therefore, other considerations aside, it is preferable to shop-retube.

If there is no impediment to removing the exchanger or bundle from its site to a transport vehicle, and you can do it expeditiously and inexpensively, and if reinstallation costs and difficulties are acceptable, it is usually advantageous to shop-retube.

Another advantage to shop retubing is that, since it frees space, facilities, and maintenance personnel and supervisors for other tasks, it reduces the assets an operating company must devote to heat exchanger maintenance.

General Retubing Techniques and Procedures

Most techniques and procedures for retubing are pretty much the same regardless of whether the work is done in a repair shop, the plant maintenance facility, or in place. In the following we first review the ones that apply universally and later discuss those that apply specifically to shop and in-place retubing.

Preparations

We discussed safety reasons for cleaning and plug removal above. A non-safety-related reason to remove plugs is to enable you to extract and replace the plugged tubes. This assumes that the failure or plugging did not damage the tubesheet. Although it is pretty easy to knock out ordinary taper plugs, it may be extremely difficult to remove welded-in ones and some other types. After removing all plugs, inspect the condition of the tubesheets. If the tube-to-tubesheet joints are welded, further inspection will require grinding or trepanning out the welds.

Extracting and reinstalling bundles

It is very important when separating bundles and shells to maintain the longitudinal axes of both in alignment. You may use a combination of manual methods for extracting removable bundles from their shells. These depend on the experience, skill, and ingenuity of the workforce. Some of the ways are as follows:

1. Anchor pairs of come-alongs (ratcheted chain pulls) to a structure that will provide reaction to the resistance of the bundle to separation from the shell. Similarly anchor the shell.
2. Use jackscrews, flange pries, and hydraulic jacks to effect the initial separation.
3. You may attach vibrators to the shell during bundle pulling to break the bundle free of any minor hangups.
4. Strike the shell with sledge-hammer blows to shake the bundle loose. Interpose a stake between the shell and the sledge hammer to avoid damage to the shell.

In using these manual techniques it is essential to pull the bundle straight out of the shell or to pull the shell straight off the bundle. Even a small amount of misalignment can damage the cage and may cause the baffles and/or supports to gouge the shell.

Commercially available for-the-purpose extractors require less time for pulling bundles on site than do combinations of come-alongs, pinch bars, hydraulic jacks, winches, and cranes. The Hydroextractor* uses hydraulics for in-place and shop bundle pulling and reinserting.[1,2] The Bundle Wagon† uses an air-mechanical system for the same purpose.[3] Both devices have mechanisms for providing anchors to resist the extraction force and for finely aligning the bundle and shell to deal with alignment problems.

You can pull and reinstall removable bundles with the shell remaining in place. Most of the time it is less troublesome to leave the shell in place if it does not need rehabilitation. However, the heads must accompany the removed bundle to permit hydrostatic testing after retubing or construction of a replacement bundle and before reinstallation.

Bundle handling

If you do not handle the bundle properly, you may damage tubesheets, tube supports, baffles, tie-rods and spacers, dams, slide bars, seal bars, zone enclosures, and appurtenances. It requires great care and thought about how to bring a bundle to the ground or lift it from the ground to reinstall it into the shell. Nylon web slings that bear against tube supports or baffles are less likely to cause damage than are chain slings. If you must use chain slings, make sure they do not bear directly on the tubes, but on dunnage that traverses tube supports. For long bundles, use spreader bars, cradles, and pairs of cranes to limit flexing that can damage the tubes and tube-to-tubesheet joints.

An excellent way to handle bundles is to use the tube bundle lifting device‡ shown in Fig. 12.1. This device consists of parallel sets of curved arms that wrap around tubed fixed-tubesheet cages, U-tube bundles, pillbox bundles, and floating-head bundles. The arms open and close in a scissor action. When closed, they gently hug the bundle. The tube bundle lifting device is then picked up, bundle and all, and gently lowered to where it will be worked on or raised to where it will be installed.

Disassembly and inspection

After the exchanger or bundle is accessible, detube and disassemble the unit and inspect the condition of its various parts. This will determine if it is worthwhile to proceed with retubing or if the unit should

---

*Registered trademark of Hydro-Extractors, Inc., Beaumont, Tex.

†Registered trademark of M & H Manufacturing Co., Vidor, Tex.

‡Available from Carey Consulting, Inc., Edmonton, Ky.

(*a*)

(*b*)

Figure 12.1 Carey Lift tube bundle lifting device: (*a*) general view; (*b*) close-up view. (*Carey Consulting, Edmonton, Ky.*)

be replaced. In addition, inspection will disclose which parts may have to be replaced.

Unbolt and disconnect removable bonnets and channels. Restore these parts to as-new condition. If the channels are integral with the tubesheet, remove the covers. Build up any eroded surfaces with weld metal and remachine as required.

Detubing and tube-stub removal.[4] Extracting the tubes from or driving them out of the cage without making cuts between tubesheets and baffles and between adjacent baffles is the most efficient way to detube when the condition of the tubes will permit it. When possible to do so, it is cost-effective to detube fixed-tubesheet exchangers by extracting or driving without cutting into the shells. Detubing very large exchangers, such as steam surface condensers, is almost always done this way.

It is preferable to extract rather than drive for these reasons:

1. Under elastic conditions, the Poisson effect reduces the tube diameter when the tubes are pulled and increases it when they are driven.
2. For inelastic tube deformation, pulling causes necking down, whereas driving may cause the tubes to buckle, rendering them almost impossible to remove.
3. Pulling attachments grip the interiors of tubes to be discarded. Driving tube ends can cause tubesheet damage at the holes.
4. Tube pullers may be designed so that the tubesheet provides reaction to the pull, or, if that is undesirable, two pullers can work simultaneously on the same straight tube to break it without loading the tubesheets.

You can pull tubes out of an exchanger or bundle and stubs out of a tubesheet with hydraulic tube extractors, or by mechanical means.

Hydraulically pulling straight tubes and tube stubs. When there is not much shell-side fouling or it is soft and the tubesheets are of moderate thicknesses, a quick, clean way to extract straight tubes is with a hydraulic tube extractor. Some hydraulic tube pullers are shown in Figs. 12.2 and 12.3.

In straight-tube units, if the tubes metal is soft enough, the tubesheet is relatively thin, and the tube diameter and gage permit doing so, use an internal tube cutter to part the tubes behind one tubesheet before pulling. However, tube cutting is unsatisfactory when the tubes are of a material that is not readily machinable.

Hydraulic tube and stub extraction is straightforward and simply requires following the puller manufacturer's directions. It is very cost-effective for these purposes:

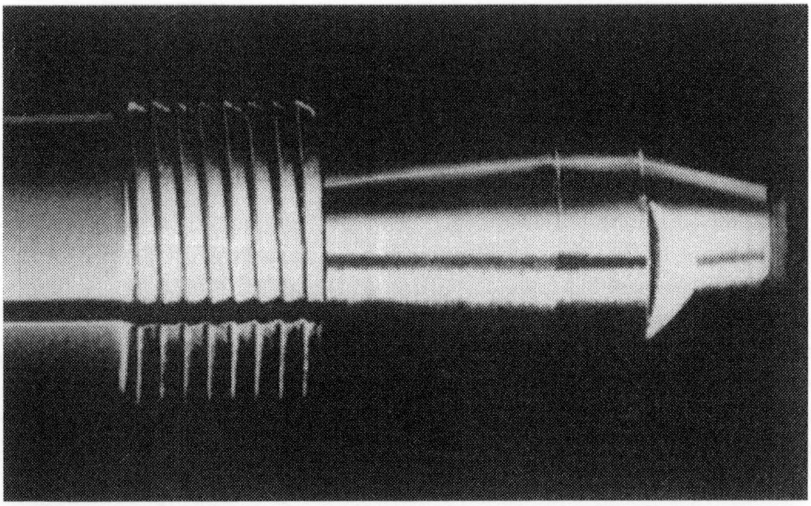

(a)

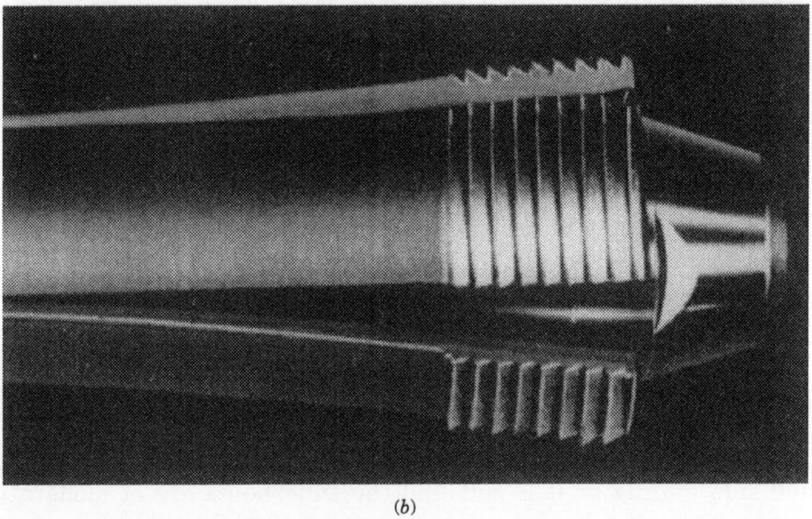

(b)

Figure 12.2 Elliott collet-type puller drawbar (*a*) before insertion and (*b*) before retraction. (*Elliott Company, Dayton, Ohio.*)

1. To break expanded tube-to-tubesheet joints.
2. To pull whole straight tubes. (It is not feasible to pull U-tubes hydraulically. Tubes cut off behind the tubesheet can be extracted from the U-bend end, leaving the skeleton intact.)
3. To pull stubs cut off behind the tubesheets. Stubs can be pulled faster hydraulically, with less damage than by thinning and driving them out.

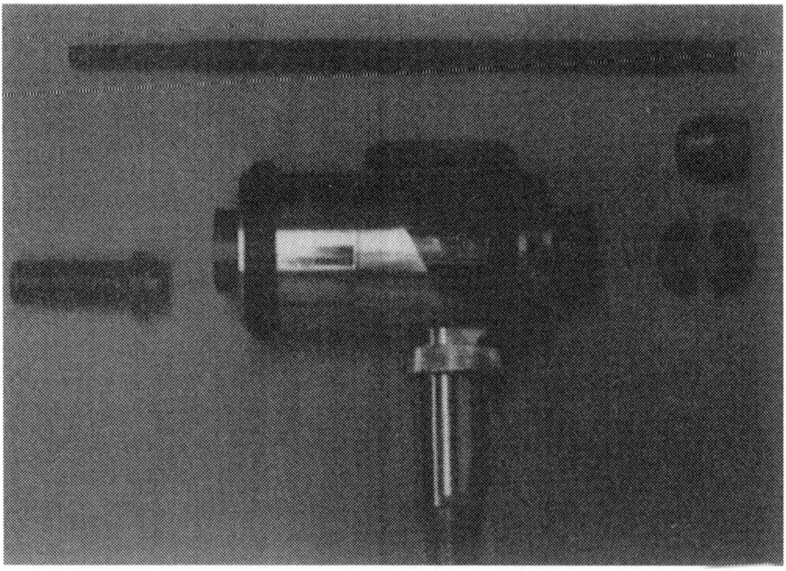

(a)

(b)

Figure 12.3 Hydropull 500 HTP spear-type tube puller (a) showing spear and jaws and (b) in use. (*Torque & Tension Equipment, Inc., Santa Clara, Calif.*)

If the tubes are covered with hard scale, applying enough hydraulic force to pull them out after they have been freed from the tubesheets will irreparably damage the baffles. A way to try to get around the problem is to simultaneously pull both tube ends with enough force to neck down and break the tube. The shock when the tube parts may crack off scale and corrosion products and the overall diameter reduction due to the Poisson effect may permit pulling out the two pieces. If this does not work, you will have to use other means to detube.

Mechanically pulling straight tubes and stubs. Most mechanical pulling systems require the whole unit to be anchored. However, they can be designed to use the tubesheet as an anchor. Extract tubes this way by thinning one end about 50 percent and collapsing it with a collapsing tool. Chuck an "easy out" (bolt stub extractor) into a slow-speed high-torque air drill. Insert it into the unthinned tube end and spin and pull. The spinning tends to flake off scale and corrosion products, permitting the tube to screw its way out of the cage.

Alternatively, if the tubes are not encumbered by scale and corrosion products, when the front end is out of the tubesheet, clamp it to a tugger hoist and pull the tube out of the bundle. Be sure to pull straight out so as not to cause the tubes to bear excessively on the tube-hole surfaces from being pulled at an angle.

Combining hydraulic and mechanical tube extraction. The Atlantic Group of Norfolk, Va., noted for field retubing steam surface condensers and other power-plant exchangers, has developed proprietary machinery for tube extracting, flattening the tubes into ribbons, chopping the ribbons into small pieces (Fig. 12.4), and feeding the scrap into trucks or gondola cars. Atlantic Group first loosens the tubes with a hydraulic tube puller, doing just enough extraction to feed into the ribbon-flattening machine. They then feed the ribbons into the choppers. The chopped tube scrap falls onto a conveyor that loads a dumpster, truck, or railcar for removal. One of their interesting variations is a modification of the flattening machine that slices the tube axially before the halves are flattened to permit decontamination.

They have also developed machines for loading replacement tubes into condensers. Another U.S. firm, specialized in rapid retubing and repair turnarounds in the oil refining, petrochemical, and chemical industries, uses hydraulic bundle extractors to pull bundles. They transport them to their factory on trucks specially outfitted to permit working on the bundle without unloading it. A third organization has outfitted oversized, closed truck trailers as traveling workshops that plug into the job-site power supply. They equip some as portable machine shops, set up others for burning and welding, and adapt others for assembling bundles and complete exchangers at the site.

Figure 12.4 Flattened and chopped tubes after extraction. (*Atlantic Group, Norfolk, Va.*)

Driving tubes and stubs. You may also drive tubes and stubs out of the unit. For this alternative, you must anchor the exchanger. To drive tubes, drill the tube ends to reduce the tube wall just enough to enable the collapsing tool to easily dislodge the tube away from the tube hole. Do not allow the drill to wobble in the tube because it might distort the hole and create a problem when trying to remake the tube-to-tubesheet joint with the new tube. If the old tube joint is welded, it will be necessary to grind the weld at the tube wall to be able to insert the collapsing tool between the tube hole and the thinned tube end.

Tap the collapsing tool so that it parts the tube in one end from the hole and collapse the tube end. Drive the other end with a power hammer, or use sledge-hammer blows and a driving tool. A thin or severely corroded tube may buckle, making it impossible to either drive it further or pull it out. It is well-nigh impossible to remove such a tube without performing major surgery on the unit. To avoid this prospect, insert a snugly fitting rod of the approximate tube length into the tube before driving. Anchor the rod at the tube end opposite the driver. Drive both legs of U-tubes with a power hammer or sledge-hammer blows.

Cleaning and inspecting the detubed bundle. After detubing the exchanger, clean all tube holes in tubesheets, baffles, and supports to ensure that new tubes will not suffer scratches when tubing and prevent making sound tube-to-tubesheet joints. Take care to clean the tube ends before tube insertion so as not to contaminate the clean holes.

Baffles and tube supports need to be cleaned for thorough examination. When possible, they should be sand- or grit-blasted. When a bundle is stabbed into or extracted from a shell or a shell pulled off a bundle, friction between baffles or supports and the shell interior causes resistance to the passage. In extreme cases the baffles or supports may be bent and the tubes crimped. The damage occurs most often at the cut line near the edge of the baffle or support. Study the failure maps for clues to damage created during installation. After cleaning, scrutinize the support or baffle edges to check for damage caused by inserting or pulling the bundle.

Examine the baffles and supports for loss of thickness attributable to erosion and corrosion. Look especially at the tube holes. Any enlargement beyond the tolerances listed in the various manufacturers' standards, or hole size greater than shown on the manufacturer's drawings, is unacceptable because (1) the tendency of tubes to vibrate is directly related to the clearance between the tubes and holes in the supports and (2) stream analysis used in calculating shell-side film coefficient of heat transfer is based on the maximum clearances specified in the manufacturers' standards.

Also examine for deterioration and need for renewal of such parts as impingement plates, air-removal systems, seal strips, skid bars, dams, and tie-rods and spacers.

Cleaning and inspecting the shell. The extent of cleaning and inspecting the shell interior depends on the shell-side service of the exchanger. If the fluid that flowed in the shell was clean and noncorrosive, a simple water or solvent flush and visual inspection may be all that is necessary to establish shell reusability. However, if the shell-side fluid was one from which material deposited and adhered tightly to the surface, or if it was corrosive, hydroclean the interior before making a visual inspection. When deposits or corrosion are severe, it may be necessary to hydroblast with abrasive dispersed in the water. Be extremely careful to avoid directing the stream at any spot long enough to abrade away the metal.

Examine weld seams and nozzle penetrations for localized corrosion. Look for rings or partial rings of thinning at the locations of cross-flow baffles and tube supports. Where severe corrosion is evident, measure remaining wall thickness ultrasonically or trepan out 25-mm (~1-in)-diameter sections and measure the shell thickness. Reweld the trepanned plug back into the hole after measurement. If general thickness loss throughout the shell exceeds the design corrosion allowance, you will have to derate the pressure capacity to reuse the shell.

Disassembling fixed-tubesheet units. If the unit is fitted with a shell expansion joint, weld temporary expansion joint bridges to the shell to maintain its axial dimensions intact. These are similar to shipping bars that expansion joint manufacturers use to maintain axial length. There are two basic ways to proceed:

1. Extract the tubes, rehabilitate the tube holes as necessary, and install new tubes. When the tubes have been extracted, examine the condition of the skeleton by peering through the tube holes and any large shell nozzles and by inserting a borescope through these openings. Repair shops will not perform borescope inspections unless you have specified in advance that they do so.
2. Cut through the shell and tubes just behind the tubesheet that does not have the tie-rods fastened to it. The most efficient way to accomplish this is with a band saw large enough for the purpose (Fig. 12.5). Repair organizations not outfitted with such equipment use a variety of other methods, discussed below.

Cutting through the shell and the tube nest. The ways to cut through the shell and tubes behind the tubesheet to which the tie rods are not fastened are sawing, abrasive cutting and burning, or carbon arcing. The saws used in sawing may be large automatic hack saws or band saws or adaptations. Figure 12.5 shows this operation taking place on a band saw. Skill and care are required to make a straight cut that facilitates butt welding the tubesheet and attached cutoff shell piece to the main shell. Figure 12.6 is a photograph of the saw.

In abrasive cutting, the shell is marked for cutting as close to the tubesheet as is convenient for working. An abrasive wheel is used to sever the shell at the cut line. A second cut is made farther away from the tubesheet. Longitudinal cuts are made to permit removing the two halves of the loose cylinder. The space between the cuts is deter-

Figure 12.5 Sawing off the fixed-tubesheet end. (*Hydrodyne, Inc. Massillon, Ohio.*)

Figure 12.6 Saw used in retubing operations. (*Hydrodyne, Inc., Massillon, Ohio.*)

mined by the working room required to abrasively cut the tubes. Abrasive cutting is slow work. Therefore, whenever circumstances permit, the tubes are arced to speed cutting the tube nest loose. Burning and arcing are the most frequent methods of cutting the free tubesheet end and tubes.

Hazards of parting shells. Always be cognizant of the hazard of fire or explosion when parting a shell. Ignition from heat from friction of the saw or abrasive wheel or from the burning torch or carbon arc is always possible if any flammable or explosive materials remain in the shell or tubes.

Combinations of methods. You may combine shell parting and tube pulling or driving when you want to examine the bundle cage more closely. Part the shell at the tubesheet end to which the tie-rods are attached. Remove the tubes and pull the tubesheet with attached shell stub and cage out for inspection.

Disassembling removable-bundle exchangers and bundles. Removable bundles are considerably easier to disassemble than fixed-tubesheet ones. Some of the same methods are used. If the unit is a floating-head type, you can saw or torch-cut off the tubes at the backside of the floating tubesheet and extract or drive out the tubes as described previously. You can extract or drive or drill out the stub ends in the floating tubesheet.

Inspecting and assessing reusability of parts

When the tubes have been removed from the bundle, carefully inspect the tubesheets, baffles and/or tube supports, tie-rods and spacers, and other internals to see if they can be reused or must be replaced. If you cut off the stationary tubesheet of a U-tube bundle by sawing, abrasive cutting, or burning through the bundle, you will cut the tie-rods and spacers behind the stationary tubesheet. These parts are relatively inexpensive to replace, and the cost is more than compensated by the value of the time saved.

Some suggestions follow for assessing reusability of parts.

Assessing reusability of tubesheets. Clean the tubesheets. Drill out all traces of the old tubes. Check the tube holes to see that they have not been made unserviceable by poor plugging techniques or ligament damage. Examine both faces carefully for ligament cracks. If the tubesheet is clad plate or forging, verify that the cladding has not separated from the base metal. Also make sure that the base metal has not been corroded or eroded behind the cladding. The higher the operating pressure of either side and the greater the differential pressure between the sides, the greater is the chance of wormholing and lateral erosion at the junction of a hard, erosion-resistant cladding and a softer base metal. This often happens to feedwater heater tubesheets in which the feedwater pressure is much higher than the steam pressure. If you suspect wormholing and/or lateral erosion at the clad-metal–base-metal interface but cannot verify it visually, examine the suspicious regions with ultrasound.

Tubesheets with expanded-only or grooved-and-expanded tube joints. When the replacement tubes are to be expanded into the tubesheet, do the following:

1. Plug-gage the holes to verify that the diameters are within tolerable limits.
2. Use a three-point tube-hole micrometer to check for roundness. (Out-of-round tube holes can be a problem in retubing with hard tube materials. Soft tube materials can generally accommodate some ovality. Tubes expanded by uniform pressure can tolerate more ovality than those that are roller-expanded. However, it is generally better practice to plug a hole that is too oval than to make a tube joint that will leak in a very short time.)
3. Make sure annular groove edges have not been upset. However the tubes and stubs were extracted, rehabilitate the holes by removing any metal remaining in the grooves, running a finishing reamer through the holes and remachining the grooves.
4. Dress out any longitudinal scratches.

Tubesheets with tubes welded to the tubesheet. As a rule, avoid reusing tubesheets into which the tubes have been welded unless you can make sure that you can remove all of the previous weld nugget without damaging the parent metal. This is especially important for weld-metal-clad ones. It is extremely difficulty to remove completely all foreign matter that may reside in the weld metal, and the character of the weld nugget may not be satisfactory for subsequent welding of tubes to it. Therefore, if the exchanger has welded tube joints, trepan out the welds. Remove the weld bond to the tube and far enough into the tubesheet to provide a sound joint preparation for rewelding. Fluid-penetrant-examine the hole edges to make sure that there are no cracks or porosity in the residual metal. Dress the holes and clean the weld surface. Grind the tubesheet to a uniform thickness and degrease and clean it before stabbing the tube ends back into the tubesheet.

Tube cage; baffles and tube supports, seal strips, skid bars, tie-rods, and spacers. Measure the thickness near the edges and holes. Check the hole diameters. Look for cracks in the ligaments. Replace underthickness ones, those with enlarged holes, or ones that are cracked. Use the tubesheet as a template to spot the holes in the replacement baffles. Replace worn seal strips. Reweld skid bars to the baffles or supports if the welds have deteriorated. Unless the tie-rods and spacers show no signs of deterioration, replace them.

Other internals. Inspect other internals, such as impingement plates, to make sure that they have not worn excessively or suffered from cracks or weld failures. Examine feedwater heater subcooling and desuperheating-zone enclosures and shrouds for cracked welds. These welds will be at the junctions of half-cylinders to flat divider plates and at the junctions of the divider plates to the tubesheet.

Parts Replacements, Repairs, Reassembly, and Testing

In the following, we discuss procuring replacement parts; repairing salvageable parts; reassembling bundles to shells, shells to bundles, and channels to shells or tubesheets; and testing the reassembled structure. Repairs include restoring wasted surfaces by welding.

Replacement parts

Order or manufacture replacement parts as early as possible in the production cycle. After you have decided to retube, order or have manufactured any parts that cannot be rehabilitated such as badly deformed and corroded channel covers or corroded-through bonnets and return covers. At the very least, order replacement gaskets and tubes. You will not learn of the need to replace baffles, tube supports, and tubesheets until disassembly is sufficiently complete to let you see the condition of these parts.

Some organizations, with a basis for predicting corrosion rates, order a set of replacement tie-rods and spacers, baffles and supports, and sets of gaskets and stud bolts from the original manufacturer when they purchase the exchanger. This eliminates the prospect of having a scheduling problem when inspection reveals wear and tear so severe that parts cannot be reused. If you order these parts with a new exchanger, make sure that the baffles and tube supports are produced to match the manufacturer's as-built diameters. This will ensure that the replacements are not oversized, which could require extensive rework to fit them into the shell; nor undersized, which can drastically reduce the exchanger's design performance.

If inspection discloses that baffle replacement is necessary, be sure to measure the existing inside-diameter dimensions of the shell throughout its length. Then set the baffle diameter to the diameter that will conform with the clearances listed in the TEMA Standards.

When tubesheets, baffles, and tube supports are drilled on numerically controlled drills during exchanger construction, you can be reasonably certain that the drilling templates of these parts will match each other. However, these parts are often laid out manually and stack-drilled on radial drills. This means it is possible for the positions of the tubesheet holes to deviate modestly from their theoretical locations and for the baffle holes in the stack to be offset from each other by the amount of drill drift at their position in the stack. Prudent manufacturers who do stack drilling on radial drills number the baffles in the stack. When they assemble the cage, baffle 1 is the first one behind the tubesheet that was at the top of the stack, followed by baffle 2 and so on with the bottom of the stack and the second tubesheet (if there is one) at the far end.

Therefore, before producing replacement baffles and tube supports, carefully measure the layouts of the tubesheet(s) to determine the as-built drilling template(s). This will permit manufacturing the baffles and tube supports to accommodate differences. This problem does not occur when the tubesheets and baffles are replaced at the same time.

Rehabilitating the shell

It is not advisable to attempt to repair pit-corroded shells by welding. However, you may successfully restore discrete regions thinned by erosion and general attack by welding. Grind out to sound metal any visible corrosion and metal loss in welds and parent metal at nozzle penetrations. If your inspection reveals rings or partial rings of crevice corrosion or flow erosion at baffle and/or support locations, grind to sound metal and restore the thickness with weld metal. For these purposes we recommend manual metal arc unless the sizes lend themselves to an automatic process.

If the inspection reveals corrosion at gasket surfaces, grind or machine down to the bottom of the deepest pits. Build up the thickness with weld metal and restore the finish using either a portable boring tool or a surface grinder.

Rehabilitating the skeleton (bundle cage)

Before inserting new tubes, rehabilitate the skeleton as needed. The baffles should be flat and not warped. The holes should not be corroded or oversized. Repair eroded or corroded tubesheet ligaments, or plug off the surrounding holes with welded-in plugs. Replace or recondition the impingement plates. Replace the tie-rods and spacers unless they are in nearly as-new condition.

Reassembly

Reassembly of a bundle and refitting it into the shell or, for many feedwater heaters, refitting the shell to the bundle requires even more care than building a new exchanger. Many times the methods used to disassemble the unit create small changes in dimensions that you must deal with in reassembly and reinstallation. Match marks made before the unit is taken apart must be precisely matched. New tubes must be installed into rehabilitated tubesheets and cages in ways that will not shorten their lives. The following paragraphs highlight some aspects of reassembling heat exchangers during retubing.

Changes of tube materials and thicknesses. As stated above, retubing provides an opportunity to change tube materials to more corrosion-resistant alloys and to replace them with extended- or enhanced-surface ones. Some things to consider are as follows:

Materials. Depending on the reason for tube failure, the reliability of the retubed exchanger may be improved with a material substitution. Keep in mind that a successful tube joint depends on the relative hardnesses of the tube and tubesheet material. Also consider that tubesheets, into which tubes have once been expanded, are already somewhat work-hardened. This is an important factor in setting the process for making the new tube joints. Remember that the NBIC considers a change in tube thickness, material, or type to be an alteration. However, it is a repair when a different material with identical allowable stress and thickness is substituted.

Thickness. Thickness requirements depend on tube metal allowable stress. Most of the time the thickness does not have to change when you change the material to a more corrosion-resistant one or to one that has a higher thermal conductivity. However, when the replacement material has a lower allowable stress, or if economics require using a thinner tube wall, make sure to calculate the thicknesses required to resist internal and external pressures. Procure tubes at least as thick as the required thickness, plus some allowance for handling and/or corrosion, and examine the effects on the heat-transfer rate.

Extended or enhanced surface tubing replacement. When performance demands require more capacity, consider retubing with extended- or enhanced-surface tubes. These can increase exchanger capacity at a relatively low cost. An advantage is that no structural changes are required. Most such tubes have to be procured to exact length, with just the right amount of bare ends to fit into the tube holes.

Tube handling and loading. Handle the tubes in ways that will not breed future trouble. Lift boxes and bundles of long tubes with cradles and spreader beams. Place unboxed or unwrapped tubes in safe regions out of the way of foot traffic. Make sure that austenitic tubes are not exposed to carbon-steel particles in the atmosphere produced by grinding and welding. Require workers handling tubes subject to stress-corrosion cracking (SCC) to wear clean gloves.

Before starting to load the tubes, fit the leading tube ends with tube guides or pilots to protect the tube ends from damage that might be caused by striking the baffles, supports, and the back side of the far-end tube sheet. When retubing a U-tube or straight-tube floating-tubesheet bundle, manually guide several tubes through the baffle assembly to line up the holes. You can do this easily after the holes in the baffles and supports have been cleaned. With fixed-tubesheet exchangers, use long rods or the first tubes you load to align the baffles. You don't need to insert the baffle alignment rods or tubes through the far-end tubesheet but only as far as the last baffle to ensure that the baffle holes are aligned.

With U-tube bundles, you have to load the tubes with the smallest bend radius first and successively load next larger bend-radius ones. Push the tubes through the bundle cage assembly and allow them to protrude 13 to 25 mm (~½ to 1 in) past the tube-side face of the tubesheet. Especially watch the tube ends when support distances are so long that the tube end might sag and not enter into the correct hole.

Load straight tubes into the bundle starting with the bottom row, working up to the top. For exchangers with floating tubesheets, push the tubes through the stationary tubesheet and cage until they extend to approximately the floating tubesheet position. Then, with the stationary tubesheet held in position, push the tubes back to position them in the tubesheet that will have the tubes joined first.

Tube-to-tubesheet joining. Whenever possible, follow the procedure used during manufacture for joining tubes to tubesheets. This should be available in the manufacturer's manual kept with the maintenance files. However, if the tube-to-tubesheet joints have been troublesome, consider changing to a procedure that will give stronger tighter joints. These are some of the methods used to expand tubes into tubesheets:

1. *Roller expanding.* Rolling the tube ends into the holes with a mechanical tube roller. This is the most common procedure. Manufacturers of tube rolling equipment will usually offer guidance and help in setting up a rerolling procedure or a procedure for expanding tubes into reconditioned tube holes.
2. *Explosive expanding.* Exploding charges in the tube ends. *Explosive expanding* is very effective for full-depth expanding tubes into thick tubesheets. It is not always feasible to set off explosions on the site. The work must be done by a licensed contractor.
3. *Rubber expanding.* Compressing an elastomer in the tube ends to create radial pressure. This method, developed by Hitachi, is rarely used in North America. It requires hiring a contractor who has the equipment available.

4. *Hydroexpanding.* Applying hydraulic pressure directly in the tube end. *Hydroexpanding* has come into widespread use. It has the advantage for retubing that (within reasonable limits) it can accept out-of-round holes that cannot be successfully roller-expanded.

5. *Hybrid expanding.* Hydroexpanding until the tube makes firm contact with the hole, followed by roller expanding. Use *hybrid expanding* when retubing with tubes such as titanium, Sea Cure, or Al6X. After initial light hydroexpansion at a pressure just great enough to create tube-hole-wall contact, roll at about half the torque value required for rolling only. The tubes are worked less and are less apt to fatigue.

Roller expanding. Tube rollers consist of a cylindrical cage with three to five (and for special purposes, seven) equally spaced longitudinal slots. Hardened steel rolls (also called *pins*), larger in diameter than slot width, are nested in the slots and held in place by a hardened steel taper mandrel that fits inside the nest. Current practice is to manufacture pins with a taper angle approximately opposite that of the mandrel in order to have the whole active length of the pin bear against the tube wall (parallel tool design). The slots into which the pins fit in the cage may be parallel with the centerline or set at an angle. When the slots are parallel, axial force is required to insert and withdraw the mandrel. Slots set at an angle cause rotation of the roller in one direction to screw the mandrel in or self-feed and reverse rotation to back it out.

The outer end of the mandrel is chucked into the rolling tool driver. A ball-bearing thrust collar and locking nut, screwed onto the cage's drive end, permits the cage to rotate opposite the direction in which the rolling machine turns the mandrel. The axial position of the rolls in the tube is set by adjusting the position of the thrust collar.

There are some disadvantages to self-feeding. Because the rolls are not parallel with the tube axis, they scrape across the tube surface. When the tube is substantially softer than the tubesheet, the tube end may take a barrel shape. If it is substantially harder, it may take an hourglass shape. Either shape change reduces the surface in contact with the hole wall. The thrust collar used to hold the tube from pulling out may work-harden the tube ends, leading to splitting.

Wall reduction is the most frequently used criterion of rolling adequacy. Rolled joint pullout strength increases approximately linearly with percent wall reduction, but rolling must not proceed to the point at which adjacent holes are distorted. It is not practical to measure the wall thickness of the tube end embedded in the hole to determine the amount of wall reduction. Therefore, wall reduction is estimated by difference. Measure the hole and tube dimensions before rolling and the tube I.D. after rolling. Calculate the clearance between the hole and tube O.D. Use Eq. (12.1) to calculate the percent wall reduction. This calculation does not account for radial ligament deformation. Therefore, it would be more appropriate to call the wall reduction, the "apparent wall reduction":

$$\%\ \text{wall reduction} = \frac{\text{postroll tube I.D.} - (\text{initial tube I.D.} + \text{clearance})}{2(\text{unrolled wall thickness})} \quad 100 \qquad (12.1)$$

where clearance = measured tubesheet hole diameter−tube O.D., mm (inch)
I.D. = measured tube internal diameter, mm (in)
O.D. = measured tube outside diameter, mm (in)

To establish a repeatable procedure, torque input is sensed and controlled.[5,6]

The torque control cuts off power at a preset torque. The recommended way to establish the optimum degree of rolling is to correlate wall reduction with the torque setting and then correlate push- or pullout strength and tightness with wall reduction. Establish the correlations by preparing representative mockup tubesheet specimens into which at least nine tubes are expanded and joints tested. Typical wall reductions range from 3 to 12 percent. Torque outputs that produce these reductions are in the range 1 to 20 ft•lb (~1.4 to 27 m•N). Specific tube-wall reduction and torque values depend on tubing properties and dimensions and must be determined experimentally. As a general rule, except for high-strength thin-walled tubes, consider rejecting expanded joints when wall reduction required to seal the tubes in the holes approaches 12 percent.

Torque may be correlated directly with joint strength. This is recommended for high-yield-stress, low-elastic-modulus, thin-walled tubes, where the small wall thickness makes wall reduction measurements difficult. An example is 1-in O.D. × 0.028 in wall (approximately 25-mm-O.D. × 0.7 mm wall) titanium grade II tubes.

To make sure that the torque cutoff point is maintained, rolling procedure specifications should require output torque calibration with a torque analyzer (dynamometer) every 50 to 100 rollings and every time the rolling tool is changed. These factors affect the quality of roller expanding:

- Cleanliness of the tube, tubesheet, and roller
- Condition of the cage, rolls, and mandrel
- Lubrication and cooling
- Tube and hole dimensions

- Torque cutoff control monitoring and maintenance
- Rolling technique
- Roller rotation speed
- Number of rolls
- Angle of rolls relative to the tube axis
- Shape of the rolls
- Worker fatigue

These are recommended cleaning steps before rolling:

1. Wipe the tubesheets and tube ends with lint-free rags.
2. Flush the tubesheets and tube ends with a volatile chloride-ion-free solvent. If you use acetone, take appropriate health and safety precautions.
3. Blow-dry the tube ends and tubesheet with warm, filtered air.
4. Repeat steps 2 and 3 at least 3 times, or more as necessary after inspecting for foreign matter.
5. Disassemble and clean rolling tools similarly before using.

The cage should rotate freely against the thrust collar. Slot edges should be smooth. The rolls should be smooth and free of nicks and evidence of flaking. The tapered mandrels must not be bent, nicked or chipped, or unevenly worn. During rolling, the tool and tube surfaces should be lubricated frequently to maintain consistency in the torque being sensed. Despite the lubrication, rolling generates considerable heat that may affect the friction of the parts, and hence the torque being sensed. Good practice is to dip the tool into the lubricant frequently between tube rollings, and to keep a second tool immersed in lubricant. Change the tool when it becomes uncomfortably hot to the touch.

Full rolled joint strength is developed in an expanded depth of 40 to 65 mm (~1½ to 2½ in). The pullout strength requirement can be met, using less than the theoretically optimum maximum expanding torque, if the product of interfacial pressure, coefficient of friction, and tube/hole-wall contact surface is large enough. The general practice for strength-expanding depth in North America for power-plant and TEMA R and B designs is to expand for the lesser of 2 in (~50.8 mm) or the full tubesheet thickness, less ⅛ in (~3.2 mm); for TEMA C exchangers, minimum strength-expansion depth required is not less than the least of two tube diameters, 2 in (~50.8 mm), or the tubesheet thickness, less ⅛ in (~3.2 mm).

When tubesheets are thicker than 2 in (~50.8 mm), it is customary to strength-expand for the depth that produces full joint strength and to contact-expand the remaining tubesheet thickness. The purposes of contact expanding are to seal the crevice between the tube and the hole that would otherwise remain as a source of crevice corrosion and reduce the effective size of the tubesheet perforations to that of the tube I.D., thereby stiffening the tubesheet.

The maximum practical depth for rolling in one step is about 50 mm (~2 in). Full-depth rolling into thicker tubesheets must be performed in overlapping steps, manually or by automatic step-rolling equipment. Because the rolling causes axial tube-end extrusion, step rolling should be regressive, starting 3.2 mm (~⅛ in) in front of the rear tubesheet face and proceeding outward to allow the extruded end to move out freely through the tube hole. Regressive step rolling is desirable for removable bundle construction because it facilitates maintaining axial dimensions. In fixed-tubesheet construction, it minimizes the compressive force on the tubes that results from restraining the extrusion.

The amount of extrusion per rolling step depends on the length of the rolls. Axial extrusion proceeds from the middle of the roll about equally toward the front and rear face of the tubesheet. To minimize locked-in axial stresses in the tube, use relatively short rolls for step rolling.

Some of the effects of tube rolling on tubesheets are as follows:

- Dishing inward or outward.
- Radial deflection of ligaments.
- Overall radial growth.
- Rotation of floating tubesheet relative to stationary tubesheet due to tube twisting.
- Torsionally stressing tubes in fixed tubesheet exchangers due to restraining twist.
- Cocking the tubesheet out of perpendicularity to the bundle axis because of extrusion and improper rolling sequence (see Fig. 12.7 for recommended tack-rolling sequence).

The effects of extrusion and improper rolling technique are intensified in conventional double-tubesheet construction. In reassembling these units, take extreme care to maintain adjacent tubesheets parallel and their tube holes aligned. Expand the tubes into the inner tubesheets and then test the joints' tightness before expanding the tubes into the outer tubesheets. Here is a recommended sequence:

1. Fix the pairs of tubesheets at each end parallel with each other and with the tube holes aligned.

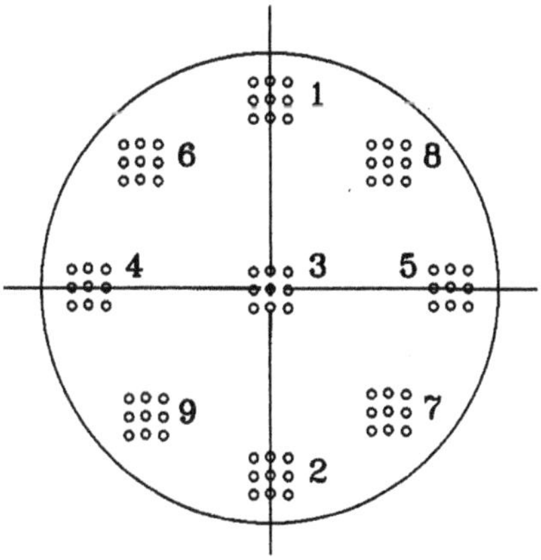

Figure 12.7 Recommended sequence of tack expanding.

2. Set the sets of tubesheets at the ends parallel with each other and with the tube holes aligned.
3. Tube the bundle. Follow the specified tube-end and tubesheet cleaning procedure.
4. Tack-expand the tubes in the front inner tubesheet in a sequence similar to that shown in Fig. 12.7.
5. Fully expand the tube ends into the front inner tubesheet. Step-roll into thick tubesheets regressively.
6. Repeat steps 4 and 5 at the rear inner tubesheet.
7. Pressurize the shell side. Inspect for leaks with strong lights and mirrors. Reexpand as necessary.
8. Repeat steps 4 and 5 for the outer tubesheets.
9. Pressurize the gaps if possible. Inspect the tube ends for leaks and reexpand as necessary. If the gaps are open to the atmosphere, pressurize the tube side. Inspect for leaks behind the outer tubesheets with strong lights and mirrors.

Near-contact explosive (kinetic) expanding. This process requires licensed contractors. It is used for the following: (1) tubesheets thicker than 50 mm (~2 in), (2) contact-expanding tubes previously strength-welded to the front faces of high-pressure feedwater heaters, and (3) expanded-only joints in low-pressure feedwater heaters.

Explosive expanding is done by setting off an explosion in the tube end.[7] It may be used to expand tubes for part or all of their depth in the tubesheet. Because the shock wave generated by the explosion attenuates rapidly with radial distance, when properly done, the surrounding ligaments are not damaged. This may require performing the explosive expanding in two stages.

A unique feature of explosive tube expanding is that with only one explosion, full expansion pressure may be applied in the vicinity of annular grooves and contact-only pressure in the rest of the tube end.[7] This is how it works:

1. The tube is the workpiece.
2. The tubesheet and air gap between the tube and the hole wall make up a forming die.
3. The explosive charge insert package consists of primacord contained in a polyethylene cylinder. The insert must be closely fitted into the tube so as to leave no air gap.
4. The polyethylene cylinder transmits the explosive force to the tube in a controlled way.
5. There are no process effects:
   a. The tubesheet is not deformed, remaining elastic.
   b. There is no measurable extrusion.
   c. There are no surface effects.

Points of attention in reviewing the explosive-expanding procedure are

- How the charge is established and whether one charge only will be used, or the work will be done in stages
- How the connections to the charges are to be made secure and reliable
- How misfires are to be prevented
- How misfires that occur are to be corrected
- How the joints are to be tested
- How cleanup is to be done after testing

Consider also safety, blast effects, and noise containment. Charge size is established by correlation with pullout test results. The selected charge is then set off in unrestrained tubes and the resulting strains measured to verify that adequate tube deformation will take place.

Hydroexpanding. *Hydroexpanding* is the process of expanding tubes by applying hydraulic pressure directly in the tube ends, using distilled or demineralized water as the medium. It is a relatively recent innovation.[8–11] It is becoming more popular, now that its use and application have been successful and the costs of the equipment have come down.

The equipment used for expanding by uniform hydraulic pressure (hydroexpanding) is capable of maintaining the preset pressure within close tolerances. The dwell time of pressure application can be adjusted for the materials and dimensions of the joint. These are some of the advantages of hydroexpanding:

- Pressure is applied uniformly throughout the tube end.
- Expanding is done in one application of pressure.
- There are no high-contact forces that cause the tube-end mashing, extrusion, and thinning produced by rolling.
- The time required to hydroexpand tubes into a 500-mm (~20-in)-thick tubesheet is not much greater than that for a 20-mm (~¾-in) one. Its principal disadvantage is that smooth-hole joints without appropriate grooves may not be as leaktight or strong as roller-expanded joints.

The following measurements made before and after hydroexpanding illustrate the effects on the tube end. In this study, 51 titanium tubes, ⅝ in O.D. × 0.05 in thick (15.9 mm O.D. × 1.27 mm thick), were hydroexpanded into 1$\frac{9}{16}$-in (40-mm)-thick steel tubesheets. Before and after measurements showed constant tube-end volume within the precision of the measurements. Tube-end shortening was approximately 1.33 percent of the embedded length, and tube-wall thickness reduction was about 1 percent.

This contrasts with tube-end extrusion, and with tube-wall thickness reductions in the range of 8 to 12 percent that is typical for titanium tubes.

Bladder expanding. Bladder expanding is a variation of hydroexpanding in which a bladder is inserted into the tube and filled with pressurized water. This method has been used in retubing steam generators to seal the crevice behind front-face primary strength-welded joints. On the basis of a limited number of tests, bladder expanding appears to be most suitable when the tubesheet's elastic modulus is equal to or greater than that of the tubes.[12]

Hybrid expanding. Fatigue and tube-end thinning effects may be minimized by first hydroexpanding the tubes into the holes into intimate contact, then rolling at a torque considerably less than would be required for rolling only. This combination, called *hybrid expanding,* is advantageous for tube materials subject to fatigue and for thin-walled high-strength tubes.

In hybrid expanding, the hydroexpanding step brings the tube into intimate contact with the hole wall in one pressure application. There is minimal general tube metal strain hardening. The subsequent rolling strain work hardens the tube, which maintains joint strength and tightness without imposing excessive cycles of loading and unloading on the tubes.[13,14]

Effects of tube-hole surface finish on expanded joints. The TEMA Standards require tube holes to be free of burrs and to be given a "workmanlike" finish. The Heat Exchange Institute (HEI) *Standards for Closed Feedwater Heaters* require only that all burrs be removed from tube holes. The HEI *Standard for Power Plant Heat Exchangers* requires the holes to be smooth and burr-free. The *HEI Standards for Steam Surface Condensers* specifies a hole finish of 500 root mean square (rms) for carbon steel and stainless steel tubesheet holes and 250 rms for nonferrous tubesheet holes.[15]

New-tube-hole finishes range from somewhat smoother than 32 rms to the roughest permissible finish. The range of commercially supplied feedwater heater, heat exchanger, and condenser tubing surface finishes is usually 60 to 100 rms. (Boiler tube surfaces may be as rough as 125 rms.)

In roller-expanded joints, rough tube holes usually yield strong joints and smooth holes yield tight ones. For rolling, rough finishes are acceptable when the tubesheets are hard and the tubes are thick-walled and soft. But for the reverse condition, finer finishes are required for tightness. A threadlike finish increased rolled joint strength over a smooth finish by as much as 35 percent.[16] On the basis of a limited number of tests of hydroexpanded joints, rougher surface finish appears to improve joint strength and tightness.

Flaring and beading tube ends. Tube ends may be flared or beaded to enhance joint strength and tightness if pressure or thermal loads tend to push the tubesheet off the tube. They may also be beaded to provide weld-joint preparation. Welded beaded joints are strong and tight under tensile and compressive loads. Flaring increases strength by virtue of the fact that the flared ends have to be drawn down to their original dimensions in order for the tubes to pull out. Beaded ends must be sheared before the tubesheet will pull away from the tubes. Flaring and beading tighten the tube joint by interfering with the outer edge of the tube hole. Tubes are sometimes flared in order to moderate fluid flow entrance effects.

Welding and brazing. Work is continually done to improve processes and techniques by such organizations as the American Welding Society (AWS), headquartered in Miami, and the New York–based Welding Research Council (WRC), by manufacturers of welding equipment, and by fabricators. In addition to literature supplied by the AWS and manufacturers, the WRC issues progress, interpretive and final reports in bulletin form on their work on new processes.[17,18,19] Welded and brazed joints are used for these conditions: (1) the joint operating metal temperature is high (welding only), (2) operation is cyclical, (3) joints must be very tight, and (4) the dimensions and properties of the joint are unsuitable for expanding.

In setting up a procedure, address these questions:

1. Can the metals be successfully joined by welding or brazing?
2. What is the most suitable process?
3. Will preheat or postweld heat treatment be required? If so, how can it be accomplished, and how will it affect the exchanger?
4. Has a procedure been established, qualified, and used successfully?
5. How will the tube and tubesheet dimensions and hole layout affect the welding or brazing?
6. Will proximity of the channel barrel or pass partitions to tube holes interfere with producing sound joints?
7. Can the joints produced by the welding or brazing be examined nondestructively? If so, what nondestructive examinations can be used?

These considerations are not independent of each other. Whether the metals can be brazed or welded is a question of not only metallurgy but also the available processes and the tube and tubesheet dimensions and tube layout. For example, titanium tubes cannot be successfully fusion-welded to steel tubesheets, but can be explosion-welded with good results. However, explosion welding is not feasible when the tube sheets are thin and the ligaments are narrow. Nor is it recommended at sites where flammable or explosive vapors are present.

The process used for new-equipment welding may not be suitable for retubing and repairs. New tube-to-tubesheet joints are most often made by gas tungsten arc welding (GTAW, also called TIG)—often automatically. However, repair and retubing welding is probably best done by manual shielded metal arc welding (stick) because it is much more forgiving of impurities in the parent metals and faying surfaces.

Brazing is hardly ever used for joining tubes to tubesheets during retubing. In brazing, the tube, hole, and braze-metal assembly is heated to a temperature above 425°C (~800°F). The nonferrous filler metal melts and flows into the space between the tube and the hole by capillary action. The base metals have higher melting points than the filler metal and therefore do not melt. The shell-side and tube-side fluids must both be compatible with the braze-metal deposit.

The maximum permissible operating temperature for brazed joints depends on the filler metal as well as the base metals. The temperature limit in Section VIII, Division 1 of the ASME Code for brazing procedures qualified under the ASME Code's Section IX is 200°F (~95°C). However, for some classifications of brazing filler metal, the limit may be raised to ~300°F (~150°C) if certain additional qualifications are met. Section VIII, Division 1 rules prohibit using brazed joints for lethal service and in unfired steam generators. Section VIII, Division 2 prohibits using brazing for strength joining tubes to tubesheets.

Brazing requires a suitable flux, atmosphere, or flux-atmosphere combination to exclude atmospheric gases that can oxidize or embrittle the braze metal. Torch or furnace brazing processes may be used. Furnace brazing may be open-flame or closed-furnace. The joint design and brazing technique must ensure flow of brazed metal into the space between the tube and hole wall.

It is hard to see if braze metal has fully penetrated from the front tubesheet face into the space between the tube and hole wall. When braze-metal rings are preplaced behind the tubesheet (secondary face), braze metal that has flowed to the front face (primary face) can be visually examined. Brazing from the secondary face forward also eliminates the crevice between the tube and the hole. The shell-side fluid must be compatible with the braze metal.

Cleanliness of the parts is essential for successful tube-to-tubesheet brazing. Any oil, grease, oxide scale, or foreign matter on the surfaces will cause porosity. Therefore, a cleaning procedure should be included as part of the brazing procedure specification.

Fusion welding. Review the welding procedure to be used with the Repair Organization. It is also prudent to get the exchanger manufacturer's recommendation. In fusion welding, a molten pool forms at the junction of the base metals. It consists of the base metals, any added filler, and elements that may be added in the welding flux. On cooling, the base metals and components provided in the welding flux form complex solid solutions.

In metal arc welding, voltage is impressed across a small gap between an electrode and the tube-tubesheet assembly. Current flow results from the voltage drop across the gap and produces an arc that generates enough heat to melt and fuse the tubesheet, tube end, and any added filler metal. The current flow across the arc may be direct, with the electrode on the positive or negative side of the line, or it may be alternating. Alternating-current flow may be continuous or pulsed.

As the joint is traversed, the electrode is maintained at an appropriate arc-gap distance to produce a uniform weld deposit. The electrode may be consumable or nonconsumable. Consumable electrodes melt into the weld puddle with the parent metals. Nonconsumable electrodes are used only as terminals for the arc. Unintended inclusions of metal from nonconsumable electrodes produce inferior welds.

Nonconsumable electrodes may be used for *autogenous welding,* in which there is no filler metal and only the base metals are fused, and also when filler metal is added to the welding puddle. Filler metal may be fed continuously from spools of wire, at rates controlled to match welding speed, or handheld wire rod may be fed manually into the arc. For some joints, it is advantageous to preplace filler-metal rings before striking the arc.

Metal arc welds can be made with unshielded bare wire, but atmospheric gases tend to cause oxidization, embrittlement, and porosity. Therefore, fluxes or shielding gases are used to exclude air from the molten-metal pool. Fluxes may also contribute metal elements to the weld. Manual and automatic processes are used. However, for retubing, most metal arc welding is done manually and frequently in position.

Manual welding quality depends on the welding procedure and the welder's skill in following it. Skilled welders, performing manual welding, may be very successful in compensating for variable welding conditions. These could be current-supply fluctuations, stray magnetic fields in the work area; sloppy fit-up, and variations in metal compositions.

Shielded metal arc and gas tungsten arc welding are used for joining tubes to tubesheets during retubing. In *shielded metal arc welding* (SMAW), the welder manually traverses the tube joint with the end of a flux-coated consumable metal electrode clamped in the jaws of a welding handle. Ligament widths, tube diameters, and tube thicknesses must be adequate for successful stick welding.

Gas-tungsten-arc welding or GTAW (or TIG, for *tungsten inert gas*) may be performed manually or automatically. For retubing it is done manually more often than in new construction. The nonconsumable electrode is usually thoriated tungsten. The weld is shielded by a blanket of inert gas (helium, argon, CO_2, or mixtures). The gas flows at a controlled rate through an annular space between the electrode and the surrounding nozzle. The back side of the joint may also be blanketed with shielding gas. Great care must be taken not to let the electrode touch the weld puddle because tungsten inclusions embrittle the weld.

GTAW welds may be made autogenously or with filler metal. For retubing, a process that uses filler metal is desirable. A variety of systems is available for automatically welding tubes to tubesheets by the GTAW process. Be cautioned that most of these are unsuitable for retubing. However they are made, the welded joints must have the following characteristics:

- They must be fully fused, resistant to corrosion, and free of porosity, nonmetallic inclusions and cracks.

- They must resist hot and cold cracking, cracking resulting from restraint, and porosity due to atmospheric gases in the weld puddle.
- The thermal expansion coefficients of weld metal, tube, and tubesheet must be similar to prevent developing high thermal stress.
- The heat-affected zone in the surrounding ligaments must be ductile after cooling.
- The notch toughness of front-face filet welds must be high.
- The joints must be mechanically reproducible, repairable by simple techniques and capable of being inspected.

To meet these requirements, the tube ends and holes must be completely free of foreign matter. If the tube and tubesheet are not scrupulously cleaned before welding, and maintained clean during welding, there will be weld porosity at best, and complete joint failure at worst.

Be sure that any weld procedure for joining tubes to tubesheets requires mechanical cleaning of the tube ends, followed by thorough washing with a volatile, chloride-ion-free solvent and filtered-air drying. To assure the quality of the joints, make this a quality control hold and witness point.

Ordinarily, clean, untreated atmospheric air is a suitable environment for welding. Some metal pairs require the work area to be surrounded by uniform-temperature, controlled-humidity, filtered air. The flux on coated rods and shielding gases must be dry. Before a cylinder of shielding gas is connected to the welding apparatus, it should be tested for dryness with a moisture-sensitive paste.

Gases generated by welding must be able to escape the weld puddle. If the gases produced by fusion welding cannot freely escape from the weld puddle, the tube joint welds may be porous and predisposed to cracking.

However, the tubes must be fixed in place before front-face welding. They may be tack-welded, or a drift pin may be used to expand the tube ends into line contact with the hole edge. Tack welds must be capable of being fused into the finished welds. Drift-pin setting may split the tube ends. The ends may also be lightly expanded into the holes. Preexpanding reduces the weld-root gap, making filet welds less subject to root-bead cracking. It may be done by rolling or expanding with hydraulic tube locking tools, such as the ones offered by system suppliers, that bulge an elastomer in the tube end to fix it in the hole without making an airtight seal.

Two disadvantages of prerolling are (1) it does not allow an escape path for welding generated gases at the weld root, and (2) it may introduce foreign matter to the surfaces to be welded. (If a lubricant is used in rolling, it causes weld porosity; if rolling is done dry, the rolls, cage, and taper mandrel may flake, providing a source of inclusions.) The hydroexpanding tube-lock tool and similar tools do not have these disadvantages.

If tubesheet temperature varies widely during welding, the size of the root opening will vary somewhat from joint to joint because of different amounts of expansion. As welding heat spreads throughout the tubesheet, the holes deviate from roundness. The amount they recover depends on the change in radial temperature gradient. As the welds solidify, they shrink, further distorting the holes near the joint being made. To reduce the amount of distortion, keep the tubesheet at a uniform temperature.

Moisture must be excluded when welding is interrupted. When welding is suspended, it is advisable to warm the tubesheet before restarting. Restart warming is separate from preheating and postweld heat treating, which serve different purposes. The restart warming helps to dispel condensation moisture.

When welding is halted for a long time, condensation moisture may settle in the unwelded joints. When the moisture vaporizes, it can make subsequent welds porous. Good practice is to put a cloth bag of desiccant in a plastic wrapping around the assembly.

The voltage and current output of the welding source must be kept constant. Variations in line voltage may change current and arc-gap length, and cause unseen changes in weld penetration and metal deposits. For best results, use a voltage regulator on the supply to the welding machine.

The welding area must be maintained free of stray magnetic fields. Work areas are full of stray magnetic fields (often generated by neatly coiled leads of welding cable). External magnetic fields add to the effect of arc blow (a deflection of welding current in direct-current welding that is caused by variation of magnetic flux in the workpiece). Arc blow may cause skips and uneven tube joint welds.

Explosive welding (EXW). The AWS defines explosive welding as "a solid-state welding process wherein coalescence is effected by a high-velocity controlled detonation." Explosive welding is desirable under the following conditions:

- Tube and tubesheet metals are not compatible for fusion welding, but the tightness and strength of welded joints are wanted.
- The tubesheets are thick.
- The tubes are on a spread pitch.

The procedure will be set by the licensed contractor.

Combining expanding and fusion welding. Preweld expanding reduces weld-root gap. The heat of welding may relax the equilibrium stress at the tube/hole-wall boundary. When the joints are heat-treated after welding, pre-weld-expanded tubes will also relax. Postweld expanding primary joints made by welding at the front tubesheet face closes the crevice between the tube and the hole. For the purpose of eliminating crevice corrosion, this is preferable to joint configurations that have enlarged holes in the tubesheet behind the part welded to the tubes.

Replacement shells and shell courses

If inspection revealed seriously corroded or eroded shells or sections of shells that cannot be restored by welding, preserve the pressure-retaining integrity of the unit by replacing the whole shell or unusable section. If the shell is a rolled plate shell, be sure to tape the circumference to enable manufacturing a replacement that will have the same diameter as the original. Make certain that the interior crowns of the joining welds are ground flush with the shell interior. Use spiders on either side of the girth seams and the minimum voltage that will give full penetration to minimize shell shrinkage that could impede joining the bundle and shell.

When fitting the replacement course(s), use levels, measuring tapes, and chalk strings to make sure that joined adjacent courses form a continuous cylinder perpendicular to the tubesheet that was not cut off. Also use straight-edges and measuring tape to make sure that when the cutoff tubesheet is fitted to the other end, it will be parallel to the one not cut off. Requirements for heat treatment and nondestructive examinations of the welds are the same as for the original construction.

Rejoining bundles and shells after retubing

Properly rejoining bundles to shells is critical to achieving the goal of extending an exchanger's life. Improperly stabbing the bundle into the shell or pulling the shell onto the bundle can damage the baffles. If you deal with baffles and supports hanging up on the shell simply by applying more force, you can bend them and cause the tube-hole edges to cut into the tubes. Applying excessive force can also damage tie-rods and spacers that hold the cage together. If there is misalignment between the impingement protection and the shell-side inlet connection, replacement tubes will fail in the vicinity of the nozzle.

Rejoining is somewhat, but not quite, the opposite of disassembly. The equipment to use is essentially the same. However, whether the shell or bundle remains stationary, it must bear reaction to the thrust force applied to mate the bundle and shell. Maintaining alignment is more critical during mating the bundle and shell than in the disassembly because any damage caused by misalignment will be hidden.

Be very careful to place gaskets of flanged, removable-bundle exchangers properly in the site. Follow the bolting-up recommendations described in Chap. 9 very carefully. If either the channel-to-tubesheet or tubesheet-to-shell joint leaks, do not try to fix it by more intensively tightening the bolts. Instead drain the unit, separate the bundle and shell, and investigate the cause. Look for crimped or misplaced gaskets, foreign matter caught in the mating parts, improperly reconditioned gasket surfaces, or distorted flanges.

Roundness retention rings should have been emplaced before cutting windows into shells that were subsequently cut apart. These may have been structural angles or tees or bars rolled on edge and welded to the shell to either side of the cut or the shell and skirt on either side of the window. However, if this precaution was not taken, the parts on either side of the cut line may not fit well enough to permit making acceptable rejoining welds. You may be able to reround the adjacent parts by pulling together angles wrapped around the shell or shell and skirt to either side of the cut line. If this fails, you will have to use a combination of heat, blows, and fitting clips to fit the parts. It is not acceptable to beat carbon-steel shells and skirts into place if doing so impairs their properties. This implies that you should heat these parts and beat them into shape while they are hot.

How you rejoin a fixed-tubesheet shell and cage depends on how you separated them. If you cut the shell instead of arcing or burning a tubesheet off its end, you may have to rejoin the parts using a consumable backing ring or a nonconsumable one that remains in place. Most of the time this closing weld cannot be radiographed. Therefore, if the code design was based on spot or full radiography, the AI will probably require an ultrasonic examination. If the shell was parted away from the tubesheet, take the same alignment cautions for replacing shell courses when refitting the parts. Depending on how the parting was done, it may be desirable to fit in a backing ring to produce a sound girth weld. In any event, magnetic-particle-test welds of shells to tubesheets when the materials can be magnetized, and fluid-penetrant-examine such welds when the materials are not magnetic.

If the fixed-tubesheet exchanger is very large, it may be more efficient to reconstruct the cage inside the shell. This requires following an assembly sequence in which the tie-rods are attached to the tubesheet that remained fixed to the shell when the unit was taken apart and carefully fitting the spacers and baffles or tube supports. If the design is one in which the impingement protection is welded to the spacers, the spacer/impingement-plate assembly at the location of the inlet will have to be made up in advance and slid onto the tie rods. If the impingement protection is welded to the shell under the inlet nozzle, it will have to be placed and welded in the appropriate sequence.

Exert care when you hang the far-end fixed tubesheet that its 0–180° and 90–270° centerlines are in precise alignment with those of the shell and near-end tubesheet. Otherwise, installing the tubes into the tubesheets will be difficult and may cause inferior tube-to-tubesheet joining.

Nondestructive examinations and tests

Nondestructive examinations and tests have been discussed in Chap. 5. The NBIC requires nondestructive examinations and tests like those used in the original construction to verify the integrity of the retubed exchanger or bundle before it is placed back into service. For the most part, they consist of the examinations discussed below.

If no special leak testing is specified, test welded-and-expanded tube to tubesheet joints in the following sequence: (1) air- and bubble-test the welds; (2) expand the tubes; and (3) after the tubes have been expanded, examine the welds with liquid penetrant to make sure that none were cracked during the expansion.

Bubble testing tube-to-tubesheet joints. Bubble-test welded tube-to-tubesheet joints after welding to determine their leaktightness. If the joints are to be subsequently expanded, perform the bubble test first.

Fluid penetrant examination of welded tube-to-tubesheet joints. Fluid-penetrant-test welded and expanded tube-to-tubesheet joints after the tubes have been expanded into the tubesheet to ascertain that the expanding operation did not cause any welds to fail.

Examining closing welds. *Closing welds* are the girth welds used to reattach tubesheets to shell courses and shell skirts to shells. The NBIC requires the welds to undergo the same kind of nondestructive examination as that which was used in building the exchanger. However, it may not be possible or practical to do so. Resolve with the AI in advance of doing the work what substitute method will be accepted as supplying the same degree of quality assurance.

Eddy-current tube testing to establish a baseline. If it is customary plant practice to monitor tube-wall loss periodically throughout the life of the tubes, the Repair or Alteration Organization (or its NDE subcontractor) should perform a baseline examination and record the results in usable form for subsequent comparison and wall-loss trending described in Chap. 5.

Pressure testing

Whenever a heat exchanger has been cut into and rewelded, the inspection code will require a hydrostatic test. The 1995 edition of the NBIC requires pressure testing repaired pressure-retaining items 80 percent of the maximum allowable working pressure (MAWP) stamped on the pressure-retaining item or operating pressure, whichever is greater, using water or another liquid medium. Each side of the exchanger must be pressure-tested with the other side at atmospheric pressure.

The NBIC requires the item under test to be maintained at the test pressure for at least 10 min before the Inspector examines the part and for as long as it takes for the examination. When pressure testing by liquids can contaminate the pressure item or is not practical, you may either (1) pneumatically test with concurrence of the Inspector, the jurisdictional authority (where required), and the owner, following precautionary requirements of the applicable section of the original code of construction; or (2) subject to the advance concurrence of the Inspector and the jurisdictional authority, use NDE methods that verify the integrity of the repair.

Address safety concerns in the safety plan. It should state that no one should be in the proximity of the exchanger under test where a part can fail catastrophically and lead to death or dismemberment.

Pressure-test temperature. The NBIC requires metal temperature during pressure testing to be not less than 60°F (~15.6°C) unless the owner provides information on toughness characteristics of the material indicating that a lower test temperature is permissible. Furthermore, during close examination, the metal temperature may not exceed 120°F (~49°C) unless the owner specifies a higher temperature and it is acceptable to the AI.

Pressure-test temperatures must be high enough to avoid causing brittle failure. The relationship of ductility and thickness of the construction material determines failure temperature. Although it is unlikely to occur at the operating pressure, which is usually 90 percent or less than that of the MAWP, a thick ferritic steel section may crack at the specified minimum test temperature. For thick ferritic steel shell and channel materials, it is prudent to apply Paragraph UG-100(c) of Section VIII, Division 1 of the ASME Code and Fig. UCS66.2, which plots *minimum design metal temperature* (MDMT), °F, against *nominal thickness,* in, for four groups of carbon steels. The ASME Code recommends that air-pressure tests be conducted at a temperature 30°F (~ 54°C) higher than the MDMT for a given thickness. For the purposes of pressure testing the tube and shell sides of a retubed exchanger, the MDMT will be the test temperature related to the thickness for each curve.

Water is the usual medium. NBIC permits alternative inert fluids. When liquid testing might endanger the structure, it allows air or inert-gas pressure testing that meets specified requirements. The test medium should be noncorrosive and free of slime-producing biological substances and suspended solids that may settle. Test water should be filtered and treated with an algicide that will not harm the unit. Do not test austenitic stainless-steel exchangers with water that has a chloride ion concentration high enough to cause stress-corrosion cracking. Most potable-water supplies are chlorinated, making it necessary to use demineralized or distilled water for hydrostatic testing.

Before you prepare a repair or alteration procedure for an exchanger to be retubed in place, discuss the pressure-testing requirements with the Inspector responsible for in-service inspection.

For many exchangers, a pressure test is all that is done to verify the tube joint tightness. However, when leaktightness is critical, consider using one of the leak-test methods described in Chap. 5 in addition to the pressure test.

Hydrostatic testing. Under shell-side hydrostatic pressure, leaks through the tube joints are readily visible. When the tube side is hydrostatically tested, leakage is directly visible only in removable bundles. The specific locations of leaking joints deep within the tube field may not be discernible. The only way to find a leak in a fixed-tubesheet unit is by pressure loss.

For removable bundles, it is best to install the heads and perform the tube-side hydrostatic test. Testing in the shell when test rings are available is the quickest way to perform the shell-side hydrostatic test.

Documentation

Depending on the nature of the retubing, more or less documentation is required. If the bundle was simply retubed, make a plant record that indicates when the retubing was done. Requirements for documenting the qualifications of Repair and Alteration Organizations and their personnel, for preparing Repair and Alteration Certificates and registering them with the jurisdiction and National Board were covered in Chap. 6.

In-place Retubing

Retubing in place is not for amateurs. Unless the plant maintenance force and supervision is trained in retubing in place and has practiced the necessary skills, it is advisable to have a retubing contractor do the work. Be very careful to investigate the Repair Organization's qualifications and hiring methods. Try to arrange to have key people from the plant maintenance department work with the contractor during in-place retubing. The knowledge they gain will be invaluable for future maintenance. Here are some considerations and techniques that apply to in-place retubing.

Space availability

There must be sufficient working space to retube in place. In front of the tubesheet, room is needed to pull out and reinstall tubes. There must be room in front of the exchanger end into which the tubes are to be loaded to insert the replacement tubes. If the unit is a U-tube one, the U-tubes will be supplied either in a box that separates the tubes of each bend radius from the others or in tied-together bundles tagged with the bend radius. Either way, there must be enough room behind the back end of the bundle cage to load the tubes through the rearmost baffle or tube support into the stationary tubesheet at the front end.

Consider also how the tubes will be dislodged from the tubesheet and how the new tubes are to be attached whether by roller expanding, hydroexpanding, or explosive expanding, or by fusion welding or a combination of these techniques.

Job specifications for in-place retubing

Write a specification that details the work to be done to satisfy the NBIC and very often to meet the requirements of the local jurisdiction. Refer to the original manufacturer's drawings on any sketches you make to describe what is proposed. Include the sequence of disassembly, replacements, and refurbishing to be done and the procedures to be used, and hold points for both owner's and Inspector's inspection. Describe all NDE and any proposed changes to NDE that were performed when the exchanger was built. Often it is cost-effective to employ a consultant who regularly deals with such work to prepare the job specification. Some items to include are described in the following paragraphs.

Welding procedure specifications

Have all the welding procedure specifications associated with the retubing on hand before the work starts. Depending on whether the retubing is to be done by the in-house maintenance group or by some outside service group, some workers may need to be qualified in using the procedures.

Procedures, techniques, and practices specific to in-place retubing

Following are some techniques, procedures, and practices that apply specifically to retubing in place.

In-place testing. The tube-to-tube sheet joints must be examined and tested after retubing. With the face of the tubesheet open to view, subject the tube joints to a test to verify that they are leaktight. Depending on the service of the exchanger and the type of joint being tested, a number of test methods are generally used.

In-place pneumatic pressure testing. It may be difficult to observe leaks when circumstances require pneumatic testing. Therefore, use an acceptable bubble former to check welds and tube-to-tubesheet joints. Based upon relative viscosities, air will leak about 52 to 53 times as much as water at the same pressure. Therefore, a pneumatic pressure test performed as permitted in the NBIC will more effectively disclose leaks than will a hydrostatic test.

Many people are reluctant to test pneumatically because of the hazard of catastrophic release of the stored energy. This is a valid concern if you are testing a new exchanger in accordance with ASME Code procedures. However, a structure that has already been in service for a period of time at its design operating temperature and a pressure about 10 percent below its MAWP is very unlikely to fail when tested as described above. The prejudice against air testing is so strong that some feedwater heater manufacturers will not pressurize shells designed for high steam pressures in the range of approximately 3450 to 5200 kPa (~500 to 750 psi) with shop air. This is so even though the usual shop air pressure is approximately 690 kPa (~100 psi) maximum. For this reason most tube-to-tubesheet bubble testing with air in the shell is usually done at pressures about 345 kPa (~50 psi) or less. However, if the shell operating pressure is far higher than that of the shop air, and the shell fluid is a vapor or a gas, the prejudice is unjustified.

Pneumatic pressure testing at pressures above the normal shop air supply requires bleeding in a source of higher pressure inert gas such as commercially available nitrogen.

In-place extracting and replacing of U-tubes. Removing U-tubes from the existing bundle requires removing the shell if the channel and bundle stays in place or extracting the bundle and channel if the shell stays in place.

Expose the tubesheet(s) and remove any plugs to enable gaining access to all tube ends and permit detubing. The options for removing the tubes are as follows:

1. Cut off the tubes behind the tubesheets with an internal tube cutter or externally with an abrasive wheel. Pull the tubes out manually or with a with a small tugger hoist located behind the bundle. Extract the stubs with a hydraulic stub puller, or drill or drive them out.
2. Drill and thin out the tube ends; pull the tubes out from the U-bend end with a tugger hoist. Once the tubes are loosened, you can easily pull them out. It may expedite driving tubes to cut or burn off the return bends just beyond the last tube support.

Place tube guides into the tube ends before loading the tubes. Discard any guides that do not fit snugly, because it may be impossible to retrieve a guide that falls out when the tube is part way down the cage. Replace the U-tubes from the smallest bend radii outward, loading all the tubes bent to a given radius before loading tubes bent to the next-larger one. Have a worker travel with the tube as it is loaded to guide the legs into the proper holes in the supports or baffles.

Partial retubing by replacing peripheral U-tubes uses the methods of cutting off behind the tubesheet and stub pulling or driving out the tubes. Be sure that tubes that are to remain are protected with sheet-metal shields when extracting the peripheral tubes.

In-place removing tubes and tube stubs and replacing tube. The work of removing tubes and stubs and replacing the tubes is as described previously. If you intend to extract, make sure that there is enough space behind the tubesheet from which the tubes will exit, or arrange the setup so that the emerging tubes can be cut into sections that fit into the available space. When you need to remove a single tube or a small group of tubes, use hydraulic pullers. When you can, avoid driving tubes when retubing exchangers in place, especially in fixed-tubesheet exchangers because of the danger of collapsing a tube in an exchanger placed where it is difficult to work. However, if you must drive the tubes, arrange for sufficient space at the far end or for cutting the pushed-out tubes into lengths that the space will accommodate.

The equipment is the same as that described for U-tube removal. This may be a drill, to reduce the tube wall; a collapsing tool, to loosen the thinned tube away from the tubesheet hole; a tube puller; a tugger hoist; a jig, to hold the drill straight when reducing the tube wall; and for welded-in tubes, a grinder to break the weld at the tube wall.

Start by removing any the tube plugs.

In-place stress relief. In general, on-site stress relief must be done after a part has been cut into when the welded part would require stress relieving under the ASME Code or prevailing construction code rules. Unless the maintenance workforce or the Repair or Alteration Organization has the equipment, controls, and knowledge required to properly perform the heat treatment, it should be done by a heat-treating contractor. In-place torch heating is difficult, requires knowledge and skill, and may be difficult to control; therefore, we recommend using heat-treating blankets.*

Shop Retubing

As stated in Chap. 6, the NBIC requires work on ASME Code–stamped heat exchangers that are going to be repaired or altered by welding, or in which pressure parts will be replaced or in which the tube gage or alloy will be changed, must be done by a Repair Organization in possession of an R stamp and have a certificate of authorization to apply it. U stamp holders are not authorized to do work in accordance with the NBIC. However, they are capable of doing work under the API rules. If you represent the owner, inform yourself of the jurisdictional code requirements for repairs to unfired pressure vessels, before permitting a U stamp holder to bid on the work. The following discusses these aspects of retubing that apply specifically to shop work:

- Owner's responsibilities for preparing the exchanger
- Shop safety
- Coordination between owner and repair shop
- Maintaining critical dimensions
- In-shop separation of shells, and bundles and cage
- Production of replacement parts
- Shop refitting and reassembly
- Factory testing retubed exchangers and bundles

Owner's responsibility for plug removal and cleaning before shipment to shop

When an exchanger or bundle is to be retubed in an off-site facility, the owner is responsible for providing the shop with the MSDS and either ensuring that all hazardous materials have been removed beforehand or negotiating prework cleaning with the shop. Complete

*The authors have had successful experience with the Piscataway, N.J.–based Cooperheat Company, who have offices and facilities throughout the United States.

cleaning requires unplugging any plugged tubes to permit removing entrapped hazardous substances.

Similarly, unless the repair shop agrees to undertake the responsibility for removing asbestos insulation or for handling the unit with asbestos insulation remaining in place, owners must arrange to remove asbestos insulation before sending the equipment to the shop. As stated previously, well-run repair and exchanger shops require owners to provide certification that the unit to be retubed is free of harmful materials.

Shop safety

Before dismantling, apprise the shop workforce of any fluids that the exchanger handled and if any hazardous liquids, gases, or particles can become airborne during the work. Provide them with the applicable MSDS. Owners should discuss the shop's safety plan with the shop management because they have a joint responsibility for the health of the people who work on the unit. Some questions the safety plan should address are as follows:

- Are hazardous substances that may have accumulated in service in the heat exchanger been cleaned or flushed out—for example, do the tubes have to be cleaned before they can be pulled out through the baffles? Or was all the asbestos insulation removed before the exchanger was dismantled?
- Have all friction-fit plugs been removed before burning or cutting the tubes? Are there any potential hazards when drilling out welded-in plugs when the tubesheet holes are being prepared for the replacement tubes?
- Is the ventilation where the exchanger is being dismantled satisfactory for any anticipated airborne releases?

Coordination between owner and shop

The organization doing the off-site retubing and the owner have to work together closely. The purpose is not only to keep the owner aware of the status of the work but also to enable making timely decisions when unexpected developments occur during the work. For example, if the shop discovers after detubing that a baffle needs to be replaced or an impingement plate needs to be modified, the owner's prompt attention will permit the work to progress quickly.

Shops and owners must also agree on and coordinate inspection hold points and the timing of inspections at the hold points. Here are some of the hold points useful for owners' monitoring retubing:

1. Witness the bundle extraction and inspect the shell and detubed bundle to become aware of problems with shell deterioration, baffles, tie-rods and spacers, impingement plates, shrouds, zone enclosures, and subcooler end plates of feedwater heaters.
2. Inspect the baffle cage assembly prior to retubing to verify that
   a. Baffles are in a satisfactory condition
   b. Baffle holes are clean and have no burrs
   c. Tie-rods are straight and spacers not corroded
   d. The impingement plate is adequately sized, properly located, and firmly fixed in place
   e. Tubesheet holes are clean and free of burrs and all traces of the old tubes are removed, especially in the grooves of the tube holes
3. Witness tube loading to note potential damage that can reduce the life of the replacement.
4. Witness bundle insertion to make sure of baffle-to-shell fit.
5. Witness the tube joint tests and examinations to verify the integrity and leaktightness of the tube-to-tubesheet joints.
6. Perform a dimensional check to make sure that the unit will fit back into its site or to learn in advance of shipment of any changes required to site piping and exchanger setting.
7. Witness hydrostatic test and preparation for shipment.

Maintaining critical dimensions

Prior to performing any disassembly work, the owner and shop should verify all dimensions that must be maintained. The shop should match-mark by center punching or with paint sticks every place where the unit is cut apart or unbolted to permit correct reassembly. It is essential that the repair organization weld temporary bridges, similar to the original manufacturer's shipping bridges, across shell expansion joints of units fitted with joints. Stretching or compressing a joint beyond its design room-temperature position will invariably shorten its cycle life. It is a prudent precaution for the shop to install a protective cover over thin-walled bellows if one is not already in place.

Separating shells and bundles and cages in the shop

It is always necessary to separate removable bundles from shells and often necessary to pull the cage-and-tube assembly out of fixed-tubesheet units. Few repair shops are equipped with the hydraulic or mechanical bundle extractors described previously. The usual procedure is to anchor the shell in place and withdraw the bundle or cage-

and-tube assembly from the shell, or pull the shell off the bundle using pinch bars, hydraulic jacks, cables, and winches. The methods and equipment used are generally like those described previously. After anchoring the shell or channel and initiating separation, you can use overhead cranes and a combination of sheaves and cables to pull the exchanger apart.

Shops equipped with large band saws can cut completely through a fixed-tubesheet exchanger just behind the tubesheet at the end away from the tie-rod attachment. You can then separate the shell and tubesheet-and-cage assembly in the same way you separate removable bundles and shells.

Production of replacement parts

When inspection after detubing shows worn or corroded baffle edges and thin or oversized tube holes, baffles need to be replaced. If the unit was constructed with two or more tubesheets, lay them one on top of the other in the order they were set in the drill, front to rear, to assess the amount of drill drift. Instead of making a manual layout for replacement baffles, or using a tape-controlled or computer-controlled drill, the shop should use a tubesheet as a template. Variations in the original drilling pattern from the theoretical should be repeated in the replacement parts and the difference in drill drift between channel and return-end tubesheets accommodated.

Generally, if detubing requires cutting or burning the tubes from the shell side, you require new tie-rods and spacers. Therefore, replace impingement plates that were fastened to the tie-rods or tie-rods and spacers. If inspection before detubing reveals tube erosion in the vicinity of the impingement-plate edges, enlarge or modify the impingement plate before retubing.

If a subcooling zone endplate of a horizontal feedwater heater requires replacement, make sure that the replacement end plate is approximately 76 mm (~3 in) thick even if the original design used a thinner one. This will prevent damage from two-phase flow into the subcooler in the retubed heater.

Shop refitting and reassembly

The purpose of match marking is to make sure that reassembly produces no change from the original dimensions. Whenever a shell is parted, there is some loss of length. Either the shortfall must be made up with a rolled plate or pipe spacer welded to the tubesheet stub or shell, or the owner must accept changes to the positions of nozzles and exchanger supports. Under no circumstances is it permissible to deflect shell expansion joints to make up for metal lost in cutting or burning.

Seam welds of reattached pressure parts must be full-penetration welds. Since the interior is not accessible, seam welds should be in the form of a single vee or J butt weld. The shop should use either (1) a backing strip that remains in place to make sure that there are no internal "icicles" and weld discontinuities or (2) a consumable backing strip. Most Inspectors will not accept a shop's use of a gas tungsten arc weld deposit in place of the backing strip.

Removable-bundle units are more easily assembled than fixed-tubesheet ones. However, if repairs or alterations were made to the shell, it may have distorted. Any distortions should be jacked out, using heat if necessary to make sure that the bundle can be reinstalled without excessive force, blows, etc.

Be sure that all tubes are fitted with tube guides before tube loading. If it is necessary to force the tubes through tight holes in replacement baffles, the tube end may be tapped with a leather or rubber-headed mallet. If considerable force is required, slip-fitted round bars should be set into the tubes to prevent their buckling. It is acceptable to lubricate the tube exteriors only if the lubricant will not harm the tubes, predispose them to corrosion, and pollute the shell-side stream when the unit is restored to service. The shop is responsible for discussing proposed lubricants with the owner.

The ends of straight tubes being tapped into the bundle should be protected with close-fitting plugs. The mallet can strike the plug rather than the tube end, to avoid mushrooming the tube. Under no circumstances is it acceptable for the assembly crew to force replacement U-tubes into the bundle by striking the U-bend end. If the tubes will not pass through the baffles or supports and enter the tubesheet holes without extreme force, the cause must be investigated and cured.

Heat treatment in the repair shop

The inspection codes discussed in Chap. 6 require heat treating if the original ASME Code or other construction code rules would have required it. This means that if a section of shell is cut into and rewelded, and the original shell was required to be heat-treated, that part of the shell on which the work was done must be re-heat-treated. Although some heat exchanger works are equipped with heat-treating furnaces, it is generally undesirable to put a whole exchanger into the furnace. Therefore, heat-treat the same way as in place. Either wrap the parts with electric-blankets and use appropriately placed remote-reading thermocouples for temperature control, or use torch heating with an appropriate temperature-verifying method.

Factory testing retubed exchangers and bundles

Testing in the shop is done to satisfy the repair code requirements and to verify that welds and tube joints have been correctly made and are leak-tight. Permissible media are as discussed previously. The atmosphere in the shop may unacceptably pollute test media. Therefore, control of shop test media is extremely important, especially for metals subject to stress-corrosion cracking (SCC), which require testing with demineralized or distilled water to make sure that chloride ion concentration is acceptable. Verify that the ambient and test medium temperatures are within the guidelines talked about above. This is especially important in unheated shops in cold climates and seasons. Any specified leak testing must conform with the approved procedure. Always witness leak tests in the shop.

Packaging and shipping retubed and replacement bundles

Most of the guidelines for packaging and shipping that apply to retubed bundles also apply to replacement ones. Therefore, they are included here.

You can ship a factory-retubed bundle or a replacement installed in its own shell if it is available. After the final hydrotests, all the water must be completely drained from the exchanger and the unit blown down with filtered, heated dry air. If the plant is awaiting its return to reinstall immediately, then, before it is shipped back to the site, if possible have bags of desiccant placed in the largest nozzles and have the nozzles sealed after the exchanger has been drained and blown down. If there is going to be a delay of 4 to 6 months or more before the retubed unit is reinstalled, then, after the water has been drained, have the exchanger hot-air-dried until all water is removed and then blanketed with nitrogen before shipping.

Retubed bundles that have been shipped to the shop without their shells and replacement bundles require careful boxing, protective wrapping, and skidding in ways that will protect them from damage by the prongs of a forklift truck during return shipment.

If a retubed bundle or replacement is to be stored on site as a spare, it is a good idea to have the shop fabricate a reusable steel shipping shell to which the bundle can be assembled easily. This applies especially to replacement bundles that come with complete new channels. A properly designed shell can be inerted to 20.7 to 34.5 kPa (~3 to 5 psig) with nitrogen and fitted with gages to indicate that there had been no pressure loss. Bundles stored this way can last indefinitely without weather damage.

A spare bundle or a bundle with a channel, which can be installed in an existing shell, requires special considerations for shipping and storage. If the spare is supplied with a channel and is hydrostatically tested on the tube side, be sure that the outside of the tubes is blown clean with shop air while the water is draining out of the channel and tubes. Require external tube surfaces of carbon-steel-tubed bundles to be covered with a suitable rust inhibitor. Have the shop place desiccants in carbon-steel channels.

Require all connections to be plugged or covered to prevent dirt and debris from entering through connections. Threaded connections that are installed for future use should be plugged with bar plugs. Those that are to be piped when the exchanger is installed can have bar plugs or threaded plastic caps. Threaded connections in channels shipped with a nitrogen blanket will be plugged.

The machined surfaces of tubesheets and flanged connections should be covered with a preservative and protected with a wooden cover to ensure that the gasket-bearing surface will not be damaged in transit. Weld-end connections must be covered with suitable caps that will protect the prepared weld end.

Skidded exchangers and bundles are usually transported by truck unless their size or the plant destination requires rail shipment. Bundles should be protected from the elements especially during shipping. Skidded ones should be wrapped with heavy plastic or canvas tarpaulin during the journey between the manufacturer's shop and the owner's plant to keep the bundle as clean as practical. Skids must protect the bundle against forklift-truck damage.

We highly recommend crating in preference to skidding for both shipping and storage of spare bundles. As a minimum, a waterproof tarp or plastic should be wrapped around the bundle and the bundle crated in a box so that it can be handled without damage.

Crates for bundles to be shipped via ocean freight should be lined with a heavy-tarpaper vapor barrier. The box should be solid so as not to allow any moisture to enter easily.

Consider shipping exchangers and bundles sensitive to mechanical shocks in vehicles fitted with vibration monitors and which travel at specified maximum speeds.

Retubing Closed Feedwater Heaters

You generally get the best results in retubing closed feedwater heaters by working with the original manufacturer. However, some manufacturers are no longer in business. Others may not have field service organizations that meet the station's requirements for in-place work. If the exchanger or bundle is to be refurbished in the shop, we strongly recommend not dealing with one that has not either built or rebuilt a considerable number of feedwater heaters. The authors' practice in recommending consideration of a shop not so qualified is to evaluate the facilities and personnel of the proposed shop and the quality of similar work. If these are generally excellent, we then suggest that the owner limit early work to low-pressure, relatively small heaters. As the shop builds an experience record, the owner can then judge when to permit the shop to bid on retubing and constructing higher-pressure and larger heaters and bundles. Descriptions of what is involved follow.

Removing and reinstalling shells of closed feedwater heaters

In contrast to how most other kinds of heat exchangers are disassembled, the shells of feedwater heaters not installed in condenser necks are removed from the bundle, rather than extracting the bundle from the shell. This is because the feedwater piping is almost always welded to the channel. Feedwater piping is thick and large, and cutting and refitting it is very costly. Shells may also be removed for shell-side repairs of impingement plates, desuperheating-zone shrouds, and subcooling-zone enclosures. Here are some suggestions for removing the shell.

Most feedwater heaters are constructed with the shell welded to the tubesheet, thus eliminating the need for a shell flange. In this construction a protection band is welded to the interior when the shell is severed. The band located under a line marked by the manufacturer as the *cutting line* is usually 100- to 150-mm (~4- to 6-in)-wide stainless steel. It protects the tubes that would be exposed to arcing or burning off the shell and the welding done to reinstall it.

In order to minimize weld spatter on the tubes when removing the shell, do not burn or arc-air completely through. Leave about 1.5 to 3 mm (~1/16 to 1/8 in) thickness. Then hand-chisel or abrasive-wheel-cut through the remaining shell thickness. In areas where the cut line penetrates the shell and there are internal shrouds, the burn can continue all the way through since there will be no spatter on the tubes. Usually the shell cut line is located so that shell-side nozzle connections do not have to be cut in order to remove the shell.

When nozzle connections have to be cut, take care to minimize the weld spatter that might fall into the heater and bounce off the impingement plate onto the tubes nearby.

With the shell cut through, pull it off of the bundle. Vertical head-up heater supports are usually located on the tubesheet or near the channel head, allowing the pulling of the shell off with the bundle and channel remaining in place. With head-down ones, guy the channel and bundle. Horizontal installations are made more difficult because the weight of the bundle is carried by the support plates in the condensing zone which bear heavily against the shell interior. As much as possible, lift the bundle to relieve this weight from the supports to assist in removing the shell. This is relatively easy with smaller bundles, under 760 mm (~30 in) in diameter, but with larger bundles, over about 1000 mm (~40 in) in diameter, the supports sometimes bind in the shell, making the bundle difficult to remove. This condition is not prevalent with bundle designs that include skid bars that connect the supports where the bundle weight bears on the shell.

To enable pulling the shell off, the manufacturer usually provides two lugs to which cables can be attached to a tugger hoist to exert the force necessary to remove the shell. As a minimum, the hoist should be rated for 52.5 N per mm of shell (~300 lb per inch of shell) diameter. Feedwater heater bundle-zone enclosures may have to be cut into, repaired, or rebuilt.

When replacing the shell, prepare the edges to be welded and add a new backing strip, fixed to the part of the shell that is being pulled over the bundle.

A Typical Shop Repair of a Three-Zone Closed Feedwater Heater

Figures 12.8 to 12.13, provided by Hydrodyne, Inc. of Massillon, Ohio, show a typical sequence for repairing a condensing and subcooling-zone closed feedwater. Figures 12.8 and 12.9 show the bundle on the shop floor ready to be worked on. Figure 12.10 is a close-up of the channel and tubesheet at the feedwater outlet pass. Figure 12.11 shows windows cut into the pass partition of another heater preparatory to cutting it away to permit tube removal and tubesheet face rehabilitation. Figure 12.12 shows the channel after rehabilitating the tubesheet and before tubing. Figure 12.13 shows the retubed bundle with windows in the subcooling zone, opened to assist in retubing.

Rebundling

Some refineries and chemical plants try to standardize exchangers to permit having common spares for several services. Sometimes an exchanger service network will be made up of 4 or 8 identical exchangers. Multistage flash distillation installations can have as many as 20 or more identical removable-bundle exchangers. In such instances it is ordinary practice to maintain on hand one or more common spare bundles. This allows the operator to change out a bundle that shows sign of deterioration in a timely way to minimize lost

Figure 12.8 View of feedwater heater channel and bundle on repair-shop floor. (*Hydrodyne, Inc., Massillon, Ohio.*)

Figure 12.9 Channel end of bundle on shop floor ready to be worked on. (*Hydrodyne, Inc., Massillon, Ohio.*)

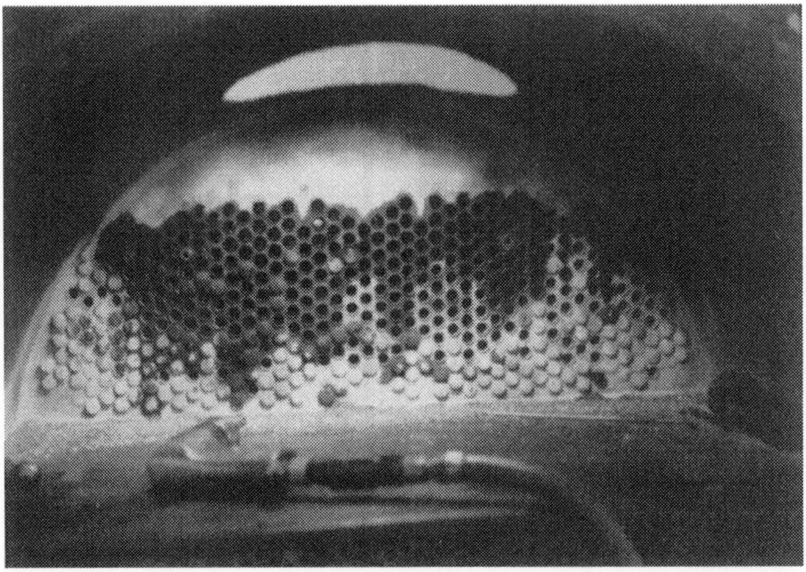

Figure 12.10 Close-up of channel and tubesheet at feedwater outlet. (*Hydrodyne, Inc., Massillon, Ohio.*)

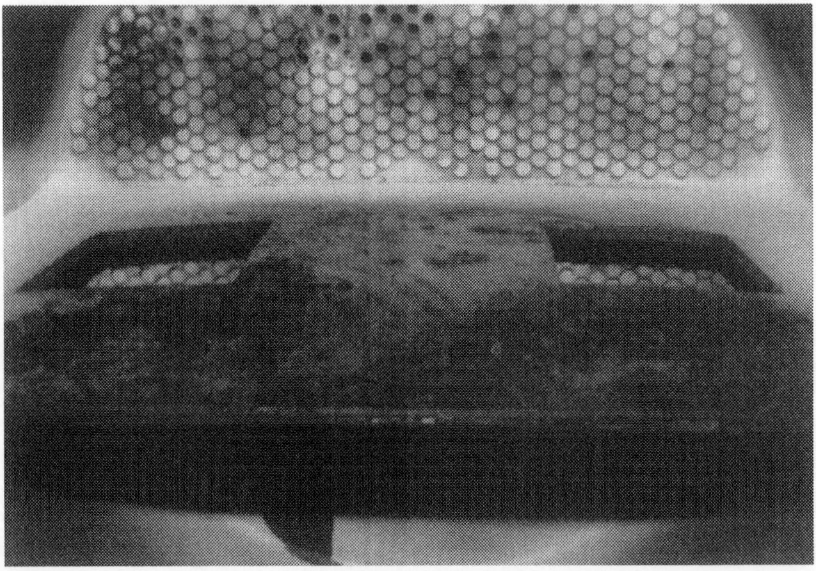

Figure 12.11 Windows cut in pass partition. (*Hydrodyne, Inc., Massillon, Ohio.*)

Figure 12.12 View of the channel after rehabilitating the tubesheet and before tubing. (*Hydrodyne, Inc., Massillon, Ohio.*)

Figure 12.13 Retubed bundle with windows in subcooling zone opened to assist in retubing. (*Hydrodyne, Inc. Massillon, Ohio.*)

production. Plant maintenance can then refurbish the bundle on a routine schedule. Alternatively, the plant can institute a regular program of repair-shop retubing and rebuilding.

Power plants have little call for spare bundles since almost all the tubular equipment is one of a kind. Furthermore, power-plant design does not normally lend itself to the quick bundle changeout possible in process plants.

Specifying bundle requirements to the manufacturer

You can purchase a spare bundle or new replacement for an existing shell by specifying the design-basis performance requirements and providing the outline drawing of the existing heat exchanger. You would think that purchasing a replacement bundle from the original exchanger manufacturer should be more economical, and this is true with small exchangers. However, when the bundle cost exceeds $40,000 to $50,000 (1997 dollars), the original manufacturer loses the competitive bidding advantage because the engineering and design effort required is only 3 or 4 percent of the cost.

As in retubing, if you send drawings to other than the original manufacturer, make sure that they are ones that have the as-built dimensions. There is nothing more embarrassing than to receive a replacement bundle that cannot be stabbed into the shell without grinding down the baffles and/or tube supports. Excessive bundle-to-shell clearance is also undesirable. It may not be crucial to heat-transfer when shell-side bypassing does not markedly affect the shell-side film coefficient of heat transfer, or when the shell-side coefficient is much larger than that of the tube side. But it is unacceptable to get a bundle in which excessive baffle-to-shell clearance reduces exchanger performance.

Without having exact bundle dimensions, the manufacturer will build the bundle to dimensions based on the maximum TEMA-allowed clearance for the nominal shell diameter. If the manufacturer's examination of the service requirements indicates that shell-side bypassing around the baffles is not critical to performance, the manufacturer may build the bundle with even slightly larger baffle-to-shell clearances to ensure that the bundle will slide in without difficulty. Typical service requirements that would permit this option are condensing vapor in the shell or a very highly viscous fluid flowing in the shell.

Extracting existing bundles before rebundling

Refinery and chemical plant layouts generally provide bundle pulling areas to permit removable bundles to be quickly changed out with reconditioned spare bundles. Bundle extraction rigs described above are available to access 6- and 9-m (~20- and 30-ft) levels above the approach to the bundle pulling area and are very useful in bundle changeouts. Power plants and paper mills, on the other hand, have minimum access, and bundle extraction is usually a basic rigger job.

Cleaning and inspecting the shell before rebundling

When a bundle is pulled during a plant outage, there is usually time in the schedule to clean and inspect the shell before stabbing in the new bundle. Inspect the shell in the same way as you would during retubing and look for the indications of corrosion and erosion described above. Keep a copy of the API *Guide for Inspecting Refinery Equipment* handy and review it before you inspect.

Correcting past mistakes

When rebundling, you may be able to eliminate some of the things that caused the original design to fail by providing the replacement bundle designer with enough information to design the new one to function correctly. The tubing first chosen may not have been suitable for the corrosive service and therefore may have corroded or failed sooner than anticipated. Failure may have been caused by an overlooked mode of operation not addressed in the original design. It may be possible to rectify the omission in the replacement bundle. Making manufacturers aware of the principal causes of early failures permits them to correct many such design flaws.

Upgrading with replacement bundles

Often in older exchangers, the current operating basis has changed so much from the originally specified one as to warrant revising the replacement bundle. You can make many changes without changing the fit of replacement bundle into the existing shell. If current shell-side flow conditions have changed from the original ones, consider changing to a baffle arrangement more appropriate than that provided in the original bundle. You can change a four-tube-pass flow bundle to a two-tube-pass one. And you can replace tubes of one diameter with those of another, adjusting the tube count accordingly. Another upgrade option is replacing plain tubes with low-finned ones, corrugated tubes, or other enhanced surface. If space is available, you can make the bundle length longer and modify the shell to accommodate the new length.

Verifying replacement-bundle dimensions

A spare U-tube bundle, of the kind usually seen in a refinery, will consist only of the tubesheet, bundle cage, and tubes. When you check these out in the factory, make the dimension check before the bundle is tubed. At this time you can check the baffle diameters, match them with any measurements of the shell interior, and verify that the impingement-plate locations will be under the inlets.

If the replacement bundle is for a floating-head exchanger, check both the floating- and stationary-tubesheet outside diameters to make sure that they will mate with the existing flanges on the shell, channel, and floating-head cover. If these details are missing, the gasket drawings may provide a clue. Otherwise you may require a complete floating-head cover and backing flange to be part of the replacement.

Replacement-bundle code status and documentation

If the pressure and size of the exchanger require code stamping, buy the bundle as a pressure part with appropriate stamping and partial data report form. Documentation is required for plant equipment insurance purposes, and the jurisdiction in which the plant operates may also require it. It is also very useful if the exchanger is reassigned to another service. Nuclear plants with safety-related exchangers have additional licensing requirements that make documentation necessary. These are some requirements.

1. All pressure-bearing material, including the tubing, must be provided with Certified Mill Test Reports (CMTRs).
2. Tube heat maps should be provided showing the location of each heat of tubes in the layout of the replacement bundle.
3. Partial data reports should be provided and the replacement bundle registered with the National Board. The National Board number of the original heat exchanger should be noted to enable the National Board to maintain an updated file. Registration is especially important for rebundled exchangers, such as feedwater heaters that are furnished complete with a new channel and channel cover.

Preparing replacement bundles for shipment and storage

Preparing replacement bundles for shipment and storage has been addressed above.

Installing replacement bundles

When cranes are available to rig spare bundles into place, installation is relatively easy. The Hydroextractor and Bundle Wagon can also be used for installation, or tugger hoists may be used to pull the bundles into the shell. When the bundle is in place before the exchanger is released for service, retest the tube joints one more time to make sure that none of the joints loosened up in transit.

References

1. W. Beckom, "Refinery Experience with Hydraulic Tube-Bundle Extractor," paper presented at 1976 NPRA Refinery Maintenance Conference, National Petroleum Refiners Association MC-76-5, Washington, D.C.
2. "Tube Bundles Are Pulled in 30 Minutes," editors, *Oil Gas J.*, June 8, 1979.
3. J. Singletary, "Bundle Wagon Eases Maintenance for Mobil Chemical Co.," *Oil Gas J.*, Jan. 8, 1979.
4. R. Clapper, "Your Guide to the Repair and Retubing of Tubular Heat Exchangers," *Power Eng.*, undated (ca. 1963). (Facts of publication from *Power Engineering Magazine,* Barrington, Ill.)
5. F. F. Dudley, "Electronic-Control Method for the Precision Expanding of Tubes," *Trans. ASME* 577–584, May 1951.
6. F. F. Fisher and E. T. Cope, "Automatic Uniform Rolling-in of Small Tubes," *Trans. ASME* 53–60, 1974.
7. I. Berman and J. W. Schroeder, "Detnaform at Power Plant Installations," paper delivered at Sixth International Conference on High Energy Rate Fabrication, Essen, West Germany, Sept. 12–16, 1977.
8. H. Krips and M. Podhorsky, "Hydraulic Expansion—a New Method for the Anchoring of Tubes," *VGB Kraftswerktechnik* **56**(7) (Federal Republic of Germany).
9. H. Krips and M. Podhorsky, "Hydraulic Expansion of Tubes," *VGB Kraftswerktechnik* (1), 1979 (Federal Republic of Germany).
10. S. Yokell, "Hydroexpanding: The Current State of the Art," paper 82-JPGC-Pwr-1, presented at the Joint Power Generation Conference, Denver, Colo., Oct. 1982.
11. D. A. Scott, G. A. Wolgemuth, and J. A. Aiken, "Hydraulically Expanded Tube-to-Tubesheet Joints," *Trans. ASME,* **186,** 104–109, Feb. 1984.
12. S. Roy, "In-Situ Retubing of Steam Generators," *Conf. Proc., Maintenance Welding in Nuclear Power Plants/III,* Knoxville, Tenn., November 6–8, 1985.
13. M. H. Jawad, E. J. Clarkin, and R. E. Schessler, "Evaluation of Tube-to-Tubesheet Junctions," paper presented at the 1986 Pressure Vessels and Piping Conference and Exhibition, Chicago, Ill., July 20–24, 1986.
14. K. Uragami, M. Sugino, S. Urushibata, T. Kodama, and Y. Fujiwara, "Experimental Residual Stress Analysis of Tube-to-Tubesheet Joints During Expansion," ASME Paper P82-PVP-61, 1982.
15. *Standards for Steam Surface Condensers,* 8th ed., The Heat Exchange Institute, Cleveland, Ohio, January 1984.
16. C. A. Maxwell, "Practical Aspects of Making Expanded Joints," *Trans. ASME,* 507–514, July 1943.
17. W. H. Kearns (Ed.), *The Welding Handbook,* 7th ed., 5 vols., The American Welding Society, Miami, Fla., 1982–1996.
18. *Procedure Handbook of Arc Welding Design & Practice,* 12th ed. The Lincoln Electric Co., Cleveland, Ohio (reprinted annually).
19. *Welding Journal,* The American Welding Society, Miami, Fla. (published monthly).

Chapter

13

Feedwater Heater Autopsies

Introduction

Few grass-roots power plants are currently planned or under construction. Energy conservation, reliability, and life extension are today's watchwords. Reliable feedwater heaters in existing plants are essential for conserving energy. Feedwater heater reliability depends on appropriate design, construction, installation, and operation. In another work, we pointed out that although feedwater heaters should last 40 years, of approximately 250 heaters replaced annually, some are replaced after less than 10 years, most last only 20 to 30 years, and some have operated more than 40 years.[1] The authors have surveyed many feedwater heaters to try to find the reasons for performance deterioration and the character and causes of mechanical failures. Mostly we examine operating heaters during planned and unplanned outages to try to determine the causes of operating problems. We attempt to learn if the reasons are faulty operating procedures, operating mode changes from those originally specified, improper design and construction, inadequate feedwater treatment, metallurgical grounds, or normally anticipated metal deterioration.

To establish failure patterns, we examine tube plugging records and records of regular eddy-current and IRIS inspections, view borescope tapes, and scrutinize whatever maintenance records are made available to us. More often than not, records are incomplete or inconsistent, especially of tube inspection and plugging in older stations.

Well-run power stations try to identify what caused the demise of heaters. Usually the effort requires a postmortem examination. This chapter discusses autopsies of heaters that have had long lives and those that came to an earlier end.

Why Conduct an Autopsy?

It is expensive and time-consuming to conduct postmortem feedwater heater examinations properly. You may require the services of a consultant to prepare specifications for performing the work and to render expert opinions about what the autopsy discloses. Station personnel may be reluctant to participate, assuming that installing new heaters will solve the problems they encountered. Why, then, perform autopsies? The reasons are (1) to train plant personnel, (2) to reveal the causes and mechanisms of the failures that led to the heater's demise, (3) to avoid repeating past specification mistakes, (4) to adjust procedures and practices to reduce the rate of deterioration of operating heaters, and (5) to dissect prematurely failed heaters for forensic purposes.

For autopsies to be worth their cost, station management must commit to investigating why their heaters did not endure for their specified life spans. They need to accept the fact that current operating practices and maintenance procedures may cause shorter-than-planned heater life and be willing to make indicated changes. Having committed to postmortem examinations and accepted the need for change, they must be willing to implement making the changes and bear the costs of personnel training and continuing education.

Who Should Participate in Feedwater Heater Autopsies and Why

The following people should participate in feedwater heater postmortem examinations: (1) maintenance personnel, (2) operators, (3) metallurgists, (4) engineers responsible for specifying and inspecting new and replacement heaters, (5) engineers concerned with extending heater life, (6) engineers concerned with operations, (7) safety engineers, and (8) consultants.

It is especially important that consultants you hire to serve as experts in legal matters witness the complete autopsy and documenting of findings. All participants should have access to the manufacturer's assembly and detail construction drawings for study before any disassembly. Maintenance personnel must be able to visualize what, where, and how cuts must be made to enable fulfilling the autopsy's purposes.

Role of maintenance personnel in autopsies

Maintenance workers are not always trained to understand construction drawings, which are abstract representations of spatial arrangements of physical objects. However, given copies of the drawings and a well-written dissection specification, a consultant familiar with heater construction can quickly show them how a heater was assembled. This trains them to understand the structure, permits them to consider the feasibility of making repairs to similar heaters, and makes clear the requirements for successfully carrying out repair procedures.

Role of operators and engineers concerned with operations in autopsies

Operators are even more unlikely than maintenance personnel to see detailed heater construction drawings. Most are taught how to set levels and maintain drains cooler temperature approach (DCA) and terminal temperature differences (TTD) for various operating modes. They learn various tricks of the trade such as keeping watch on the position of the valve stem of the drains control valve, which can give an early warning of tube leaks. When the operation deviates from nominal for the current mode, they know what adjustments to make to restore the correct values. When these corrections do not improve the situation, they suspect that the heater has suffered some mechanical trauma.

Engineers concerned with operations then become involved. With the operators, they speculate on possible kinds of failure, consider strategies for fixing the problem without bypassing and shutting down the heater, and determine the temporary and/or permanent repairs that may be needed. Their primary goal is to restore operational status as quickly as possible. Although the station may discuss the difficulty with the manufacturer or a consultant, economics usually dictate reestablishing service and deferring major repairs to the next scheduled outage *even when doing so may cause further deterioration.* Little time is allowed to explore the roots of the problem.

Usual practice for unplanned outages is to access the channel and identify and plug any leaking tubes. So little time is allotted for repairs that intact surrounding tubes are frequently plugged as insurance against the possibility that they were damaged. Indeed, the PowerFect-patented removable temporary plug discussed in Chap. 10 was invented to enable insurance-plugged tubes to be examined during a subsequent planned outage. After opening the channel for an unscheduled outage, the maintenance crew makes only repairs that they can do quickly. In addition to tube plugging, these are mostly to pass partitions, pass-partition covers, and tube-to-tubesheet welds. They may include reexpanding expanded-only tube-to-tubesheet joints of low-pressure heaters.

Maintenance, operating, and engineering personnel are all familiar with and understand what they see when the channel is opened for examination. However, they may not have a clear mental picture of the arrangement and construction of the shell-side internals of the various heater configurations. Having operators and operating engineers study the construction drawings and participate in autopsies helps them understand the relationship between channel- and shell-side mechanical design and construction and heater performance.

Another benefit of having operating personnel participate is that it may provide graphic evidence of the results of failing to follow recommended practices. Examples are not following the manufacturer's recommendation for starting up and shutting down, not following the manufacturer's recommendation for cooldown rates before opening the channel, failing to properly vent the heater, and not using the nitrogen or steam padding system when shutting down. It is noteworthy that, although most tube failures result from occurrences in the shell side, during unplanned outages time is rarely available to investigate what has been happening there. Often this results in exacerbating damage that could have been limited if the situation had been corrected in a timely way. Many heaters fail prematurely because a few minutes were not spent to find the axial position of tube damage. Locating the failures may reveal a trend that can be stopped by a prompt repair. Examples are tube failures at subcooler inlets, suggesting level-control difficulties and midspan breaks indicating vibration problems.

Role of metallurgists in autopsies

It is essential that metallurgists participate in heater autopsies. They should decide where to extract or cut out tube specimens for metallurgical examination from each zone and representative U-bends. Where inspection on disassembly reveals multiple failures within the tube holes, in the tubes just at the rear face of the tubesheet or within about 25 mm (about an inch) beyond the rear face, they should examine specimens for evidence of mechanisms such as stress-corrosion cracking (SCC) and vibration-induced fatigue. Where tubes were plugged because of failures in welds of tube-to-tubesheet overlays, they should examine the overlay weld metal, the tube-to-overlay weld nugget, and, if possible, the condition of the roots of the tube-to-tubesheet welds. The metallurgist can determine whether it is desirable to take a boat sample of the tubesheet, and if so, instruct the work force about location and sample size. By analyzing the surface of tubes that failed in desuperheating zones, the metallurgist can ascertain if there was chloride ion entrapment in the oxide layer of austenitic tubes.

Role of engineers concerned with life extension in autopsies

By participating in autopsies, engineers concerned with extending heater life may be able to learn how to prolong the lives of similar heaters that are still in operation. For example, when many welded tube-to-tubesheet joints or ligaments between tubes exhibit cracks and the metallurgist suggests thermal shock as a probable cause, this arms the life-extension engineer with a basis for insisting that cooldowns follow the manufacturer's recommended procedure, no matter how inconvenient.

Observing that many tube failures spread out from the steam inlet, and further noticing that the failures are in the vicinity of the impingement-plate edges, the life-extension engineer can recommend cutting into the shells of similar heaters and modifying their impingement protection to stop further deterioration.

A life-extension engineer who observes extensive creep-fatigue cracking in shell skirts of heaters that had been in high-temperature service will be alerted to investigate whether improper material selection was made and to avoid repeating such an error.

When the autopsy discloses cracks in the welds of the part-cylinder-to-flat-plate of subcooler zone enclosures, life-extension engineers may examine operating records to see when the DCA began to deteriorate as a result of steam in leakage. This can indicate similar problems for comparable heaters. These welds may be repaired if done soon enough, and an alert to timely action may prevent a heater's early demise.

Role of engineers responsible for specifying and inspecting new and replacement heaters in autopsies

Autopsies reveal the physical effects of specification errors and the results of poor work and practices that inspectors did not uncover in the factory. The graphic nature of shell-side deterioration in each zone provides an unforgettable education to specifiers and inspectors. Removing the bundle from the shell and cutting loose the desuperheater wrapper from its flat plate dramatically exposes what happens when insufficient wet-wall protection for low-load operation was specified.

Examining the bundle at the U-bend end can show if the tubes were hammered into the bundle (see Fig. 13.1). The latter emphasizes the importance of specifying that the purchaser's inspector be present during tube loading. Viewing the tubes between supports can reveal tube wear

Figure 13.1 Damage at U-bend ends due to hammering tubes into bundle. (*MGT Inc. photograph.*)

from tube-to-tube collisions resulting from vibration and erosion of tubes, and enlargement of holes in tube supports and cross-flow baffles attributable to vibration (Figs. 13.2 and 13.3). Such damage is usually so shocking and attention-getting that it conditions specifiers to include overload conditions in their specifications. The condition of the tube supports and shell in the vicinity of the tube supports may show, in striking fashion, the results of allowing higher-than-recommended steam velocities.

Role of safety engineers in autopsies

It is essential that safety engineers be included in the autopsy team. The team leader should familiarize the team with the proposed site of the work, the sequence of dismemberment and examinations, and the tools and equipment to be used. The safety engineers should provide appropriate text to be incorporated in the specification for performing the autopsy.

Documents to Gather in Preparation for the Autopsy

Beforehand, gather the following documents to prepare for postmortem examination: (1) operating logs or plots that show the performance deterioration that led to replacing the heater, (2) heater main-

Figure 13.2 Tube vibration damage at baffle penetrations. (*Source unknown.*)

Figure 13.3 Enlargement of holes in tube supports due to tube vibration. (*Source unknown.*)

tenance records, (3) nondestructive examination reports, (4) plug maps and associated information, (5) the setting plan, (6) the original purchase specification and exchanger data sheet for the heater, (7) any available heater assembly and detail drawings, and (8) the maintenance manual supplied by the manufacturer.

Until about 1985, many manufacturers furnished setting plans and location drawings, but not construction details, which they considered proprietary. (The authors, who have seen hundreds of feedwater heaters during construction, maintenance, and disassembly, believe that there is little difference between the several manufacturers' designs and construction and do not think that the variations significantly affect performance capability or price.) Manufacturers' reluctance to provide details stemmed from their desire to have owners buy repair and retubing services and replacement bundles and heaters from the original source. Currently, well-written purchase specifications require construction detail drawings as part of the documentation package.

Although manufacturers prefer to provide only drawings useful for routine maintenance, when specified as part of the purchase contract, most will supply construction drawings if the purchaser agrees to maintain their confidentiality and not use them to obtain competitors' bids for duplicates, replacements, or replacement components.

Assembly and detail drawings, hidden away in engineering department files, are often not made accessible to the maintenance workforce. When assembly and detail drawings are not available or cannot be obtained, use the maintenance drawings provided in the manufacturer's maintenance manual.

Use the operating logs, maintenance records, and plug map to guide the search for causes of the heater's demise. For example, if the logs display an increase in the terminal temperature difference (TTD), the cause could be a broken pass-partition-to-tubesheet weld that permits feedwater to bypass the heater. If the maintenance records and plug map show a pattern of increasingly frequent plugging of tubes that failed in the vicinity of the steam inlet, direct the postmortem examination to examining the condition and adequacy of the steam inlet impingement plate. Compare the operating logs with the conditions shown on the exchanger data sheet for clues to deviations from the originally specified operating conditions and mode of operation that may have hastened the heater's end.

You can use the autopsy to establish the validity of eddy-current test results in currently operating heaters. Do this by comparing records of gauged tube-wall thicknesses of specimens taken at random locations throughout the bundle with thicknesses indicated by the most recent eddy-current test results.

The setting plan and feedwater heater data sheet usually have the weights of the complete heater and bundle. Most of the time it indicates the structure's center of gravity and lifting points and locates the shell-to-skirt cut line (which may also be marked on the shell itself). The assembly and detail drawings show the positions of welds, the sizes and locations of zone enclosure parts, and the locations and construction of seal rings (close-off plates) welded to the inside of the shell and outside of the desuperheater and subcooler enclosures. Although the drawings provided for maintenance purposes do not have all the construction details, they may have enough information to prevent disassembly in a way that obscures the causes of the heater's demise.

Scheduling Autopsies

The engineer directing the autopsy should establish the schedule for its performance. This includes timing when to place the heater on the site, arranging the acquisition of required equipment and tools, scheduling the workforce selected to perform the work, booking NDE contractors if they are to be employed, and planning when participating parties are to be present.

Space Requirements, Facilities, and Tools

Ideally, postmortem examinations should be conducted in a facility in which the heater will be protected from the weather and where the workforce can perform its functions without weather-related impediment. The site should be readily accessible to the participants, equipped with lifting and winching devices to handle and pull apart the heater, and power and compressed air to operate the required tools. Most of the time the deceased heaters are moved as soon as possible to an outdoor yard to which all the required equipment must be brought and to which participating personnel must come.

Space

The site must be spacious enough and have dimensions appropriate for performing all disassembly tasks, making inspections and examinations, and photographing and videorecording the work. The surface on which it is placed should be flat, level, hard, and capable of bearing the heater's weight. There should be enough space to accommodate pulling the shell off the bundle from the cut line and leave walking room at the U-bend end to permit examination. Either existing structural members, adequate for anchoring the heater against the loads required to pull the shell and other parts free of the bundle, must be available at the site, or you must provide for installing a temporary structure. To prevent damage as the shell slides off the bundle, movable props must be placed under the bundle's tube supports and reset appropriately as the tube supports are exposed. There must be enough room around the heater to permit doing these tasks safely and conveniently, and the surface on which the movable props rest must be hard and flat.

Handling facilities

All handling facilities must meet current governmental and plant safety requirements. Winches and cranes must be fitted with safety hooks. Crane cables should be properly lubricated and not be stretched, frayed, or unraveled. Crane and winch controls should be in good working order, and their brakes should have been recently inspected and tested. Do not permit unqualified persons to operate handling and lifting equipment such as cranes and winches. The following should be on site prior to beginning disassembly:

1. Crawler crane, A-frame, or other lifting devices with capacity adequate to handle the channel cover and channel end of the bundle and to lift the whole heater. As stated above, the setting plan has the weight of the heater and bundle. The channel weight is included as part of the bundle. The channel cover weight can generally be found in the maintenance instruction book or on maintenance drawings. It is important to establish the bundle's center of gravity and to provide means for supporting it without inflicting damage before parting the shell and bundle.
2. A winch of capacity at least 50 percent of the combined bundle and shell weight to pull the shell off the bundle.
3. Structural members embedded in the ground or other structures capable of anchoring the bundle against axial movement when the shell is pulled off the bundle. At least two anchors are required, spaced behind and to either side of the channel end. The angle between each anchor and the channel centerline should be approximately 45°. Each anchor should be capable of providing reaction of about 70 percent of the bundle weight.
4. Cable or nylon slings and spreaders for lifting the heater, heater shell, and bundle, and for attaching the channel to the anchors. If cable or nylon ones are not available, use chain slings only if they have recently been tested and inspected. Do not permit the slings to bear directly against the tubes. They should be positioned to bear against a tube support or on dunnage that bridges two or more tube supports.

5. Clevises appropriately sized for the loads to be borne.
6. Steel I-beams, channels, and shims, 200×200-mm (~8×8-in) wooden beams, and similar materials to be placed under the bundle's tube supports to prevent sagging as the shell is pulled along the bundle length.
7. Steel wedges to be fitted between the bundle supports and beams on which the supports rest to prevent the bundle from rolling off the supports in case of an untoward event.
8. Hydraulic jacks. 90-kN (10-ton) jacks are desirable for supporting and adjusting the position of the channel after the bundle has been set in position for the autopsy. Pairs of smaller jacks are useful for starting the shell on its axial travel after it has been parted from the shell skirt.
9. Pinch bar(s).
10. Saddles, support beams, dunnage or other material to support the bundle in the horizontal position.

Note that heater support saddles (item 10, above) welded to the shell generally cannot be used for supporting the unit during disassembly unless the heater is set where the saddles can slide or roll with the shell when the shell is being pulled off the bundle. Initially there should be one support under the channel, one under the shell skirt adjacent to the cut line (or shell flange if the unit was constructed with one), and three supports approximately equally spaced between the support on the shell at the cut line or flange and at the U-bend end of the shell.

Equipment and tools

In addition to the handling facilities and equipment, the following equipment and tools should be on site before the work begins:

1. Carbon arc outfit consisting of a 200-amp or greater amperage portable DC welder with cables and carbon-stick holder.
2. Gas-burning outfit—oxyacetylene, oxy-mapp gas, or oxypropane are suitable.
3. Source of compressed air at pressure in the range 500 to 700 kPa (~75 to 100 psi) with an output of approximately 3 m^3/min (~100 ft^3/min).
4. Edge grinder with appropriate personnel wheel guard adequate for handling approximately 300-mm (~12-in)-diameter resin-bonded abrasive wheels suitable for safely abrasive cutting steel shells.

5. Pencil grinder for removing tube-to-tubesheet welds; chipping guns with chisels suitable for parting steel.
6. Hydraulic tube-stub puller.
7. Nut runners, wrenches, and miscellaneous small tools as required.
8. Miscellaneous plate, angles, channels, and other steel structurals for fabricating lifting lugs, seats for prying apart for initial separation of the shell from the shell skirt, etc.
9. Steel shimming plates of various thicknesses from 6 to 25 mm (~¼ to 1 in).
10. Hydroblaster.
11. Safety equipment, including but not limited to vision and hearing protection; protective clothing for all participants; grinding shields, welding helmets, burning goggles, and gloves; safety shoes, chaps, and vests for personnel protection during burning and welding; isolation tapes; and whatever else is recommended by the station's safety department.
12. Inspection and examination equipment consisting of high-intensity low-voltage lights; scrapers, 30-power (or higher) portable illuminated microscope, borescope, portable videocamera, 35-mm camera, and light source and other equipment recommended by the station's nondestructive examination personnel.
13. Marking crayons or litharge-lead marking refillable marking sticks.

Disassembly Procedures

Disassembly procedures must be adapted for each configuration. The process will differ for a full-opening channel heater and a manway access one. The approach to removing the shell may depend on the diameter. The procedures for dissecting partial-pass full-length heaters and vertical head-up ones will be similar to each other, but will differ in some respects from the procedure for horizontal ones. Different requirements will apply to single-zone and multizone units. The procedure for cutting apart dual heaters is distinctly different from the dissection procedures for other types of heaters.

The authors have found that including clip-art icons and other graphics at appropriate locations in the specification is a very effective way to alert the workforce to the requirements of the job. Following are typical specification headings. Some headings are followed by applicable comments. Others are followed by suggested text appropriate for performing a postmortem autopsy of a horizontal, three-zone, medium-sized, high-pressure manway-access heater.

1. Introduction

   Suggested text: There are two ways to disassemble feedwater heaters—dismantle them in (*a*) the shortest time and least expensive way required to recover their scrap value or (*b*) in a way that will permit establishing the modes and deterioration that caused its demise. Since the purpose of dismantling this heater is to find out why and how it expired, this procedure specification is for the second mode. It is essential that the techniques and methods used not obscure the kinds of deterioration and failures that caused the heater's death. The heater to be autopsied may contain foreign matter in the shell, zone enclosures and tubes. Collect and identify where specimens of this material were found and turn them over to the metallurgical laboratory for analysis. The analysis may provide clues to the mechanism or mechanisms of deterioration.

   1.1. Icons

   Suggested text: Throughout this specification icons are used to flag points of special interest. The hand ✌ indicates places where the work should be vidoegraphed and photographed, or where it should be interrupted for videography and photography. The magnifying glass ❍ is shown when the disassembled parts should be examined and inspected.

2. Autopsy team

   Comment: Name the autopsy team members and their departments at this point in the specification.

3. Drawings and documents

   Comment: When available, list and include in the specification copies of the drawings and documents that underpin the heater construction. The specification should make it clear that the autopsy team should study these documents before any further work is done. We have found it helpful to include manufacturers' exploded-view posters showing general construction of their heaters among the documents. These graphically show the various parts that will have to be cut loose for the postmortem autopsy. (One manufacturer now has available a video showing how to open its full-opening channel and another general video of its heater manufacturing operations. As other manufacturers make such videos available, they will be important predisassembly training tools.)

4. Maintenance records

   Comment: Specify the type of records to be studied and the period the records span.

5. Operating logs
   Comment: Specify the beginning and ending dates of the operating logs to be studied.
6. Site
   Comment: Describe the site where the work will be done. When it is an outdoors one, include paragraphs that describe how the heater should be protected from the elements before work commences and during its performance.
7. Required equipment and tools
   Comment: List the equipment and tools required to do the work. Prepare the list together with the maintenance workers who will perform the work.
8. Safety precautions
   Comment: Include text provided by the safety engineering department.
9. Work sequence
   9.1. Set the heater in place
   Suggested text (if suitable trunnions and lugs are not already on the unit): Prepare and weld to the shell lifting lugs adequate to carry the heater's weight. If the saddles originally welded to the shell cannot be arranged to permit sliding as the shell is pulled off the bundle, place structurals or adequate wooden dunnage under the channel and at four equally spaced places along the shell. Set the heater on these members with the saddles on top. This means that the desuperheater will be on the bottom and the subcooler on top. Set wedges in place between the shell and the support dunnage to prevent the unit from rolling. Then fabricate and weld anchoring lugs to the channel just behind the joint of the channel to the shell skirt parallel with the ground 180° apart. Each lug capacity must be equal to 70 percent of the heater weight. Lug holes must be large enough to receive clevis pins. Fabricate a pulling lug capable of handling half the bundle weight and weld it to the shell head crown at the U-bend end of the heater.
   9.2. Gain access to the channel barrel and tubesheet face
   Suggested text: Remove the cover. It is not necessary to take special precautions for protecting the cover-to-stud-pad surfaces. Enter the channel through the manway. (*Caution!* Breathing air, cooling, and ventilation must be provided. A safety watch must be established. Observe all OSHA, other governmental, and company requirements for entering closed spaces. Wear protective clothing and safety gear!)
   9.3. Unbolt and remove the pass-partition covers

Comment: If the pass-partition covers are welded to the frame, the specification should describe how to burn the covers off.

Suggested text: Videotape and photograph the tubesheet face in the inlet and outlet passes for record purposes and subsequent study. Examine the channel and tubesheet for evidence of leaks between the pass-partition and tubesheet or bolting frame, or through holes in the pass partition or pass-partition cover. Scrutinize the pass partition and tubesheet for distortions, broken welds, worn-away pass-partition cover to frame bolts and nuts, eroded-away tube ends, washed-away tube-to-tubesheet welds, and channel-barrel-to-tubesheet transition corner cracks. Then carbon-arc or burn away the attachment of the pass partition to the tubesheet and the pass-partition cover flange to the channel barrel. Remove these scrapped parts from the channel. Examine the region of the feedwater inlet and outlet nozzles for erosion and cracking. Videotape and photograph evidence of deterioration and failures for further study. Then draw a cut line completely around the head either at its knuckle or where it is attached to the channel barrel neck. Burn or arc away the head from the channel. Again videotape and photograph the tubesheet face to enable establishing the plugging pattern. Wire-brush the tube-to-tubesheet welds. Use a 30-power (or higher) portable illuminated microscope to examine a representative number of welds. Examine the tube-to-tubesheet welds, especially of plugged tubes. If required by the metallurgist, take boat samples of the tubesheet.

9.4. Remove the shell

Suggested text: Anchor the channel to the structural reaction mass by inserting clevises into the lugs welded to the channel. Attach cables to the channel lugs and reaction mass. Align the cable-to-reaction structures at 30 to 45° angles from the heater's axial centerline. Rig the winch cable and hook to the lug on the head at the U-bend end of the heater. Insert clevises attached to each end of a cable sling in the shell lifting lug holes. Insert the crane lifting hook into the sling clevis and take a strain so as to take the major portion of weight below the cut line off the supports under the shell.

Comment: If the heater is very long, a spreader bar may be used with two cable or nylon slings.

Suggested text: Then burn the shell free from the skirt at the cut line. When the shell has been burned through, weld two pairs of opposed angle clips to the shell and skirt, with the pairs 180° apart on the sides of the heater. Allow sufficient space between the opposed clips of each pair to fit a hydraulic jack. Jack the clips apart at the same time to initiate moving the shell axially. Engage the winch and commence pulling the shell off the bundle. Set wooden beams or structurals under the bundle supports as they are exposed to prevent the bundle's sagging. As the shell moves, relocate the shell supports to maintain its position parallel with the ground. Make sure to support the shell with the hanging cable(s) and sling(s) as it slides free. Then, in stepwise fashion, pull the shell about the distance between the bundle supports and relocate shell-and-bundle-supporting wooden beams or structurals as required to maintain the shell in its horizontal position and to prevent the bundle's sagging as it emerges from the shell. Finally, pull the shell completely off the bundle and set it on the ground in a location convenient for inspecting its interior before it is sent off to the scrap yard.

9.5. Cut through the close-off plates

Suggested text: The shell skirt cylinder is welded to the tubesheet at one end and to the shell at the other end. Inside the shell skirt, seal-off plates welded to the exteriors of the approximately half-cylinder zone enclosure wrappers and the interior of the shell skirt should be visible. The typical three-zone heater graphic shows about where these seal plates are located. Their function is to make sure that all the superheated steam enters the desuperheater and all the condensate enters the subcooler. Torch-burn or carbon-arc through the welds of the seal-off plates, leaving as little of the weld on the shell skirt as is convenient. A wide burn is desirable.

9.6. Part the shell skirt from the tubesheet

Suggested text: Torch-burn the welds of the zone flat plates to the shell completely through. Leave as little of the weld behind as is convenient. It is necessary to cut windows in the shell on both sides of the divider plates to gain access. Cut, carbon-arc, or burn to within 3 to 1.5 mm (~⅛ to ¹⁄₁₆ in) of the inner surface of the skirt cylinder and chisel through or abrasive-cut through. The tubes in this vicinity must not be impinged on by the torch's burning flame or covered with slag. The weld that joins the shell skirt to the tubesheet appears to be a fillet weld, but it is a multipass full-penetration one. Carbon-arc it away to approximately the root pass. Use the edge grinder to completely sever this weld so as to avoid filling the space between the shell skirt, tube bundle, and tubesheet with slag. Then remove the freed shell skirt from the bundle in a fashion similar to removal of the main shell. Use hydraulic jacks, dunnage, or structurals to adjust the position of the bundle and support it as the freed shell skirt is moved toward the U-bend end. Set the shell skirt on the ground near the shell. Cover the ends to protect the interior against foreign-matter intrusion.

Comment: It may be acceptable to remove the shell and skirt in sections instead of in the way described above. The pieces should consist of top and bottom halves, and segments between the shell cut line and the dished head end that can be conveniently handled by the available lifting device. This alternative requires fabricating and welding lifting lugs to the sections. Part the sections by burning or arcing almost, but not completely, through their thickness. Complete the cut with the abrasive cutting wheel. This procedure will prevent slag from obscuring the parts that are to be observed.

9.7. Inspect the longitudinal zone-enclosure-to-flat-plate welds

Suggested text: The divider plates are flat plates welded to the shell-side face of the tubesheet and extend to the ends of the enclosures. Examine for cracked or broken-away welds. Photograph and videotape any damage.

9.8. Inspect the subcooler end plate

Suggested text: Examine the penetrations of the tubes through the end plate for enlarged holes and broken or eroded tubes. Photograph and videotape any such damage.

9.9. Examine the open end of the desuperheater

Suggested text: Inspect to determine the actual clearance and condition of the holes in the last baffle. Photograph and videotape if the holes are enlarged.

9.10. Expose the desuperheater and subcooler divider plates and tubesheet inner face

Suggested text: The half-cylinder zone-enclosure plates are welded to the flat divider plates and the tubesheet. In the subcooler they are also welded to the end plate. To ascertain the condition of the tubes, cross-flow baffles in the desuperheater and subcooler, and the backface of the tubesheet, cut large windows into the half-cylinder enclosures as follows. On each side, burn full-length axial slots completely through the half-cylinders as close as possible to the flat plates. Carbon-arc a groove parallel with the tubesheet in each zone enclosure approximately 25 mm (~1 in) from the tubesheet. The groove should be as deep as possible without cutting through. Follow the same procedure at the end plate of the subcooler. Abrasive-wheel through the grooves to release the two partial cylinder windows. Then inspect the exposed tubes and flat plates. Scrape and collect any deposits on the tubes and tubesheet and accumulated on the flat-plate dividers for laboratory analysis. Photograph and videotape the exposed tubes. Examine the welds of the flat separator plates to the back face of the tubesheet. Look for gaps or separations between the edge of the flat plate and the tubesheet. If any are present, photograph and videotape the scene. If the plate has pulled a substantial distance away from the tubesheet, lay a steel tape on the plate, perpendicular to the tubesheet so that the photographs and videos indicate the size of the gap.

9.11. Examine the shell and skirt interior
Suggested text: Following all OSHA, governmental, and station safety requirements for entering closed spaces, enter the shell and inspect it for the full length. Look especially for corrosion and erosion-corrosion in the vicinity of the tube supports and all shell inlet connections. Record your observations by photography and videography. Examine the shell skirt in similar fashion.

9.12. Inspect the tube bundle
Suggested text: Progress with visual examination, photography, and videography from the tubesheet end to the U-tube end. Especially survey tubes that were plugged. Where possible, cut sections of plugged and unplugged tubes at various locations along the bundle for laboratory examination. Give special attention to the places where the tubes penetrate the tube supports and the degree of tube-support erosion. Examine the support edges for erosion and erosion-corrosion and the tube holes for enlargement that may have been caused by vibration, erosion, and corrosion. Then scrutinize the tubes midway between the tube supports for flat abrasions and, where they penetrate the supports, for fretting that indicates excessive vibration. Address the condition of the U-bends, using jointed inspection mirrors to see if the inner parts of the innermost bends were crimped. Look for shiny surfaces or flat, smooth surfaces on the crowns of the U-bends for indications of collisions and rubbing due to vibration. Also see if there are dents on the U-bends that indicate that the tubes were hammered into the bundle. Cut out cracked and corroded sections for laboratory examination.

What to Look for When Conducting Autopsies

Most tubes fail because of erosion, corrosion, SCC (stress-corrosion cracking), fatigue, or vibration. These phenomena may be interactive.

For example, vibration-created tensile stress may add to residual stress in the transition from unexpanded to expanded tubes, thereby increasing the likelihood of SCC.

Following is a guide to what to look for when conducting autopsies.

What to look for in the tube side

Channel barrels. Look for cracks in the pass-partition-to-tubesheet welds and tubesheet deformation. Figure 11.4, previously referred to, illustrates damage attributed to pressure surges or water hammer in the feedwater system.

Look for fractures at the transition of the channel barrel to the tubesheet. Cracks are most likely to be present in older heaters with small transition radii. Use a portable 30-power (or higher) microscope and fluid penetrant examination to reveal small cracks not obvious to the unassisted eye. Seeing corner cracks shows specifying engineers the importance of requiring manufacturers to provide a generous corner radius. To resolve differences of opinion about what constitutes "generous," we recommend the tubesheet-to-channel intersection radius be no less than 25 mm (~1 in) for the first 6900 kPa (~1000 psi) of design pressure, with an additional 13 mm (~½ in) for each additional 6900 kPa (~1000 psi) of design pressure thereafter with interpolation between increments.

We caution readers *not* to repair corner cracks at the transition of the channel barrel to the tubesheet of currently operating heaters without first analyzing the effects of continued operation in the cracked state and the effects of grinding out and welding the cracks.

Check the tube ends for washed-away tube-to-tubesheet welds and tube ends worn to knife-edge thickness. These indicate high localized feedwater inlet velocity. Blow out a random number of tubes with compressed air and catch what comes out the other end with a strainer to permit examination of any loose corrosion products. Copper-nickel tubes may have snake skins that blow out as discrete cylinders. Analyze the collected material for copper content. It is not unusual to find copper content of 90 percent or more. Such findings are a serious matter and should prompt investigating where copper may have deposited elsewhere in the system. Substantial copper deposits on turbine blades, which reduce turbine efficiency, may have to be removed by sand or grit blasting.

Examine the inside of the tubes either by cutting short specimens or probing with a borescope. This part of the autopsy can reveal various types of corrosion and the presence of various corrosion products. If operating logs show a slow decline in performance, cut sections of tubes in various locations and examine the specimens for internal fouling. Determining the nature of the deposits will provide information valuable for correcting similar problems in operating units.

Examine nozzle-to-channel-barrel welds. Cracks observed in these welds should raise a cautionary flag about currently operating heaters in service similar to the one on which the postmortem is conducted.

What to look for in the shell side

Following is a guide to making observations after the shell has been removed from the bundle.

Shell. Examine the shell interior after the shell has been cut loose from the shell skirt and pulled off the bundle. If there is substantial sediment on the bottom of a horizontal heater, scoop out samples for analysis by the metallurgical laboratory. The removed shells of all but the smallest heaters can be entered for inspection. Follow all the procedures required for safely entering a confined space during this examination. Use high-intensity low-voltage lights to investigate the inside surfaces for rings of erosion where the tube supports were situated and axial scoring by wet steam that may have exited from desuperheaters with insufficient wet-wall margin.

Desuperheating zone. Inspect the desuperheater before and after the shroud has been removed. Look for signs of vibration at the midspan of the longest unsupported tube spans and the baffle holes and the tubes where the tubes pass through the baffle plates (Fig. 13.4).

Look for indications of tube and baffle plate erosion near the outlet of the desuperheating zone (see Fig. 11.2). Such erosion results from steam conditions that permit initial condensation to take place where the bulk steam velocity is high enough to entrain condensate droplets.

Some tube materials, prone to corrosion when wet and exposed to air, do not usually corrode in the desuperheating zone where the tubes are surounded by dry superheated steam. Examples are carbon steel and 70-30 copper-nickel. Carefully inspect for SCC in materials such as austenitic stainless steels and copper-nickel and nickel-copper alloys near the back side of the tubesheet and sometimes near the steam inlet.

Cracked welds in desuperheating zone shrouds can lead to tube failure (Fig. 10.5). Eroded, broken-off, or inadequate impingement plates can expose the tubes to very high impact forces generated by the incoming steam. Figure 10.6, taken from a borescope examination of a high-pressure heater, shows how an impingement plate can be broken off and completely destroyed by incoming superheated steam.

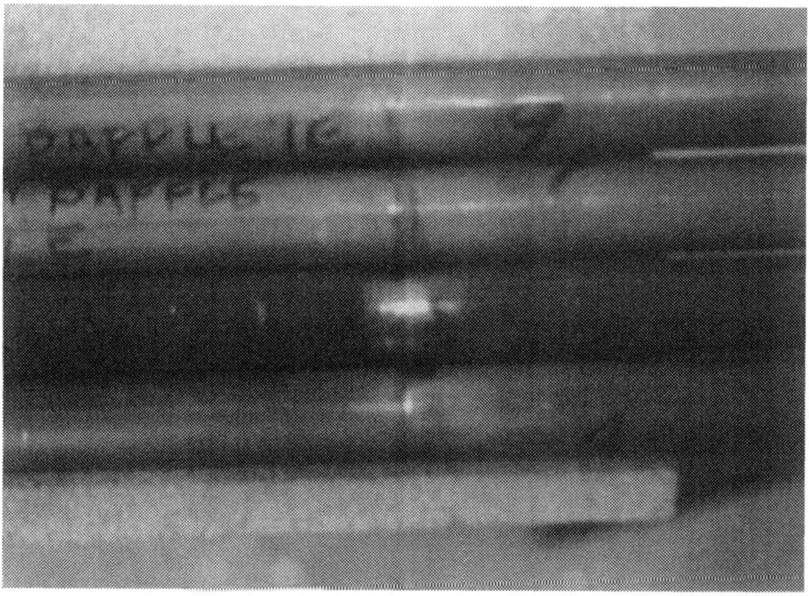

Figure 13.4 Shiny appearance of vibrating tubes passing through tube supports. (*MGT Inc. photograph.*)

Therefore, scrutinize these places. If desuperheating-zone shroud welds fail in vertical head-down heaters, water can collect in the bottom of the desuperheater during shutdown. Accumulated water not drained during start-up can be the source of two-phase flow that eventually erodes the tubes in the bottom of the desuperheater. Accordingly, if examination of the shroud welds before shroud removal showed failures, examine the tubes in this region.

When autopsying heaters in which the superheated steam inlet temperature approached 425°C (~800°F), survey the regions of the steam inlet nozzle and the shell skirt for parent metal, weld, and nearby tube failures. Examine the construction drawings for possible use of inappropriate materials and insufficient flexibility to accommodate the wide range of service temperatures.

Condensing zone. In condensing zones, look for evidence of tube vibration where tubes penetrate the support plate under the impingement plate(s) (Fig. 13.5). Explore the midspans of the U-bends. Manufacturers try to separate adjacent bends enough to avoid vibration-induced collision damage. However, when ligament spacing is insufficient or when the U-bends of successive rows are too closely spaced, the U-bends may collide. The physical aspect of such damage is obvious, as shown in Fig. 13.1.

Figure 13.5 Severe erosion of tubes and support plates in vicinity of perforated impingement plate. (*MGT Inc. photograph.*)

Also examine the tubes for erosion just beyond the impingement-plate edges. Such erosion is caused generally by impingement plates too small to completely block the path of the incoming steam from striking the tubes and by too narrow a space between the longitudinal edges of the impingement plate and the adjoining support plate. Some older heaters were designed with perforated impingement plates. The authors have observed severe erosion in tubes and support plates in the region of such plates (Fig. 13.5). Therefore, if perforated-plate construction was used, examine the impingement region and downstream support plates for erosion.

Examine condensing-zone tube supports and the surrounding shell closest to the exit from the desuperheater. Erosion may cause the edges of these supports to become very thin and the clearance between tubes and holes to become enlarged by wet steam rushing axially along the bundle (Fig. 13.6). The shell may also have been eroded where wet steam passed between the sections of the supports that have small support-to-shell clearance.

When cascading drain inlets are located within the bundle space—as, for example, in vertical heaters—examine the bundle for tube and support-plate erosion in the regions near the impingement-plate edges, especially near the top longitudinal edge.

Figure 13.6 Tube support erosion at exit of desuperheater in vertical head-down feedwater heater. (*MGT Inc., photograph.*)

Insufficient purging of air from heaters during start-up or on the shift from low to high load of units in peaking service, and failing to blanket with steam or nitrogen during shutdowns can lead to corrosion of carbon steel and 70-30 copper-nickel tubes. When the autopsy exposes them, look for a mottled appearance characteristic of exfoliation (Fig. 13.7). In this process, copper is extracted from the tubes and forms complex compounds on the tube surfaces. As the heater works, these compounds flake off and may be carried out with the condensate or accumulate somewhere in the heater.

Insufficient venting is a principal cause of SCC in innermost U-bend rows of tubes sensitive to SCC. Before the autopsy commences, it will probably not be known if air removal and blanketing were properly done over the heater's life. Because the noncondensibles tend to migrate to the coldest part of the heater, examine the tubes in the region of the coldest part of the bundle for corrosion and the inlet pass tubes nearest the thermal centerline for SCC.

Explore for tube and support-plate wear in the region of peripheral tube penetrations of support plates in the condensing zones of split-pass full-length heaters.

Subcooling zone. Almost all feedwater heaters are built with subcooling zones. Postmortem examination of subcoolers should be done before and after removing the subcooling-zone enclosure. Most subcoolers are constructed as full-pass short-length, but some horizontal

Figure 13.7 Exfoliation. (*Heat Transfer Consultants Inc., photograph.*)

heaters and all vertical channel-head-up ones are split-pass full-length. Scrutinize split-pass full-length heaters for vibration damage midspan between baffles and at the locations where the tubes pass through the baffle holes.

Cracked welds in drains cooler enclosures permit steam to rush into the subcooler, causing water hammer. The appearance of the resultant damage is very similar to that of erosion. When the enclosures have been removed, look for tube erosion in the first couple of rows above the inlet to horizontal units and, in vertical channel head-down designs, at the inlets to the subcooler. This erosion results from operating with low levels that permits steam to enter the subcooler with the drains. The two-phase stream is very erosive.

In heaters with full-pass short-length drains coolers, look for excessive wear on the top row of tube holes where they pass through the end plate and cracked welds of the drains cooler divider plate to the shell-side face of the tubesheet. If the tube plug maps and records are complete, they may reveal whether the failures in the inner rows of tubes at the top of the drains cooler were at the tubesheet or the end plate.

Seals of zone enclosures to shells. In multizone heaters, seal rings (also called *close-off plates*) are welded to the inside of the shell and outside of the desuperheater and subcooler enclosures. In heaters fitted with desuperheaters, they prevent steam from bypassing the desuperheater before it enters the condensing zone. In those with subcoolers, they prevent drains from bypassing the drains cooler before leaving the heater. Remove the shell parts so as to expose the seal rings. Examine them for plate failures and failed or washed-away welds of the seal rings to the shell and enclosures. Operating logs that show marked performance deterioration indicate that such failures have probably occurred.

Duplex (dual) heaters. Look for steam inlet erosion of the tubes in the subcooler of the low-pressure side of duplex (dual) heaters. Dual heaters consist of two units combined in a common shell with a divider plate sealed with a gasket to separate the two steam pressure stages. The extraction stage pressure to the lower pressure unit is usually about 0.5 to 1.0 kPa ($\sim$7.5 to 15 psi) below that of the higher-pressure one. A breach in the divider-plate gasket seal permits higher pressure extraction steam to rush into the lower-pressure side. This may cause difficulty controlling the condensate level. If the low-pressure side includes a drains cooler, the level-control problem may bring on steam inlet erosion.

Tubes and U-bends. The paragraphs on disassembly procedures suggest things to search for in the straight-tube lengths and U-bends.

Flat, shiny abrasions midway between the tube supports and fretting where tubes penetrate supports and baffles indicate excessive vibration (see Fig. 13.4). Shiny surfaces and flat, smooth surfaces on the crowns of the U-bends indicate collisions and rubbing due to vibration. Dents on the U-bend ends indicate that the tubes were hammered into the bundle as shown in Fig. 13.1.

Recording Observations

The best way to record the observations made during postmortem examinations are photographs and appropriately edited videos. Number, date, and label each photograph with a label that indicates the part it shows. The videocamera should have a microphone to record the statements of the observers. Use its time and date features at the start of each videotaping sequence. Relatively inexpensive systems are available for editing out extraneous material and adding text and sound. When borescopy is performed, text and sound can be added to explain what the video output disclosed.

Carefully label specimens of tubing taken for analysis to identify the tube from which the specimen was cut and its location in the bundle.

Reports

The authors have found a dual report to be most effective for communicating the information and conclusions derived from performing the autopsy. Part 1 incorporates prints of the photographs in a written text report which refers to part 2, a report made from edited videotapes. Text and video should use simple, clear language suitable for training. To get the greatest benefit from the expenditure of time and money, we recommend review and discussion by the complete autopsy team. The team should summarize their conclusions in an executive summary to assist management in understanding their recommendations for improvements to current specification, procurement, and operating practices.

Reference

1. S. Yokell and C. F. Andreone, "Feedwater Heaters Should Last Forty Years," paper presented at the October 1992 Joint Power Generation Conference, Atlanta, Ga.; published in *Practical Aspects and Performance of Heat Exchanger Components and Materials, PWR,* vol. 19, 1992, pp. 97–107.

Glossary

ASME The American Society of Mechanical Engineers.

ASME Boiler and Pressure Vessel Code The boiler and pressure-vessel safety code of the ASME (ASME Code). The ASME Code has sections with rules for specific types of boiler and pressure-vessel construction and sections with general rules applied by reference in the construction rules. The construction sections that apply to shell-and-tube heat exchangers are : Section VIII, Division 1, "Rules for Construction of Pressure Vessels," for most units; Section VIII, Division 2, Alternative "Rules for Construction of Pressure Vessels," for exchangers that undergo severe cycling, have unusual configurations, or have design pressure between 3000 psi and 10,000 psi; Section VIII Division 3, Alternate Rules for High-Pressure Vessels, "Rules for Construction of Pressure Vessels," for equipment that operates at 10,000 psi and above; Section I, "Rules for Construction of Power Boilers," or Section VIII, Division 1 for unfired steam generators; and Section III, "Riles for Construction of Nuclear Pressure Vessels," for units in nuclear service in the primary loop or essential for the safe shutdown of the unit. The division of Section III to which nuclear units must conform depends on their positions in the loop. In addition, several mandatory and nonmandatory appendixes apply.

Sections that the construction rules apply by reference are Section II, "Material Specifications"; Section V, "Nondestructive Examination"; and Section IX, "Welding and Brazing Qualifications." The *National Board Inspection Code* (NBIC) and the *API Inspection Code* apply the ASME Code's rules by reference.

Editions of the ASME Code are published at 3-year intervals. Before the 1986 edition, Addenda were published twice a year. Since 1986, Addenda have been published annually. Code Cases, which deal with rules for materials, design, and construction that are not incorporated within the main body of rules and addenda, are issued for limited periods and discontinued, reaffirmed, or incorporated into the rules.

axial nozzle An inlet or outlet connection in the head of a heat exchanger that is aligned with the tube axis.

baffle One of a series of drilled or perforated plates, slightly smaller than the shell I.D., located between the front and rear tubesheets or stationary tubesheet and return end through which the tubes pass. The functions of baffles are to direct the flow of the fluid in the shell and to support the tubes. See **single segmental baffles, double segmental baffles, triple segmental baffles** and **disk and doughnut baffles.**

baffle bypassing Leakage between the baffle O.D. and the shell I.D.

baffle clearance The space between the design I.D. of the shell (or in closed feedwater heaters, the zone-enclosure I.D.) and the baffle or support plate O.D. The TEMA Standards have tables of recommended clearances between shell design I.D. and baffle O.D. for various shell sizes.

baffle cut The fraction of a segmental baffle or support plate that is cut off to allow passage of the shell-side fluid, expressed as a percent of the shell or closed feedwater heater-zone I.D. Cuts may be single segmental, double segmental, or triple segmental. See also **percent cut.**

baffle cut overlap The number of tube rows in horizontally cut baffles and the number of tube tiers in vertically cut baffles common to adjacent baffles or tube supports. As a minimum, there should be one row or tier of tubes common to adjacent supports of baffles.

baffle ear A horizontally projecting uncut portion of a baffle at the end of the cut line of a vertically cut segmental baffle. Baffles are provided with ears to (1) receive tie-rods, (2) prevent vapor bypassing in horizontal condensers when tubes have been omitted from the layout to accommodate impingement protection, and (3) provide a dam to maintain a liquid level in the subcooling zone of a horizontal process condenser-subcooler that condenses on the shell side. See also **dam baffle.**

baffle edge distance The distance between the outermost tube-hole O.D. and the baffle O.D.

baffle hole clearance The baffle or support hole diameter minus the tube O.D.

baffle notch A drainage notch in the bottom of a segmental baffle that has a horizontally cutoff upper sector. Baffle notches are not required for nonviscous fluids because the drainage space between the baffles and shell is adequate.

baffle pitch The center-to-center distance between adjacent baffles or tube supports.

baffle spacers Spools of pipe or tubing between adjacent baffles to maintain baffle spacing and alignment. The spacers are slid over tie-rods that run from the rear face of the front or stationary tubesheet to a point just beyond the last baffle.

baffle spacing The distance between adjacent segmental baffles. The baffle spacing is equal to the baffle pitch minus one-half the baffle thicknesses of the adjacent baffles.

baffle span See **support span.**

bare tubes Heat exchanger tubes with smooth inside and outside surfaces.

beaded end An expanded tube-to-tubesheet connection in which tube ends that are extended beyond the outer face of the tubesheet are beaded over. Beading enhances the strength and tightness of the joint. It may be combined with welding the tubes to the tubesheet.

bellows expansion joint A thin-walled expansion joint with a flexible element no thicker than 3.18 mm (~⅛ in). The element consists of one or more corrugations shaped like a U, V, omega, toroidal, or similar configuration. Corrugation roots may be fitted with root rings to prevent blowout under pressure and to distribute deflections equally between multiple corrugations. Bellows expansion joints are characterized by relatively low spring rates and high cycle lives. The deformations when the joint deflects may be elastic but are usually inelastic.

bend interference Metal-to-metal contact between U-bends in the same plane. Construction in which there is bend interference is unacceptable.

bend line The point of tangency of the straight leg centerline to the bend radius of a U-tube.

bend radius The radius of the centerline of a U-tube bend. The bend radius is half the bend spacing. See **bend spacing.**

bend spacing The center-to-center distance between the legs of a U-tube.

bend tangent See **bend line.**

bend tangent line See **bend line.**

bent tube A straight tube with a small bow created by forming the tube to an arc height approximately 2.5 percent of the tube length. Bent tubes are used in water evaporators to remove scale buildup by shocking the hot tubes with cold water. They may also be used to accommodate differential thermal expansion between tubes within a bundle and between the bundle of tubes and the shell. See **shock tubes.**

bimetallic tube A tube made by inserting a tube of one metal into a second tube of another metal with a close fit, then drawing the pair together to create an interference fit between the inner and outer tubes. A short length of outer tube is pared away at the tube end and replaced by a ferrule of inner tube material in order to present only inner tube metal to the channel side of the tubesheet. Bimetallic tubes with a metallurgical bond between the inner and outer tubes have been produced experimentally by explosively cladding and reduction drawing a hollow, but are not commercially available.

blowdown (1) A connection in an evaporator for removing accumulations of solids, sludge, and scum by partial draining during operation; (2) continuously bleeding liquid from the bottom of a boiler, evaporator, vaporizer, or kettle-type reboiler.

bolster A structural or fabricated plate spacer set between stacked heat exchangers or heat exchanger shells to maintain a desired separation. Spool pieces are installed between connecting pipe nozzles.

bolt spacing The bolt-circle circumference of a flange divided by the number of equally spaced bolts.

bullets See **tube guides.**

bull plug A round or hexagonal bar-stock or forged-pipe plug, 65 to 75 mm (2½ to 3 in) long.

bundle See **tube bundle.**

bundle entrance area Unrestricted flow area for shell-side fluid to enter a bundle from the inlet connection. See **shell entrance area.**

bundle entrance velocity The average linear velocity of the fluid entering the space between the tubes and the shell.

bundle exit area The unrestricted area for shell-side fluid to leave a bundle to the outlet connection. See **shell exit area.**

bundle exit velocity The average linear velocity of the fluid leaving the space between the tubes and the shell (or zone enclosure).

bundle-in-column reboiler A reboiler in which the tube bundle is inserted into a horizontal nozzle at the bottom of a vertical column. The bundle may be U-tube or floating-head.

bundle tracks See **rails.**

capacitance The volume of water in a closed feedwater heater required for successful drain-control-valve operation. Control-valve operation should be available in the heater design in the operating range that lies between the normal liquid level and the low liquid level. Drain-control valves operate successfully when they respond to load changes without causing the liquid level to drop so low that the drains cooling zone inlet is exposed to erosive steam-water mixtures.

cascading drains Any water that enters a closed feedwater heater shell from higher-pressure sources.

channel-down The position of a channel when it is at the bottom of a (1) closed feedwater heater or (2) vertical multipass process heat exchanger that has an even number of tube passes.

channel-up The position of a channel when it is at the top of a vertical (1) closed feedwater heater or (2) multipass process heat exchanger that has an even number of tube passes.

circumferentially finned tubes Finned tubes that have the fins perpendicular to the tube axis, or on a helical pitch. See **extended-surface and low-fin tubes.**

cleaning lane Space between adjacent rows of tubes in removable-bundle heat exchangers large enough to permit cleaning devices to enter. Cleaning lanes are provided by arranging the tubes on a square pitch or on a spread triangular pitch.

cleaning turbine An air-, hydraulic-, or electricity-powered tool for cleaning inside heat exchanger and boiler tubes. It consists of a drive motor, flexible drive cable or hose, and head. The heads may be arrangements of blades on hubs, modified drill bits, burrs, or brushes. Cleaning-turbine heads may be constructed to allow water to flow through passageways to wash away deposits removed by turbining. Cleaning turbines are sometimes adapted for cleaning tube exteriors when there are cleaning lanes.

cleanout An access opening for cleaning out scale and sludge from the bottom of an evaporator, kettle reboiler, or vaporizer.

closed feedwater heater A shell-and-tube heat exchanger used for heating boiler feedwater or reheating condensate. The feedwater or condensate flows in the tubes. Turbine extraction steam enters and its condensate leaves the shell.

code See ***ASME Boiler and Pressure Vessel Code.***

condensate level See **liquid level.**

condensing zone The portion of a closed feedwater heater tube bundle in which boiler feedwater is heated by condensing saturated turbine extraction steam.

confined joint A flanged joint in which the gasket is retained in a recess (female flange) that receives the raised face of a mating male flange.

connection protector A device used to protect machined surfaces or threads of heat exchanger connections during shipping and handling. Flanged connections may be protected with gasketed metal, wood, or composition disks bolted to the flange. Screwed connections may be protected with pipe caps, plugs, or plastic inserts or caps.

continuous vents Connections and collection devices in the shell of a closed feedwater heater for continuously collecting and removing noncondensibles from extraction steam to prevent noncondensibles from accumulating, thereby causing capacity loss and corrosion. Continuous vents should be fitted with orifices that limit steam flow to 0.5 percent of the inlet steam quantity. They should be bypassed during start-up to allow for rapid purging of inerts.

conventional double tubesheets Double tubesheets in which each tubesheet of a pair of double tubesheets is an individual plate or forging. See **double tubesheet.**

cores Devices inserted in tubes for their full length to constrict the cross-sectional flow area by forming an annular flow channel.

crisscross U-tubes An arrangement of the innermost rows of a U-tube bundle in which the plane of the bends is rotated alternately clockwise and counterclockwise. The bends rotated in one direction pass behind the bends rotated in the other.

cross-flow baffles See **baffle.**

crossover area The flow area in a pass compartment measured in the plane perpendicular to the tubesheet. The minimum recommended crossover area for TEMA R units (the TEMA classification for exchangers in severe service; "severe service" covers erosive and high-velocity flows) is 1.3 times the flow area of the tubes in one tube pass. For TEMA C and B units (TEMA classifications for less severe commercial and chemical services), the flow area may be smaller but never less than the flow area of the tubes in the pass.

cut line (1) The line on which segments of segmental cross-flow baffles or semisupport plates are cut; (2) the circumferential line at which the shell skirt is severed from the shell of a closed feedwater heater to permit the shell or bundle to be removed.

dam baffle (1) A vertically cut segmental baffle with a bottom ear—the ear dams the condensate flow to provide a liquid level in the bottom of a horizontal process condenser-subcooler; (2) a vertical band approximately 75 mm (~3 in) high that is welded to the support plates in the condensing zone of vertical feedwater heaters. Its purpose is to prevent condensate collected on the support from raining down past the steam inlet when a heater is constructed without a desuperheating zone or into the desuperheating zone when it is constructed with the channel head down.

davit An inverted L-shaped structure used for handling channels, bonnets, and covers. The davit may be fabricated from structural elements, or it may be a round bar or pipe bent on a radius to provide horizontal and vertical legs. The vertical member or leg fits into a socket in a bracket welded to the shell. An eyebolt, set vertically into the end of the horizontal leg, is directly connected or hooked to the part to be handled. Alternatively, the eyebolt or hook may be connected to a carriage fitting that rolls between stops along the horizontal member.

descaling tube See **bent tube.**

depth of groove In confined joint construction, the depth of the groove milled into tubesheets and/or channel covers to receive gasket ribs and pass partitions.

design maximum working pressure The pressure for which a vessel is structurally designed. See **maximum allowable working pressure.**

desuperheating zone That part of a closed feedwater heater's outlet tube pass that is reserved for transferring sensible heat to feedwater from superheated extraction steam.

desuperheating zone shroud An enclosure surrounding the tubes of the outlet pass at the outlet end of a closed feedwater heater that is designed to be supplied with superheated steam. The shroud is constructed so that superheated steam enters near the inlet end, then flows through a baffle system suitable for sensible-heat transfer. Enough superheat must remain in the exiting steam to keep the tubes dry at the zone exit.

diaphragm A thin disk of metal continuously welded to the manway or channel-flange face when there is no pass partion gasket rib. The diaphragm, which takes the place of a gasket, is backed up against positive tube-side pressure by the channel cover. Diaphragms are used when even small leaks cannot be tolerated.

disk and doughnut baffles A baffle system in which the center is removed from a full circle and located one baffle pitch away from the remaining annulus. The center is the disk; the annulus is the doughnut. The baffle span is two pitches long.

dismantling clearance Clearance required at the (1) stationary end of a removable-bundle process heat exchanger or feedwater heater installed in a condenser neck to permit bundle removal, (2) front end of a fixed-tubesheet heat exchanger to permit tube removal, or (3) return end of a closed feedwater heater to permit shell removal.

distributor belt An enlarged section of a shell into which the shell inlet connection enters. The bundle is wrapped with a 360° impingement baffle under the enlarged section. The impingement baffle may have several large holes offset from the inlet nozzle to provide access to the bundle for the incoming fluid, or it may be perforated by smaller holes except under the nozzle. An alternative is to allow space for flow into the bundle between the end of the impingement protection cylinder and one end of the enlarged section.

double grooves Two annular grooves machined into the holes of a tubesheet, provided to enhance strength and tightness of expanded tube-to-tubesheet joints. The standard TEMA and HEI configuration is two grooves $\frac{1}{64}$ in deep × $\frac{1}{8}$ in wide (~0.40 mm deep × 3.18 mm wide). Only one groove, more than two grooves, and configurations other than the TEMA ones have been used. Wider grooves are desirable when tubes are to be hydroexpanded or explosive-expanded.

double segmental baffles A cross-flow baffle system that consists of alternating A- and B-section segmental baffles. A sections are opposite sectors of a circle; B segments are sections of a circle that remain after opposite sectors have been cut off.

double tubesheet Two tubesheets at the tube ends, of which the outer face of the outer tubesheet is exposed only to the tube-side fluid and the inner face of the inner tubesheet is exposed only to the shell-side fluid. Conventional double tubesheets are independent plates or forgings separated by a discrete air gap. The gap may be sealed by a peripheral shroud, or the outer edges of the tubesheets may be welded to each other or to a spacer cylinder. Integral double tubesheets consist of a single plate or forging into which annular grooves, approximately 3 to 6 mm (~$\frac{1}{8}$ to $\frac{1}{4}$ in) wide, are cored into the tube holes until adjacent grooves intersect. This creates two separate tubesheets, joined at the peripheral, undrilled edges and by metal not removed in coring out the annular grooves.

drains subcooling zone See **subcooling zone.**

dry pipe A perforated horizontal pipe located above the upper tube row of a horizontal steam generator. A vertical length of pipe connects the dry pipe to the steam outlet nozzle. The increase of velocity of the wet steam flowing through the perforations occurs at the expense of pressure. The reduced pressure causes some of the moisture to vaporize. The two changes of direction of the steam passing through an adequate-sized dry pipe, in which the steam velocity is below the entrainment velocity, allow the water droplets to drop to the bottom of the pipe. The droplets collect in the bottom of the dry pipe and are piped back to the boiling pool from drain connections at each end.

drying surface Tube area at the top of an evaporator or vaporizer used to provide a small amount of superheat to the vapor to ensure that it is dry.

drying tubes Tubes at the top of a horizontal evaporator or vaporizer bundle used to provide drying surface.

dual bank An arrangement of tubes, partitions, and (1) nozzles in a surface condenser in which each of the two halves of the tube count may be operated independently; (2) and connections in a refrigeration-system condenser in which each of the two halves of the tube count may be operated independently.

dual-gage tubes Tubes in which the wall of the region to be bent into U-tubes is two gages thicker than in the rest of the tube. The extra thickness is added to the O.D.; therefore, there is no change in cross-sectional flow area.

Bending a tube stretches the metal beyond the neutral line (bend radius centerline) and compresses it at the inner part of the bend. Some flattening occurs, causing the bend region cross section to become ovoid. The extra thickness in the section to be bent allows the minimum thickness in the bent portion to be at least equal to the minimum thickness in the rest of the tube. Because the severity of thinning and deformation depends, among other things, on the bend radius, it is customary either to require dual-gage tubes or tubes that are two gages thicker than the rest for the inner two rows of the tubes of a U-tube bundle. When thicker-walled tubes are provided, the reduction in flow cross section is ordinarily neglected. Dual-gage tubes are usually available only in nonferrous material.

dummy tubes Tubes not used for transporting fluids, but that occupy space in a tube bundle. Dummy tubes are installed to prevent fluid bypassing around or through a bundle. They are usually welded to the inside face of the front or stationary tubesheet and end a short distance in front of the rear tubesheet or U-bend. See **sealing device.**

duplex feedwater heater Two feedwater heater stages contained in a common shell installed in the neck of a steam surface condenser. The stages receive steam from the two lowest extraction points of the low-pressure turbine.

eddy-current examination A means of examining tubes for flaws and approximate wall thickness by measuring the variations in electrical characteristics induced in the tube by alternating current coils or probes. Variations and discontinuities in the tube cause changes to the eddy-current strength at the location of the variation or flaw. The eddy current and its variations are detected by an electronic device and displayed on a cathode-ray terminal.

egg-crate support A tube support in which a grid or grating of interlocked flat bars arranged to provide square or triangular openings is contained within a circular band. The tubes pass through the openings in the grid.

enhanced-surface tubes Tubes in which the (1) walls have been deformed or shaped on a pattern that increases turbulence in the fluid stream or (2) surfaces have been treated to permit high flux rates in boiling.

equalizer connections In a closed feedwater heater, connections provided for installing pipe lines to equalize the pressure between the top and bottom of liquid-level controllers and gage glasses. The function of equalizer lines is to avoid false readings due to siphoning.

expanded tube joints Heat exchanger tube-to-tubesheet connections made by expanding the tubes into the tube holes. The tubes are deformed elastically or plastically until they make contact with the hole. Expanding force application is continued, enlarging the tube and the hole until the tube metal approaches its plastic limit. The holes may deform, elastically or plastically, but when expanding force is released, the hole must recover more than the tube to create an interference fit between the tube and the hole.

The magnitude of the interference and the fit pressure it generates depends on the relative yield strengths of the tubesheet and tubes and other metal properties, the tube diameter-thickness ratio, and the maximum expanding force applied. The methods used to expand tubes into tubesheets are roller expanding, hydraulic expanding (hydroexpanding), and near-contact explosive expanding (high-energy expanding). Expanded-only joints comprise a large majority of process and power-plant heat-exchanger tube-to-tubesheet joints. They are frequently used on low-pressure closed feedwater heaters but are not suitable for high-temperature service or for services where a very high degree of tightness is required. Tubes may also be expanded when the primary tube-to-tubesheet connection is made by welding the tubes to the front face of the tubesheet. This is done to seal the crevice between the tube and hole, isolate the welded joint from vibration, and stiffen the tubesheet if contact between the tube and hole can be maintained under all conditions of loading.

extended-surface tubes Tubes in which the outer- or inner-wall bare surface has been extended by adding metal projections. The added metal may be in the forms of (1) continuous or discontinuous longitudinal high fins, (2) transverse helical or annular high fins, (3) continuous thread-shaped low fins, or (4) pegs or studs. High fins extend beyond the bare tube surface; low fins do not. Longitudinal and transverse high fins may be attached by shrink-fitting, fitted into scored grooves and held there by peening the upset groove metal against the fin, or by resistance welding. Longitudinal high fins that are welded to the tube are first formed into a channel shape to permit the web of the channel to be welded to the tube. Low fins are formed from bare tubes without adding metal. The fin O.D. is the same as or slightly less than the bare tube O.D. Therefore, the tubesheet and baffle drilling is the same for low-fin and bare tubes.

Ferrule (1) A short cylinder of inner-tube material of a bimetallic tube that fits the tube end where the outer tube material has been peeled away (the end of the tube presents only inner-tube material to the tube-side tubesheet face when the tube is expanded into the tubesheet); (2) a threaded sleeve fitted over the end of the tube of a packed tube joint (the ferrule acts as a packing gland follower to compress the packing in the tube hole); (3) a sleeve of tube material inserted between a tube and an oversized hole to enable a tight expanded joint to be made; (4) the thin sleeve of material used in bridging the damaged tube area in sleeving.

fixed-tubesheet exchanger A straight-tube exchanger in which a tubesheet is fastened to each end of the shell.

flange-edge distance The distance between the bolt circle diameter (B.C.) and the O.D. of a bolting flange [$\frac{1}{2}$(O.D.−B.C.)].

flanged-and-flued expansion joint A thick-walled expansion joint that is fabricated from a pair of flat flanged heads with centrally located flued-out holes. The I.D. of the flue holes is equal to or slightly larger than that of the shell. The straight flanges of the flat heads are butt-welded to form a single corrugation joint.

flanged-only expansion joint A thick-walled expansion joint fabricated from a pair of flat flanged heads with centrally located holes. The flat heads are butt-welded to each other and welded to the shell at the hole locations.

flared end joint A tube-to-tubesheet connection in which tightness and strength are enhanced by flaring extended tube ends. The outer edge of the holes may be relieved in preparation for tube flaring.

flared nozzle A nozzle in which the entrance line diameter is increased to a larger diameter where the connection pierces the shell. Flared nozzles are used to (1) reduce pressure loss in vapors entering condenser shells and (2) enable installing impingement plates without eliminating tubes at the shell inlet.

flash chamber An auxiliary pressure vessel set next to a feedwater heater or a built-in length in the end section of a horizontal feedwater heater shell extending beyond the U-tube ends in which steam flashed from cascading drains is separated from the condensate before it is introduced into the bundle.

flat face A flange facing in which the gasket sealing surface is in the same plane as the rest of the flange face.

floating head A floating tubesheet and its attached floating cover.

floating-head cover The return or outlet cover attached to a floating tubesheet.

floating head with backing device A floating head in which reaction for bolts or clamps used to hold the cover to the tubesheet is provided by a backing ring behind the tubesheet. The most frequently used backing device designs are (1) a split backing ring that fits behind the tubesheet and that has an I.D. smaller than the tubesheet O.D. and (2) a solid backing ring with an I.D. slightly larger than the tubesheet O.D. and that bears against a split shear ring inserted into a groove in the tubesheet rim. The O.D. of the backing device is larger than the shell diameter. Therefore, the shell cover must be larger than the shell to permit it to fit over the backing device. The shell flange at the cover end and the shell cover flange have matching O.D.s and B.C.s, but the shell flange I.D. is smaller than the shell cover flange I.D.

full-circle supports Tube supports that are full circles drilled to receive all the tubes, tie-rods, etc.

fusion weld cladding An integral cladding method in which a base metal is faced by fusion welding metal to it. It is commonly used on the channel side of forged-steel high-pressure heat exchangers, such as high-pressure closed feedwater heaters, to provide a homogeneous base for tube-to-tubesheet welding.

gasket A member placed between mating surfaces to effect a seal by flowing into the irregularities in the surface when the member is compressed.

gland A cylinder that fits into a packing box and bears against the outermost ring of packing to compress it. Most glands are made with an integral bolting flange. Studs, threaded into the packing box body, pass through boltholes in the flange of the gland. Nuts are mounted on the studs to drive the gland against the packing.

grid supports See **egg-crate support.**

hole count The number of tube holes drilled into a tubesheet. In U-tube heat exchangers, the number of holes is twice the number of tubes.

honeycomb See **integral double tubesheet.**

horizontal cut Segments of cross-flow baffles cut off alternately at the top of one baffle and at the bottom of the adjacent baffle to direct stream flow up and down over the tubes. See **vertical cut.**

hotwell Space included in the shell below the tube bundle to provide liquid capacitance as required by the liquid level controller to maintain a liquid seal in the exchanger in order to prevent the vapor from blowing out of the condensing section. The space can be provided as a pot for small capacitance requirements or the shell may be enlarged to satisfy larger requirements.

hydraulic lance A hydraulic tube cleaning tool consisting of a hollow rod, somewhat longer than the tube length, fitted at its front end with a nozzle that enters the tube. High-pressure water is pumped through the lance to clean deposits off the inner-tube surface as the lance is moved through the tube. The nozzle orifices are designed to suit the nature of the deposit. When hydraulic lances are manually operated, the operator walks with the lance to feed it into the tube. This has given rise to the name "walk lance." Automatically fed lances are safer and more effective.

hydrocleaning Using high-pressure water to remove scales, gums, and other substances that foul tube surfaces.

hydroexpanding Expanding tubes into tubesheet by applying hydraulic pressure directly inside the tubes. The tube wall is not crushed, and the small amount of wall reduction that takes place is due only to the increase in tube diameter. Instead of extruding as in roller expanding, the tube end becomes shorter.

HydroPro The trademark of Torque and Tension Equipment, Inc. for a system and equipment for hydroexpanding.

HydroSwage The trademark of Haskel Inc. for a system and equipment for hydroexpanding.

impact plate A metal plate located between the shell inlet nozzle and the tube bundle to prevent fluid from the inlet nozzle from impinging directly on the tubes.

impingement baffle See **impact plate.**

impingement plate See **impact plate.**

inert-gas blanketing Purging a unit of air with inert gas, then filling it with inert gas as it is shut down. Inert-gas blanketing is used to (1) prolong the life of closed feedwater heaters that operate on peaking (cycling) service, (2) exclude air when shutting down process heat exchangers through which highly flammable or explosive substances flow, and (3) prevent stored units from corroding.

inside surface Tube heat-transfer area based on the I.D. of the tubes. Inside surface is used in heat-transfer calculations for extended-surface tubes.

insulated pass partition A shell-side pass partition made of two layers of metal separated by an air gap, and sealed at all four edges. Insulated pass partitions are used to reduce thermal leakage between the passes of a two-pass shell heat exchanger.

insurance plugging The custom of plugging tubes that have not leaked that surround tubes that have. It is intended to avoid risking a subsequent leak in tubes that have not leaked, but which may have been eroded by impinging fluid from tubes that are known to have leaked. Insurance plugging is costly in terms of its effect on the unit's capacity. Therefore, when there is sufficient outage time, it is better to eddy-current-scan the tubes that surround leakers to find out if they have been damaged. When immediate outage time is not available, it is good practice to use removable plugs to permit eddy-current or internal rotary inspection scanning the surrounding tubes at the next planned outage.

integral double tubesheet A double tubesheet made by machining annular intersecting grooves into the tube holes at the midpoint of the thickness of a single tube plate. See **double tubesheet.**

intermediate nozzles Mating shell and channel nozzles of multiple shells arranged in series. See **stacked shells.**

Internal Rotary Inspection Scan (IRIS) A means of inspecting the condition and wall thickness of tubes and pipes developed by Shell Development Corporation and licensed to various inspection services contractors. It consists of a rotating probe with a mirror that carries its own source of water. High-frequency sound is reflected radially from the mirror and reflected back from the inner and outer tube walls. The reflections are converted to an output on a cathode-ray tube that shows the precise location of the inner and outer tube walls. This permits accurately measuring the tube wall thickness precisely for its full straight length, determining wall loss and through-wall penetrations. Unlike eddy-current examinations, the output signal is not affected by the presence or absence of tube supports or baffles.

IRIS See **Internal Rotary Inspection Scan.**

lantern ring A metal cylinder equal in I.D. and O.D. to the packing in a packed floating head. The lantern ring is placed between two sets of packing rings. It is drilled with radial holes to permit leakage through either set of packing rings to drain to the atmosphere.

lap-joint flange A flange that fits loosely around a cylinder and bears against a gasket sealing ring integral with the cylinder end. Lap-joint flanges may be used with channel barrels, shell or cover cylinders, or nozzle-neck lap-joint stub ends. See **stub end.**

lifting lug A welded attachment to facilitate lifting a channel, channel cover, or shell. The lifting lug is fitted with a hole sized to receive the pin of a shackle.

ligament The metal in a tubesheet between adjacent tube holes.

ligament width The width of a ligament, equal to tube pitch distance minus the tube-hole diameter.

liquid level The level of liquid in the shell. The level must be maintained high enough to provide a liquid seal to prevent the vapor from blowing out of the condensing section. In feedwater heaters the level should not be so high as to cause water to back up into the steam extraction line. In horizontal vaporizers the liquid level should be high enough to cover all of the tubes.

liquid-level equalizer line A pipeline, external to a closed feedwater heater, that connects the bottom of the shell-level connection in the condensate region to the top of the shell-level connection in the condensing zone. Its purpose is to preclude condensate siphoning.

long baffle See **longitudinal baffle.**

longitudinal baffle A shell-side pass partition. Longitudinal baffles may be sealed to the shell of a nonremovable bundle exchanger by (1) continuous or discontinuous internal welds, (2) plug or slot welds that pierce the shell, or (3) making the shell in two halves with the longitudinal baffle welded between. In removable-bundle equipment, the baffle is made part of the bundle skeleton. It may be sealed to the shell by packing or metal-to-metal sealing strips sprung against the shell wall.

longitudinal baffle seal The means used to prevent fluid bypassing between the shell and edges of shell pass partitions.

longitudinal fins Fins arranged parallel with the longitudinal axis of extended-surface tubes.

loop seal An external, inverted U-shaped pipe loop in the condensate outlet of a condenser-subcooler, for the purpose of maintaining a liquid level in the subcooling zone. The top of the loop is vented or connected to the condensing zone by a tie-line to prevent siphoning.

low-fin tubes Circumferentially finned tubes in which the fin diameter is equal to or slightly less than the bare tube diameter. Low-fin tubes must be bare at their ends for a distance somewhat greater than the tubesheet thickness. Therefore, the mill must furnish them cut to length. Low-fin tubes may also be obtained bare where they are to be bent into U-tubes. They are mechanically interchangeable with bare tubes of the same diameter.

lug (1) A mechanical projection forming part of a breech-lock type of channel closure; (2) one of two or more supports welded to the shell of a vertically mounted unit.

lug supports See **support brackets.**

maximum allowable working pressure (MAWP) As defined in the ASME Code, the maximum pressure permissible at the top of the shell and at the top of the channel in their normal operating positions at the operating temperatures specified for the pressures. The MAWP is the pressure stamped on the nameplate for each side of the unit (in pounds per square inch) and is the pressure setting of at least one pressure-relief device on each side.

To provide a suitable margin between the operating pressure and the primary safety-relief device setting, it is good practice to set the MAWP higher than the operating pressure. Each industry has its own standards. Typical in the process industries is to set the MAWP to at least 110 percent of the operating pressure, but not less than the operating pressure + 103.4 kPa (~15 psi) for pressures up to 6895 kPa (~1000 psi) and at least 110 percent of the operating pressure above 6895 kPa (~1000 psi). For closed feedwater heaters, recommended MAWP settings are at least 115 percent of operation pressure, but not less than operating pressure + 103.4 kPa (~15 psi) for pressures up to 6895 kPa (~1000 psi) and at least 115 percent of the operating pressure above 6895 kPa (~1000 psi). For closed feedwater heaters, the shell-side MAWP is typically set to at least 115 percent of the extraction steam pressure from the turbine, based on the turbine's operating at 5 percent over pressure with valves wide open, but not less than 345 kPa (~50 psi) and full vacuum. The tube-side MAWP is typically set at 115 percent of the condensate- or feed-pump shutoff pressure.

minimum bend radius The minimum centerline radius to which a U-tube may be bent without flattening more than 10 percent or becoming crimped. The minimum bend radius that is attainable is a function of the tube-wall thickness, tube diameter, and metal properties. Typical minimum bend radii in the process industries are 1.5 to 2 tube diameters. The minimum recommended in the HEI CFHS is 1.5 tube diameters.

mixed layout A tubesheet layout in which flow is partially ribbon and partially quadrant. See **ribbon layout, quadrant layout,** and **pie layout.**

multinozzle heater A low-pressure closed feedwater heater, usually located in the neck of the steam surface condenser, in which steam enters the shell through nozzles located on either side of the thermal centerline. See **thermal centerline.**

multisegmental cuts Cuts of segmental cross-flow baffles that divide a full circle into three or more parallel segments. The baffles produced by double segmental cuts consist of a pair of equal-width segments cut off on parallel chords separated by one baffle pitch from a central portion that remains after two equal-width segments have been cut off a full circle on parallel chords. Outer, two-piece and inner, one-piece baffles alternate, with their projections overlapping. Triple segmental baffles consist of regularly alternating outer, intermediate, and inner pieces, with overlapping edges.

multitube exchanger An exchanger in which the shell consists of two legs joined by a fixed or removable-end cover. U-bent tubes are inserted with one leg of each U-bend in each shell leg before the shell cover is closed. The tube ends may be joined to a separate tubesheet with bonnet at each shell-leg stationary end, or one rectangular tubesheet and an elongated two-pass bonnet may be used. If the fixed-end cover is made from sections of bent pipe, welded to the legs, multitube exchangers can be considered to be U-tube, U-shell exchangers. Multitube exchangers may be arranged for true countercurrent or cocurrent flows.

nondestructive examination A means of examining the condition of a part by a method that does not damage the part. ASME Code, Section V, "Nondestructive Examination" defines the following methods of nondestructive examination: RT, radiography; UT, ultrasonics; MT, magnetic particle; PT, liquid penetrants; VT, visual; LT, leak testing; ET; eddy current; and AE, acoustic emission. Section V of the Code has glossaries of terms, not repeated here, for the various methods it covers.

no tubes in window (NTIW) Construction in which each tube passes through each cross-flow baffle and intermediate supports when provided. To accomplish this, tubes that would pass through the plane of a baffle where a segment has been cut off are excluded from the layout. See **window.**

nozzle A flanged pipe connection to a heat exchanger or pressure vessel. Also used to describe pipe connections that have their ends beveled for field welding.

nozzle entrance velocity The average linear velocity of the fluid in the inlet(s) to the exchanger.

nozzle losses The pressure losses in a fluid due to friction created as it expands from the inlet nozzle to the bundle and contracts when it enters the outlet nozzle.

nubbin A narrow raised portion of the gasket surface of a flange face, too thick or wide to be considered a tongue. The function of the nubbin is to apply high unit force to the central part of the gasket while permitting some compression in the remainder.

offset bend lines An arrangement of U-tubes in a bundle in which the bend lines of successive rows are offset axially by a constant amount. The straight lengths of successive rows are thereby increased by a constant increment. The axial distance between successive bends is the sum of the pitch distance and the increment.

offset nozzle A nozzle with its centerline located in a plane parallel with a plane that passes through a unit's axial centerline, and in which the distance between the parallel planes is less than the maximum possible. See **tangential nozzle.**

operating pressure The pressure for which a side of a unit is thermally designed and rated.

O-ring gasket A ring gasket in which the cross section is circular. O-ring gaskets are most often used in self-energizing closures.

orifice baffle A full-circle flow baffle in which the tube holes are substantially larger than the tube O.D.s. The area of the annuli formed between the tube O.D.s and the hole I.D.s is approximately equivalent to what would be provided if 25 percent cut segmental cross-flow baffles were to be used. Flow in the shell of a unit fitted with orifice baffles is essentially axial.

outer-tube center The diameter of the circle on which is located the center of the tube farthest from the center of the tubesheet.

outer-tube limit The diameter of the circle that circumscribes the tubes farthest from the center of the tubesheet.

outside-tube surface Tube surface based on the tube O.D.

packed floating head A floating tubesheet sealed to the shell by packing. In some versions, the floating-end cover and shell cover are one and the same. They are sealed to the tubesheet with packing separated by a lantern ring.

packed joint A tube-to-tubesheet connection in which the tubesheet holes are enlarged and tapped for approximately three-fourths of their depth. Packing is compressed in the annular spaces between the tubes and holes by ferrules screwed into the tapped holes. This seals the tubes to the tubesheets.

packing box An enlarged portion of the shell or cover machined to receive packing.

packing gland See **gland.**

packing ring A ring of fibrous or elastomeric material inserted between the tubesheet and the packing box of a packed floating-head unit. The packing is compressed by a gland to seal the floating tubesheet to the shell. Usually the packing is square, but chevron rings may be used.

pass (1) A group of tubes manifolded together in the tubesheet that receives the full flow from the inlet; (2) one transit of fluid through the straight-tube length in a tube pass.

pass-partition bypassing Flow between a pass partition and channel cover, or between a pass partition and tubesheet. In bolted assemblies, pass-partition bypassing may be due to cover or tubesheet deflection under the influence of the tube-side pressure. For units designed to the TEMA or HEI PPS, the small amount of pass-partition bypassing due to deflection probably does not materially affect thermal performance. However, if the tube-side fluid is at high pressure, or is erosive or corrosive, pass-partition bypassing may lead to erosion-corrosion in the region of the pass partition which can noticeably affect performance.

pass-partition cover A cover plate of the outlet pass compartment of a two-pass tube-side unit. When a pass-partition cover is used, the channel cover is sealed to the channel flange by a ring gasket without a rib or by a welded-on diaphragm. The seal of the pass-partition cover to the outlet pass compartment needs only to resist the differential pressure between the inlet and outlet to the tubes. When a pass-partition cover is used, the channel cover does not have to be rigid to seal the gasket rib to the pass partitions because channel cover deflection cannot cause pass-partition bypassing.

pass-partition grooves Grooves milled into tubesheets and channel covers to accept gasket pass-partition ribs when confined joint construction is used.

percent cut off The height of the segment cut off a segmental cross-flow baffle expressed as a percent of shell I.D. See **baffle cut.**

pie layout A tubesheet layout similar to a quadrant layout, but with more than four tube passes. See **quadrant layout.**

pigtail A loop of pipe or round bar welded to the top of a cover to provide a handling ring.

pipe shell (1) A heat exchanger shell made of standard pipe as opposed to one made of rolled and welded plate; (2) the term applied to a heat exchanger when the shell is made of standard pipe.

pitch pattern The arrangement of tubes that pierce the tubesheets' baffles and supports. Typical pitch patterns are (1) square, with shell flow entering the bundle perpendicular to the pitch line; (2) rotated square, with shell flow entering the bundle at the intersection of perpendicular pitch lines; (3) triangular, with shell flow entering the bundle parallel with the base of the equilateral triangle on which the tubes are arranged; (4) rotated triangular, with shell flow entering the bundle at the apex of the equilateral triangle on which the tubes are arranged; and (5) diamond.

plain face See **flat face.**

plate shell (1) A heat exchanger shell cylinder made by rolling a plate into a cylinder and welding the long seam (rolled plate shells are used for units that have diameters larger than standard pipe sizes, or when standard pipe construction is more costly than fabricating a shell); (2) a heat exchanger that is larger than a standard pipe size unit.

plug See **tube plug.**

plug map A tubesheet layout marked to show the locations of plugged tubes in the tube field. Plug maps should also show the axial positions of leaks, the radial orientation of the failures, and the dates of plugging. See **tube failure profile.**

plug welding Fusion welding an applied loose lining to a backing surface on a pattern of holes perforated through the lining.

pull-through floating head A floating tubesheet and bolted cover that fit entirely within the shell.

quadrant layout A four-pass tubesheet layout in which each pass occupies one quadrant of the tubesheet.

radial nozzle A nozzle whose centerline is located in a plane that passes through the axial centerline of the unit. Radial nozzles may be pitched within the axial plane that passes through the centerline of the exchanger..

radial pitch A tube pitch pattern in a crossflow condenser in which the tubes are laid out on arcs that have a common origin in the bottom of the condenser. The space between the tubes decreases in successive arcs as the origin is approached.

rails Guides in a removable-bundle heat exchanger, kettle-type reboiler, or closed feedwater heater for aligning and supporting removable bundles. The rails may be made from structural angles, flat bars, or other structural shapes. The bundle support plates may be notched to slide on the rails. The rails may also be used as roll bars on which rollers fastened to the bundle roll. See **skid bars.**

ribbon layout A tubesheet layout in which each pass partition is parallel with a common centerline. Inlet and outlet nozzles are on opposite sides of the channel.

rolled plate shell See **plate shell.**

rolled tube joints Expanded tube-to-tubesheet connections in which the expanding is done by a tube roller. The roller consists of a slotted cage that contains three or more hardened rolls and a tapered mandrel that fits between them. The tube roller is inserted into the tube and is rotated by a power source that turns the tapered mandrel and thrusts it forward. The mandrel simultaneously rotates the rolls in a direction counter to its rotational direction and forces them against the tube wall. As the rolls bear on the tube wall, they first force it outward until it makes contact with the hole. Up to this point tube expansion may be elastic or plastic. After contact, continued rolling stretches both tube and tubesheet. Tubesheet deformation may be fully elastic, or fully plastic or verge from plastic to elastic. However, to obtain a seal, the tube must deform plastically after contact. When the roller is withdrawn, the tube hole must recover more than the tube. The high unit force of the rolls on the tube exceeds the tube metal's plastic limit, causing some wall reduction and axial extrusion. The residual stresses in the tube wall due to tube crushing resist tube recovery and help maintain the joint tight and strong.

roller expanding The process of making rolled tube joints. See **tube rolling.**

root rings Round bars, pipes, or cast rings fitted into the roots of adjacent corrugations of bellows expansion joints. Root rings reinforce the roots of the corrugations against internal pressure and maintain equal distances between corrugation crowns.

rubber expanding Expanding tubes into tubesheets by axially compressing a cylindrical elastomeric plug inserted into the tube end. When the plug is compressed, it bulges, forcing the tube against the hole.

sacrificial anodes Replaceable metal wasting plates affixed to internal parts of an exchanger that are lower on the electromotive-force scale. The preferential corrosion sacrifices the anodes, thereby protecting the exchanger parts.

saddles See **support saddles.**

seal strip A longitudinal metal strip that traverses the length of a tube bundle, placed so as to prevent fluid bypassing around or through the bundle. See **sealing device.**

sealing device A device installed in a tube bundle to prevent fluid bypassing around or through the bundle. Sealing devices may be dummy tubes, seal strips, tie-rods and spacers, or a combination of these parts.

seal rings and seal plates Plates cut to shape to fit between the shell inside diameter and the exterior of desuperheating and drains cooling-zone enclosures to prevent entering steam and drains from bypassing the zones.

self-energizing closure A channel cover closure in which hydrostatic force due to operating pressure is used to push a self-energizing gasket into the space to be sealed.

semicircular baffle A cross-flow baffle from which a segment approximately 45 percent of the diameter is cut off.

semisupport A tube support plate from which a segment approximately 45 percent of the diameter is cut off.

separator An internal drum in the vapor space of a power-plant evaporator in way of the vapor outlet. Its function is to separate droplets of water entrained in the vapor stream and return the collected liquid to the boiling pool.

shell (1) One complete heat exchanger; (2) one heat exchanger module in a group of modules connected in series or parallel; (3) that part of a heat exchanger or closed feedwater heater into which the bundle is inserted; (4) the pipe or rolled-plate cylinder from which a shell is fabricated.

shell cover The rear head of the shell of a U-tube or floating-head heat exchanger.

shell entrance area The unrestricted area for flow between the shell inlet and the bundle.

shell exit area The unrestricted area for flow between the bundle and the shell exit.

shell plate The rolled plate from which a shell cylinder is fabricated.

shell skirt A short length of shell cylinder welded to the stationary tubesheet. The shell skirt is usually welded to the shell after the bundle is inserted into the shell, or (less frequently) the skirt is welded to a flange and bolted to a mating shell flange. It is customary to install shell skirts in U-tube closed feedwater heaters to facilitate disassembly by parting the skirt from the shell. If a shell skirt is integral with the shell, it is parted at a predetermined cut or burn line. A thin, wide metal band is set in place on the inside of the shell under this line to protect the tubes and other parts that flame might impinge on when the skirt is parted from the shell during disassembly. When the bundle and shell are reassembled, if the protective band is intact, it is used as a backing ring that remains in place. If it has been too severely damaged to serve as an effective backing ring, it is replaced. In either case, the rejoining weld is in the category of a single-butt weld from the outside.

shell-side pressure drop The pressure loss through the shell of a heat exchanger, excluding static losses. Permissible shell-side pressure drops in process heat exchangers are determined by process requirements and the economics of operation. The suggested limits in closed feedwater heaters are 30 percent of the differential pressure between stages at the lowest anticipated extraction pressure and no more than 35 kPa (~5 psi) in any one zone. When heaters are operated at low extraction pressures [140−35 kPa (20−5 psi)] or there are appreciable line losses, these limits should be reduced. See **tube-side pressure drop.**

shim An adjusting plate between (1) the matching saddles of stacked units or (2) between shell saddles and roller supports to accommodate uneven floors.

shock tube (1) A tube that is prebent to a bow to withstand the thermal impact of cyclical or sudden high-temperature heating and chilling (shock tubes are used in regenerative heat recuperators); (2) bowed tubes of boiler feedwater evaporators designed to be shocked with cold water to crack off scale. See **bent tube.**

shoes Flat bars installed on the horizontal edges of notches in support plates or baffles of removable-bundle units. The shoes ride on slide rails, reducing friction during thermal expansion and contraction and facilitates bundle removal. See **skid bars.**

shoulder bolt A stud bolt with an annular shoulder approximately half way along its length. The cross section of the shoulder may be rectangular, or the shoulder may taper from a perpendicular face to the stud-bolt diameter. Shoulder bolts are used in removable-bundle exchangers when the stationary tubesheet is extended as a flange. Their function is to maintain the tubesheet connection to the shell flange when the channel is removed. When shoulder bolts are used, the tubesheet bolthole is recessed to receive the rectangular shoulder or chamfered to receive the tapered shoulder so that it is positioned just below the tubesheet face. It is customary to install shoulder bolts in one-quarter of the boltholes. However, half or all of the bolts may be shoulder bolts.

single inlet steam nozzle A steam inlet nozzle that enters the shell of a closed feedwater heater at the plane of the thermal centerline, and through which all the extraction steam enters. See **thermal centerline.**

single segmental baffles Cross-flow baffles in which sectors of adjacent baffle circles are cut off alternately side to side or top to bottom, leaving behind baffle segments that direct the fluid up and down or side to side across the tubes.

skid bars Pairs of parallel thick rectangular bars, fastened about two-thirds of the shell diameter apart, to a heavy bundle's baffles or tube supports. The skid bars ride on the shell, permitting inserting and pulling the bundle without damaging the baffles or supports.

snap ring A machined square or rectangular ring with one cut across its cross section. The ring fits into a groove on the periphery of the floating tubesheet of a floating head with backing device. The snap ring is sprung apart so that it can be slid over the tubesheet. It springs into the groove, providing a shear ring for reaction to the clamping device.

spacers See **baffle spacers.**

split flange A lap-joint flange, cut on a diameter between pairs of boltholes. Split flanges are usually ring flanges, thicker than solid rings. They may be used in pairs with the split lines 90° apart.

split ring A floating-head backing device consisting of a split bolting flange.

split shear ring A machined, rectangular ring cut into two or more parts, used as a shear ring in a floating head. The I.D. of the split shear ring is approximately that of the depth of a rectangular groove machined into the periphery of the floating tubesheet. The O.D. extends far enough to permit a backing ring to bear against the side of the split ring. The backing-ring I.D. is slightly larger than the tubesheet O.D., which permits it to clear. The floating-head cover flange bolts to the backing ring. Multipart shear rings are used in the closures of some high-pressure full access channels. See **snap ring.**

spread pitch A tubesheet drilling pitch greater than $1\frac{1}{4}$ times the tube diameter.

spring washers Conically shaped washers installed under the nuts of bolting-flange stud bolts to ensure that enough tension will remain in the bolts to keep the gasket tight under varying temperature and pressure regimes.

stacked shells Horizontal heat exchanger shells mounted one upon the other. The bottom supports of the next-upper shell bolts to mating supports on the preceding lower shell. When an exchanger consists of stacked multiple shells in series, the intermediate nozzles are welded or bolted to each other.

standpipe A vertical extension of the liquid-return connection of a kettle-type reboiler. The standpipe extends above the level of the highest tube to ensure that the bundle is always fully immersed.

start-up vents Vent connections in the shell of a feedwater heater or condenser that bypass the operating vents during start-up to effect rapid removal of noncondensibles present before start-up.

stationary-tubesheet exchanger See **fixed-tubesheet exchanger.**

steam chest The shell side of the tubular portion of a vertical-tube calandria-type evaporator.

steam distribution zone An untubed region under the inlet nozzle of a closed feedwater heater.

string See **train.**

stub end The end of a nozzle or cylinder turned over to form an integral annular ring that serves as a gasket surface. A loose flange (lap-joint flange) fits over the cylindrical part of the stub end and bears against the back of the lap. The O.D. of the annular ring may be as small as the gasket O.D. or as large as the lap-joint flange O.D. Stub ends may be fabricated by (1) welding an annular ring to the end of a pipe or cylinder and machining as required, (2) forging integrally and machining, or (3) bending over the end of a cylinder or pipe and facing the gasket surface.

stuffing box See **packing box.**

subcooling zone In a (1) process condenser-subcooler, the region of tubes flooded to subcool the condensate; (2) closed feedwater heater, the portion of the tube surface enclosed by a drains subcooling-zone enclosure.

subcooling-zone enclosure An enclosure in a closed feedwater heater around the inlet pass tubes, into which condensate and drains from the condensing zone are fed. The surface contained within the subcooling-zone enclosure acts as a condensate-feedwater heat exchanger in which the condensate and drains surrender heat to incoming feedwater. In horizontal heaters, the subcooling zone can be either split-pass full-tube-length or full-pass short-tube-length. Subcooling zone enclosures of split-pass types envelope a fraction of the tubes in the inlet pass for their full length; in full-pass short-tube-length heaters, they surround all the tubes in the inlet pass for a short length behind the tubesheet at the inlet end.

sump An auxiliary portion of a condenser shell for the purpose of accumulating condensate. Sumps may be horizontal or vertical. Those that are horizontal may be integral with the shell and extend for part or all of the shell length. See **hot well.**

support brackets Brackets for supporting heat exchangers vertically or horizontally. The brackets may consist of loose saddles or a framework for horizontal units or loose lugs that bear against rings on the shells of vertical exchangers. Brackets are usually loose, whereas saddles and lugs are welded.

support lug See **lug.**

support ring A plate or bar ring joined to the shell of a vertical heat exchanger used to support it in a structure. The ring is stiffened by lugs or gussets.

support saddle A structural or fabricated support for supporting a horizontal unit. Support saddles are usually installed in pairs, but very long units may have intermediate saddles. The stationary or front-end saddle is usually fixed by bolting to the structure on which it rests. The other saddle(s) are free to move axially to accommodate shell length changes due to temperature variations.

support span The distance traversed by a tube between support points. When full-circle supports are used, the support span is equal to the support pitch. When segmental baffles or semisupports are used, the support span is two semisupport or baffle pitches.

tangential nozzle A connection with its centerline in a plane parallel with a plane passing through the axial centerline of the exchanger, and in which the distance between the parallel planes is at the maximum. Tangential nozzles may be rotated within their plane of entry. They may also enter the dish of a bonnet or channel. See **offset nozzle.**

taper plug A tube plug machined from a round bar to provide a locking taper when driven into a tube end (included angle of approximately 5°). Taper plugs are installed to seal off leaking tubes. See **tube plug.**

test flange A loose ring flange used to seal the tubesheet of a removable-bundle exchanger to the shell with tube ends exposed during hydrostatic testing. The tubesheet is sandwiched between the gasketed shell flange and the test flange. Test flanges are also used with U-tube units in which the tubesheets are neither welded nor bolted to the shell.

test gland A device used in place of a bonnet to seal the floating tubesheet to the shell during hydrostatic testing of an externally sealed, packed floating head. The tube ends at the floating end are exposed.

test ring and gland A device used to seal the floating tubesheet to the shell during hydrostatic testing a pull-through floating head or floating head with backing device. The front face of the test ring is machined and drilled to make a gasketed, bolted joint with the shell flange at the cover end. The back of the test ring is constructed as a packing box to receive a gland that compresses packing against the O.D. of the floating tubesheet. When a test ring and gland are used during hydrostatic testing, the tube ends in the floating tubesheet are exposed.

thermal centerline The point along a condensing-zone closed feedwater heater tube bundle where steam flow is distributed equally in both directions.

thick-walled expansion joint Heat exchanger shell expansion joints greater than 3.18 mm ($\sim\frac{1}{8}$ in) thick. They are characterized by higher spring rates and (usually) lower cycle lives than thin-walled (bellows) expansion joints of the same diameter designed for identical pressures. See **bellows expansion joint, flanged-and-flued expansion joint,** and **flanged-only expansion joint.**

tie-rods Round bars that pierce the cross-flow baffles or tube supports and are attached to the stationary tubesheet at one end and to the last baffle or tube support at the other. Usually, baffle spacers are mounted on the tie-rods to space and align the baffles and supports. In some units, the baffles are welded directly to the tie-rods.

tongue-and-groove joint Confined joint construction in which a gasket is confined in a groove and compressed by a tongue that fits the groove.

torpedo See **tube guides.**

train A group of closed feedwater heaters connected in series on the feedwater side. Two or more trains may be arranged in parallel. The words *train* and *string* are used interchangeably.

transverse fins Fins arranged perpendicular to the tube axis of finned tubes.

transverse support clearance See **baffle clearance.**

triple segmental cuts A segmental cross-flow baffle system that consists of cyclically repeating arrangements of two A segments, two B segments, one C segment, and two B segments (ABCBABCBABCB,..., etc.) The A segments are equal sectors of a full circle that together occupy somewhat more than 40 percent of the shell diameter. The B segments are made from a full circle by cutting off equal parallel outer sectors and an inner segment that has sides parallel to the bases of the sectors. Two B segments also occupy a little more than 40 percent of the shell diameter. The C segments are full circles from which sectors have been cut off on parallel chords. A single C segment takes up a little more than 40 percent of the shell diameter. The segment edges overlap to permit tie-rods and spacers to support and align adjacent baffles. Triple segmental baffles remove very little energy from the shell stream. The main flow direction is axial. Their use results in minimal pressure loss and freedom from vibration problems. See **multisegmental cuts.**

trunnions Round bars or pipes attached to a heat exchanger or feedwater heater to facilitate lifting a vertical tubular from the horizontal to the vertical position during installation.

tube bundle (1) In a removable-bundle heat exchanger, the assembly of tubesheet or tubesheets, baffles and/or supports, tie-rods and spacers, tubes, and any other appurtenances not connected to the shell: (2) in a fixed-tubesheet exchanger, the part of the unit that extends from the outer tubesheet face to outer tubesheet face.

tube count The number of tubes in a tube bundle. For U-tube heat exchangers, the tube count is half the hole count. See **hole count.**

tube failure profile A record of the dates and locations of tube failures in a unit. The record also shows the axial position and orientation of the failures. It may consist of a plug map and accompanying notes that detail the failure positions and types of failure. Tube failure profiles are desirable for maintenance, repair, and replacement planning. They may also indicate where design and operating deficiencies require correction or where additional protections must be provided within the structure to prevent continuing deterioration and failures.

tube guides Devices that have bullet-shaped heads with the same diameter as the tube O.D., and twisted, short brush-and-spring tails that fit snugly into the tube end or other means to hold the guide in place. Tube guides protect the tube ends and guide the tubes through the tubesheet and baffle holes as they are loaded into the bundle. When a tube has penetrated the hole at the far end, the guide is removed and reused.

tube-hole grooves One or more annular grooves in the holes of a tubesheet. Tube-hole grooves are used to enhance strength and tightness when tubes are expanded into tubesheets.

tube joint temperature The temperature of the tube-to-tubesheet joint. In process and power-plant heat exchangers, the tube joint temperature is estimated from the design heat-transfer calculations. For the purpose of establishing temperature limits of expanded-only joints in closed feedwater heaters, the tube joint temperature is considered to be the outlet feedwater temperature at the design-point conditions.

tube lane See **cleaning lane.**

tube metal temperature The highest tube metal temperature calculated by considering the relative heat-transfer coefficients on both sides of a tube.

tube pitch pattern See **pitch pattern.**

tube plug A plug for sealing the end of a tube when there is a leak in the tube wall. Common types are (1) single-piece metal or fiber taper plugs with a locking taper angle, (2) two-piece plugs in which the inner piece is a taper plug that fits into a matching taper inside the outer piece, (3) two-piece extension plugs for sealing the far end of a tube that is inaccessible, (4) thimble-shaped plugs that are roller or hydraulically expanded into the tube ends, (5) explosive plugs, and (6) welded plugs. Welded plugs vary from a length of round bar press fitted into the tube (or tube hole when the tube end is cut away) and fusion-welded with a fillet weld to the surrounding structure to shapes designed to prevent impurities from the backface of the tubesheet and the space between the tube end and the hole from contaminating the weld. Plugs may also be explosively welded to the tube end or hole. Proprietary types are commercially available. The tightness of nonwelded plugs depends on friction between the tube and the plug. Taper plugs make line contact before they are driven, but deform the tube to provide some surface for friction. The required interfacial fit pressure between the taper plug and the tube is high because the surface available for friction is small. Two-piece, thimble, and explosive expanded plugs provide greater contact areas than do one-piece taper plugs. Therefore, they require less interfacial pressure to seal. Two-piece plugs may have their exterior barrels serrated to enable them to cut through scale of deposits that cannot be readily removed before plugging.

tube thinning allowance Excess tube wall thickness allowed for thinning of a U-bend at its outer radius. The TEMA minimum tube wall thickness before bending is

$$t_o = t_1\left(\frac{1 + d_o}{4R}\right)$$

where t_o = original wall thickness
t_1 = minimum ASME Code straight-tube thickness
d_o = tube OD
R = mean bend radius

In the HEI standards, the minimum wall thickness before bending is

$$t = \frac{Pd(1 + d/4R)}{2S + 0.8P}$$

where t = original tube wall thickness before bending
d = outside tube diameter
P = design pressure
R = mean bend radius
S = allowable tube design stress at the temperature that corresponds with the shell design pressure

Thinning should not exceed 17 percent of the nominal wall thickness. Flattening at the bend should not exceed 10 percent of the nominal tube O.D.

tube rolling Expanding tubes into tubesheets by roller expanding.

tube-side pressure drop The pressure loss through the tubes, tube-side connections, and pass turnarounds. See **shell-side pressure drop.**

Tweeners A proprietary tube-support/baffle system formerly produced by Metals 'n Things that consists of folded and formed sections. Flow is generally axial and pressure drop is low. Therefore, spans can be short enough to minimize the prospects of flow induced tube vibration damage. The successor to Tweeners is the Non-Segmental Baffle produced by Holtec International.

U-bend exchanger An exchanger made with U-bend tubes (U-tubes). Except when there are more than two passes, horizontal U-bend exchanger bundles are preferably built with the bends in the vertical plane to take full advantage of the flexibility of the bent ends. For example, if steam is being generated in the shell side of a U-tube unit, the low water level that exposes upper tube rows can create a far more severe differential thermal expansion problem in bundles with tubes bent in the horizontal plane than in the vertical. Whenever possible vertical U-bend exchangers are arranged to operate head down to permit the tubes to drain.

unit A heat exchanger consisting of one or more shells. Shells in a unit may be arranged in series, in parallel, or in series-parallel flow combinations.

U-tube, U-shell exchanger See **multitube exchanger.**

vapor belt A cylindrical enlargement of the shell concentric with the shell cylinder of a process condenser. The vapor belt is somewhat longer than the width of the vapor inlet nozzle. The enlargement permits incoming vapor to distribute itself to the tubes. When the vapor belt extends for substantially the full shell length, the flow characteristics approach croseflow.

vapor bustle See **vapor belt.**

vapor distribution belt See **vapor belt.**

vertical cut Segments of crossflow baffles cutt off alternately at either side of adjacent baffles to direct stream flow side to side across the tubes.

wall clearance The clearance between the O.D. of cross-flow baffles or tube supports and the shell or zone-enclosure I.D.

wasting plate A plate intended to be preferentially eroded or corroded. See **sacrificial anodes.**

water box A channel of a steam surface condenser.

weep hole (1) A small hole drilled in the cover plate to indicate leakage in an applied lining; (2) a small tapped hole in a doubler plate or reinforcement fully sealed to a pressure part used as a leak indicator and vent; (3) a small hole drilled through a pass partition to permit full drainage of the compartment.

weir A transverse plate behind the tube bundle in a kettle reboiler. The top of the weir is located above the level of the highest tube. When the weir is sealed to the shell to enclose the bundle in a hydraulically tight trough, it is an overflow weir; when the seal is interrupted at the bottom by an open space, it is an underflow weir. Overflow weirs are used with viscous liquids.

Liquid that is not vaporized flows over or under the weir to the compartment behind it. The compartment empties to the column bottom or recirculating pump. The weir provides a calming zone for instrumentation and level control.

weld-end nozzle A connection with the end beveled or otherwise prepared for welding to the connecting pipe.

window The distance between a baffle or tube-support cut line and the shell. In a closed feedwater heater desuperheating or condensing zone, the distance between the cut line and the zone enclosure.

wire drawing Erosion due to high-pressure fluid on one side of a tubesheet passing between the tube and hole into a low-pressure space on the other. An example is leakage of feedwater between a tube and tubesheet. See **wormholing.**

wormholing Spiral or curved grooves on the tube and tube-hole surfaces, similar in appearance to holes made in wood by worms. The holes are initiated at the high-pressure side of the tubesheet and penetrate axially. Wormholing may become so severe that the ligaments between adjacent tubes are penetrated.

Index

*Italic page numbers are for index entries from figures and tables.

ABOUT THE AUTHORS

CARL F. ANDREONE, P.E., is president of Heat Transfer Consultants, Inc., and a consultant on power generation heat transfer equipment to Stone & Webster Engineering Corporation. An internationally recognized authority on specifying feedwater heaters and auxiliary exchangers, he has worked on more than 400 closed feedwater heater replacements in the past 15 years.

STANLEY YOKELL, P.E., is president of MGT, Inc., and a senior consultant to Hagler, Bailly Consulting, Inc., an international economics and engineering consulting firm. He is a leading expert on tube-to-tubesheet joining of tubular heat exchangers, and has been involved in the design and construction of over 3000 tubular heat exchangers. Since 1979 he has consulted on a variety of problems with process and power generation heat exchangers.

Both authors are Fellows of the ASME. The authors have presented more than 100 short intensive courses on shell-and-tube heat exchangers, closed feedwater heaters, and tubular exchanger inspection, maintenance, and repair in the United States, Canada, and Europe.

www.ingramcontent.com/pod-product-compliance
Ingram Content Group UK Ltd.
Pitfield, Milton Keynes, MK11 3LW, UK
UKHW041642031026
15587UKWH00001B/445